Palgrave Studies in Sound

Series editor
Mark Grimshaw-Aagaard
Aalborg University
Aalborg, Denmark

Palgrave Studies in Sound is an interdisciplinary series devoted to the topic of sound with each volume framing and focusing on sound as it is conceptualized in a specific context or field. In its broad reach, Studies in Sound aims to illuminate not only the diversity and complexity of our understanding and experience of sound but also the myriad ways in which sound is conceptualized and utilized in diverse domains. The series is edited by Mark Grimshaw-Aagaard, The Obel Professor of Music at Aalborg University, and is curated by members of the university's Music and Sound Knowledge Group.

More information about this series at
http://www.springer.com/series/15081

Tom A. Garner

Echoes of Other Worlds: Sound In Virtual Reality

Past, Present and Future

palgrave
macmillan

Tom A. Garner
University of Portsmouth
Portsmouth, UK

Palgrave Studies in Sound
ISBN 978-3-319-88086-0 ISBN 978-3-319-65708-0 (eBook)
DOI 10.1007/978-3-319-65708-0

Cover illustration: Fatima Jamadar

Printed on acid-free paper

This Palgrave Macmillan imprint is published by Springer Nature
The registered company is Springer International Publishing AG
The registered company address is: Gewerbestrasse 11, 6330 Cham, Switzerland

Acknowledgements

This book has *emerged* from what to me feels like a somewhat unusual set of circumstances. As I continue my nomadic postdoctoral quest, searching for a faculty to call home and moving from one unexpected project to the next, I have enjoyed learning about many fascinating things from a range of exciting disciplines. This book is about emergence and its application for better understanding virtual reality sound, and experiences like these have undoubtedly contributed to the shape this work has taken.

If we were to map out an emergent framework of the book itself, one decidedly significant component would be Mark Grimshaw, the gentleman who not only co-authored with myself the notion of emergent perception as it pertains to sound, but just so happens to be the editor of the series *Studies in Sound*, of which this book is but a humble part. Throughout the course of writing this book, I have been most fortunate to receive various forms of support from my colleagues at the University of Portsmouth. Special thanks go to Wendy Powell, who provided insight into the wider applications of virtual reality but who also, and crucially, put everything in place so that I had the opportunity to write this book as a central part of my employment. Thanks also to

Dion Willis and Matthew Higgins who provided ongoing feedback and advice, not to forget Marc Cook, Jahingir Uddin and Vaughan Powell, who were frequently and consistently interrupted from their own work to answer innocuous questions on everything from digital games culture to the finer points of biofeedback.

Lastly, I would like to thank my wife Hayley, son James and daughter Sophie, who all had to tolerate my increased absence, particularly in the latter stages of writing. Whilst this had knock-on effects (particularly since, Sophie, you decided that being born right in the middle of this project would be a good idea), the three of you managed to absorb them completely, and this book exists because of your great support. I owe you a holiday. Somewhere nice.

Contents

List of Figures

1

Introduction

In 1963, the modern concept of virtual reality was born. From then it took more than twenty years for us to give this concept a name and, since then, we continue to have trouble agreeing upon the meaning of (virtual reality)VR and how exactly the term should be defined. VR is a field of study, a synthetic experience, an interactive computer simulation, a source of synthetic feedback, a three-dimensional representation of both the concrete and the abstract, an immersive experience generated by a computer—the list goes on. The same difficulty applies to sound, but here the alternate perspectives are even less compatible with each other. Sound can be, amongst other things, physical perturbations in the air or electrical impulses triggered by displacements of the cochlea hair cells. Sound can be synonymous with a source object, an event or a listener. It can exist only with a listener and it can exist without a listener. This emphasises the great challenge in understanding the composite of the two, VR sound.

T.A. Garner, *Echoes of Other Worlds: Sound In Virtual Reality*,
Palgrave Studies in Sound, DOI 10.1007/978-3-319-65708-0_1

Virtual Reality Sound as an Emergent Phenomenon

The central position of this book builds upon several years of research into auditory perception, sound design for digital games and applications of VR. These themes that have been brought together by emergence theory. Inherently integrative and holistic, emergence theory encourages us to try and reconcile various notions and positions on a subject, in order to understand it from a broader, more ecological perspective. It discourages silos of thought and proclaims the huge complexity of things, based on the assertion that such things do not exist in a vacuum and are instead part of a huge and ever-changing 'aggregate' of potentials. Therefore, VR sound cannot be separated from either sound or VR, nor can it be separated from the numerous wider related elements that contribute to its precise nature. Such a framework may initially seem plausible but incomprehensibly dense and without clear practical application, but fortunately this is not the case. When we centre the thing we wish to understand within an emergence model, it becomes possible to map out the likely elements of its aggregate. Once constructed, we are then able to conceptually step back and observe both the potential contributors to the subject, and also the potential interactions that exist between them.

Let us take sound as an example. From the various perspectives on this subject, we can position various properties upon our conceptual map. This could include the source object, the event that is occurring, the physiology of the listener, the acoustic properties of the physical environment and the affective and cognitive state of the listener. This is itself a very cut-down list and within each of these properties we are likely to find many further sub-properties. How much detail we wish to include is at our discretion, but to collect the various elements together and draw lines between them provides us with a platform from which to further explore. For instance, our map could draw a line between proprioceptive feelings of comfort and short-term memory, raising the question of how these elements may influence each other and collectively (along with many other things) determine our perception of sound. The same can be done for sound in VR, which brings us the central argument of this book; that to better understand and create sound material for VR (and, in turn, create better VR in general) requires VR sound to be recognised as an emergent phenomenon. The intention is to present a more integrated framework of sound in VR that emphasises the importance of perceptual effects and user experience from within a wider cultural and technological domain. It acknowledges and attempts to uncover some of the hidden properties that influence our experience of sound in VR with the hope that this will encourage readers to take a step back when approaching VR design—improving both sound in VR through greater awareness of wider interrelating factors, and VR in general by way of acknowledging the vital importance of sound.

Chapter 2

Throughout the second chapter, we begin to uncover some of the key components that will constitute the emergent framework of VR sound. This includes a consideration of some of the broader cultural and philosophical issues that shape our experience in VR, including a discussion regarding how our history of associated technologies has steadily prepared us for acceptance of contemporary VR. We discuss some of the technological precursors that first made virtual experience a ubiquitous phenomenon (television, telephony) alongside contemporary associated technology such as applications of the Internet (emails, online shopping, video calls, social media, etc.) and mobile systems (smartphones, tablets). The central assertion of this chapter is that through such technology, the broader landscape has shifted to make present day a period in which we are now highly open to VR as both a concept and a technology. This chapter also addresses some fundamental issues regarding how VR is presently defined, what its relationship to digital games is and what is meant by 'mixed-reality'. How things have changed since the infamous narrative of 1990s consumer VR, that fell from grace so spectacularly, is also considered as this chapter asserts how issues of wider application, increasing corporatisation, technological ubiquity and ever-greater integration have contributed towards the much greater levels of public acceptance the contemporary VR has so far acquired. Lastly, this chapter makes preparations for the following discussion on an emergent framework of sound by discussing Deleuzean philosophy and virtuality as an organisational theory. The central argument raised here presents the virtual as a component of reality, countering the widely held assumption that a virtual world is an alternative to a 'real world'.

Chapter 3

The third chapter brings the focus promptly back to sound with an exploration of Sonic Virtuality and how sound can be conceptualised as an emergent phenomenon. As with definitions pertaining to VR, this chapter posits that conventional perspectives on sound lack comprehensive explanatory power. Perspectives such as 'sound as object', 'sound as event' and 'sound as phenomenon' are reviewed, and we question where sound is actually located (at the source, in the transmission space between the source and the listener, at the point of hearing, or within the auditory cortex—or even nowhere at all). This chapter asserts that conventional theories on sound incorrectly present its empirical components as dominant and neglect to fully acknowledge to role of the listener. We then observe some of the primary issues with such perspectives by revealing problems in reconciling them with observable phenomena such as non-cochlea sound and auditory hallucinations (tinnitus, exploding head syndrome, musical hallucinations, etc.). In response to this, some of the other existing concepts that are beginning to acknowledge sound in more holistic terms are presented. These include theories of acoustic ecology, auditory grouping, perceptual localisation effects and the Figure Ground phenomenon. This chapter then closes with a discussion on Sonic Virtuality and how our perception of sound is best understood from an emergent framework. The underlying theory upon which Sonic Virtuality is based is also discussed, and we directly explore the literature surrounding emergence theory and associated concepts including Plato's Allegory of the Cave, embodied cognition and construal level theory.

Chapter 4

Bringing together the preceding discussions on sound and VR, this chapter reviews key literature pertaining to the entity at the centre of all this, the user, discussing prominent user-experience notions. The review begins with a brief outline of physiological issues, specifically adverse effects pertaining to cybersickness, before addressing psychological components of user experience including flow, diegesis and fun. As a more contentious point of discussion, immersion and presence are discussed in more detail as we review the differences between system-side and user-side explanations of immersion. Largely originating in theory pertaining to digital games, these points of user experience are posited to be of equal, if not greater relevance to VR. Throughout this chapter, the various components of user experience are revealed to be connected to one another and, in each instance, sound is presented as a significant element, whether it be increasing immersion by way of its inherently surrounding and dynamic properties, or expanding diegesis by seamlessly transitioning its state from the diegetic world to the non-diegetic world. This chapter closes with a discussion on the implications of VR for identity and our understanding of ourselves. Again, taking notions from digital games a step further, the central assertion raised here is that VR is dramatically changing our concept of the Self, by presenting us with seemingly limitless opportunities for customising our existing identity and forming entirely new ones. Including matters of depersonalisation and even out of body experience, the increasing fidelity and convincing realism of VR are shown to be emphasising new issues that were once considered purely science fiction.

Chapter 5

'Representations' considers the various ways in which fictional representations of VR in cinema, television and literature have dramatically influenced user expectation and experience in the past, and how they continue to do so today. From the utopias of ancient classicism to the Orwellian nightmares of the twentieth century, the nature of fictional worlds is revealed to be a substantial influence upon the nature of virtual worlds. From matters of aesthetics and scale to alternate forms of literary realism, both the classic conventions of world-building in fiction and the established methods of breaking them are shown to draw noteworthy parallels to the worlds in digital games and VR. Throughout Chap. 5, the discussion is centred upon the treatment of sound in such representations, examining both how the sounds of fictional worlds have relevance to those of virtual ones, and also how the quality of sound in fictional VR is described in literature and portrayed upon the screen. This chapter continues with a review of the more contemporary representations of VR in cinema and television, as we consider how such visions shaped the expectations for 1990s VR, setting the bar so incredibly high that actual VR had an insurmountable fantasy to live up to. The chapter then expands upon these expectations by looking into the foundational desires underpinning the fantasy of VR.

Chapter 6

'Technological history' adds another collection of components to the overarching emergent framework of VR sound by tracing the history of artistic and mechanical precursors to VR sound and reviewing how the functionality and aesthetics of the past continue to influence VR design and technology today. Key characteristics of contemporary VR (including panorama, stereoscopy, parallax and multimodality) are documented and, throughout this discussion, the lesser-mentioned auditory aspects of these technologies are detailed, from cacophony of whirring film reels that sonically characterised the Kaiserpanorama to the use of binaural audio recordings in the Sensorama. The history of VR, up to and including the 1980s, reveals notably more points of interest with regards to VR research, points that are explored in this chapter before we turn our attention to the consumer VR of the 1990s. Here, the development of VR sound is reviewed, and it is revealed just how consistently underappreciated sound was throughout this period. This chapter comes to a close with a more detailed exploration of VR sound technology. Reaching back to the dawn of electronic sound and the earliest approaches to sound recording, this discussion makes two primary assertions. Firstly, that sound reproduction technology is the auditory equivalent of VR, with headphones and loudspeaker arrays essentially HMDs (head-mounted display) and CAVE systems for the ears. Secondly, because headphones and loudspeakers were not subject to the same hype as HMD/CAVE systems, their technology avoided the 1990s crash into the trough of disillusionment and instead have made consistent progress which continues to the present day—to the extent that certain properties of contemporary headphones may be indicative of future developments in VR-HMDs.

Chapter 7

'Expectations, reality and digital games' focus specifically upon VR's turbulent narrative throughout the 1990s and early 2000s with a look at how VR sound continued to progress, largely through its implementation in digital games. This chapter begins with a response to the expectations set by fictional representations (discussed in Chap. 5), revealing precisely how the realities of 1990s VR inevitably failed to match such impossible standards. Throughout the majority of this chapter, digital games are posited to have been the primary incubator of consumer VR concepts and technology, both in general and with regard to VR sound. Points discussed include the development of the first-person perspective, multichannel audio technology and new approaches to positional/3D audio, intelligent non-player game characters that responded to sounds made by the player, progress in haptics and motion tracking technologies, new speech recognition and voice-command systems and refined affective frameworks of game sound. These issues reveal how the underlying concepts and technologies of VR and VR sound were successfully retained and advanced by digital games, meaning that as the time for contemporary consumer VR rolled around, the technology and design techniques were more refined and in a much better position to deliver a quality experience.

Chapter 8

'Current status' brings us up to now with a review of how sound is currently treated within VR, plus speculations on future developments. Commencing with a hypothetical exploration of the likely integrated future of VR, this chapter asserts that, by way of technology that includes the Internet of Things, social media and collaborative virtual environments, integration is a pivotal theme when considering the shape of things to come. This incorporates integration at the technological level, in which future VR is poised to be the central hub of our daily lives, and also at a social level, with people from across the globe able to transcend physical limitations and come together in massive shared virtual spaces. Sound does feature throughout this section and is argued to be a vitally important element of our virtual integration with both things and people. The subsequent sections of this chapter address audio for VR more directly, with a review of contemporary hardware, software and design techniques. How audio hardware is treated in modern HMDs is documented, both in terms of audio outputs (headphones) and inputs (microphones). The wide range of currently available software is also reviewed, from audio source development kits and plug-ins to digital audio workstations and digital game audio middleware packages. Specific methods for designing VR sound are also addressed, from ambisonics and head-related transfer function, to environmentally modelled audio and procedural generation. The overarching position of this chapter is one that seeks to champion the value of good VR sound as a pivotal means of producing a high-quality VR experience. It is also acknowledged that several of the design methods that are discussed within this chapter have yet to be utilised in contemporary consumer VR systems. This emphasises the opportunity for improvement in this area, and it is speculated that future VR sound shall steadily begin to accommodate these features whilst also responding to the broader need for greater integration by becoming more cross-compatible from within a single user interface.

Chapter 9

The final chapter documents some of the wider applications of VR and considers the value of good sound design across such applications. Here we take a look at data visualisation (including the auditory equivalent, sonification), telepresence (inclusive of VR-based teleconferencing, multi-user shared virtual spaces and VR broadcasting), education (the role of VR in contemporary eLearning and professional skills training), creative applications (painting, musical composition, sculpture and even VR design), rapid product prototyping and health applications (including VR for pain distraction, phantom limb pain therapy, attenuating anxiety and post-stroke limb rehabilitation). Throughout this chapter, sound is positioned as an underappreciated yet vitally powerful contributor across all wider applications of VR; from increasing the ease of comprehension in VR visualisations, enabling more complex data sets to be more readily interpreted, to facilitating the retention of more information in an eLearning application by increasing users' engagement and immersion in the material. The primary assertion of this chapter is that VR extends far beyond a recreational and digital games context to draw connections with great number of further disciplines that significantly expand the emergent framework of both VR and (by way of its consistent and extensive value to such applications) VR sound.

A Few Brief Notes Before We Begin

To avoid a lot of repetition and likely inconsistency (and potentially insulting the reader's intelligence), (VR) and (HMD) are presented solely as acronyms from here on in. More than a few further acronyms have found their way into this book, explanations for all which are included within the glossary. It should also be pointed out that throughout this book, 'we' is used instead of 'I' as a personal pronoun of preference. Before any accusations of pretentiousness are flung, this is not the majestic plural (or 'Royal We') but rather a personal pronoun that includes two initials; me as the writer and you as the reader. As *we* shall get to in due course, sound cannot exist without someone to hear it, VR cannot exist unless it is being experienced, and a book cannot have a voice if it is not being read. Furthermore, the writing of this book was most certainly an exploratory adventure and acknowledging this in the writing style seemed appropriate.

With regard to terminology, this book presents quite a substantial collection of ideas and with that comes something of a barrage of terminology. Every effort is made to keep things clear but, as with the acronyms above, the following are commonly used terms that are often interchangeable in other texts but will benefit from clarification here. Firstly, 'sound' and 'audio'. How these differ from one another is a central discussion point throughout this book but, in basic terms, sound is used to describe the subjective experience and is a highly complex phenomenon, irrevocably tied to many other things. Audio by comparison is used in the book to identify a physical and objective phenomenon, relevant to terms such as 'sound wave' or 'acoustics'. This can be further elucidated by the distinction between 'sound designer' and 'audio engineer', both largely interchangeable terms that here differentiates, respectively, someone who generates material with a focus on the listener (received meaning, interpretation, emotional content, potential to influence behaviour, etc.) from someone who prioritises the acoustic sound wave (spectral composition, attack, decay, sustain, release, etc.). Another set of commonly used terms that require separating are digital games, video games and computer games. Throughout the book, 'digital games' is used most commonly as it is essentially an umbrella term that encapsulates the other two, whilst 'video games' exclusively refers to digital games played upon a home console and 'computer games' denotes those run from a personal computer. Now moving from games to players, there is also much inconsistency within existing literature over the use of 'players' and 'users'. In this book, players and play are used to describe interaction with games (specifically, software that are designed primarily as games and this excludes software with another primary function even if it has gameplay features). Outside of this instance, 'user' is the preferred term.

Finally, a couple of points concerning references. As this book cites a large number of sources across a very broad range of types, effort has been made to organise these as efficiently as possible. Consequently, at the end of each chapter you will find a bibliography featuring all the text-based sources, alongside all artefacts (films, digital games, television programmes, etc.) that can be clearly attributed to an author or authors. Internet sources with no clear author appear in the notes section at the end of the book. Any Internet-based citations, whether in the notes section or the chapter references, do not include individual 'accessed on' dates, as each of them was checked at the latest possible time just prior to submission (1 July 2017).

Despite significant investment, the future of VR remains uncertain, particularly as a mainstream technology. It is hoped that this book will be of interest and of value, both to those who wish to create VR and those who wish to experience and better understand it.

2
The Domain of Virtual Reality

Virtual technologies permeate our lives to near ubiquity, mediating our careers, social lives, finances, shopping habits and leisure activities to name a few. This is as much a result of our changing attitudes and understanding of the virtual as it is due to technological developments. This chapter introduces some of the philosophical foundations of the virtual and VR. What these terms mean and how that meaning has shifted over the years are discussed, and we explore the differences between the virtual as a technology and as a broader theory of organisation. There are two primary assertions made within this chapter. Firstly, that our perception of the virtual and VR is changing, not solely because of the technology (which is too often incorrectly positioned as the dominant aspect of VR), but because of a much wider web of interrelating factors all centred around us as perceivers of the virtual. The second position is that existing definitions and explanations of the virtual and VR are limited, but that each has value as a piece of the larger puzzle. It is bringing these pieces together that will help to create a more holistic understanding and support better VR design and implementation.

As a first step towards mapping out an emergent framework of VR sound, this chapter commences with some wider context regarding how

T.A. Garner, *Echoes of Other Worlds: Sound In Virtual Reality*,
Palgrave Studies in Sound, DOI 10.1007/978-3-319-65708-0_2

our relationship with virtual technology has paved the way for contemporary VR by steadily and subtly introducing us to both its conceptual and technological aspects. Following from this is a look at the technological perspective of the virtual, including an introduction to the various forms of virtual technologies. An additional, wider question is then asked relating to the global changes affecting contemporary VR and how things have changed (and how they've remained the same) since the 1990s. This chapter then closes with an exploration into the concept of the virtual from an organisational perspective, introducing the literature that has inspired the central position of this book—that both VR and sound are best understood as emergent phenomena.

Global Context

Typically referred to as '1990s VR', the period that commenced in the late 1980s and petered out towards the millennium represents the first generation of consumer VR. As described in greater detail later within this book, disappointment with the actual hardware and software of this generation caused great disillusionment with VR as a mainstream product and preceded something of a dark age. Subsequently, the historical timeline of consumer VR then begins to document the activity of note from roughly 2012 onwards, as an unknown start-up company, Oculus, looked to crowdsourcing to fund the development of their *Rift* HMD. Their initial designs ended the consumer VR dark age and sparked a resurgence of interest that marks the second generation of consumer VR. This section begins our exploration into the wider issues that contribute to VR as an emergent phenomenon, with a look at how our perspectives and priorities compare between the two generations of mainstream VR. Additionally, we consider some of the global changes that have been instrumental in repositioning VR in its contemporary form, as a serious contender for being one of the most widely accepted and ubiquitous technologies of the modern age.

Then and Now

How we define VR has certainly changed since its first consumer generation. For a fundamental example, we can consult the Merriam-Webster dictionary which, as of 2017, defined the term as: '[A]n artificial environment which is experienced through sensory stimuli [...] provided by a computer and in which one's actions partially determine what happens in the environment'.[1] Whilst this definition contains some issues, we can compare it to an earlier dictionary definition as a simple means of examining change in thought across the VR generations. In 1994, Heim (p. 109) quoted Merriam-Webster's dictionary definitions of both *virtual* and *reality* (the combination of these words apparently not yet accepted in dictionary lexicon). The former was defined as 'being in essence or effect though not formally recognized or admitted' and the latter 'a real event, entity or state of affairs'. There is no technological aspect to the definition, with virtual used in much more of an organisational sense. Heim interprets the two together as something that is real only in effect, it is not real, but merely appears to be so. What this reveals to us most prominently is the difference between the generations in terms of general acceptance.

For researchers throughout the 1990s, VR was largely defined by the technological hardware that facilitated it. However, the literature of the time was already raising concerns (see Steuer 1992) regarding this definition, specifically that such a perspective limits conceptual understanding and analysis, making it difficult for us to compare VR with non-VR beyond a basic dichotomy. It was also asserted that this perspective restricted the potential to comparatively examine multiple instances of VR (e.g. VR against AR or augmented virtuality, etc.). Steuer's counter to the hardware definition of VR is decidedly more phenomenological, positioning VR as an experience. Within this definition, Steuer identifies presence and, by extension, telepresence as key components of conceptualising the virtual. Comparing this to the second generation, a significant degree of similarity can be observed. VR remains a predominant technology/hardware-focussed concept but with pockets of thought describing VR as an experience. Unlike the 1990s, however, the prevalence of the technological perspective is beginning to wane, presenting us with one noteworthy example of changes in contemporary VR. Perspectives focussing upon user experience have become progressively more pronounced, and the number of publications advocating this understanding continues to increase (see Jerald 2015).

In 1990, the Special Interest Group on Computer Graphics and Interactive Technologies held a panel session with speakers from the Human Interface Technology Laboratory on the anticipated future directions of VR (Bricken 1990). Some of the notes taken from that discussion reveal that many of the research and development themes of VR in the early 1990s are strikingly similar to their contemporary counterparts. Examples of this similarity include: a research focus upon bi-directional behaviour transducers (approaches to mapping natural motions/gestures into VR systems and creating VR content that can influence user physiology and behaviour); increasing sensory immersion (virtual environments that surround the user); and forming a detailed psychological understanding of the user (user experience: cognition, emotion, meaning, etc.) to inform design of virtual environments and interfaces. A search of VR applications (see Chap. 9) also reveals that prominent ideas pertaining to how VR can be practically utilised are comparable between then and now. Prime examples include teleoperation/telepresence, physical and psychological rehabilitation therapies and military training simulations. Research focus upon multimodal feedback in VR is also comparable across the generations; 1990s research closely mirrors its contemporary cousins with much emphasis on visual feedback, a second-place position for auditory content and distinctly smaller pockets of research exploring further modalities that include haptic/tactile (Iwata 1990), olfactory (Krueger 1995) and gustatory sensations (Robinett 1994).

The hardware that facilitates VR has been refined over the generations and arguably still has great distance yet to travel in terms of further improvement. That said, the principle designs have changed relatively little over the decades. For example, HMDs of the 1990s may have been heavily criticised, but contemporary VR still positions them as *the* interface for VR experience. What is even more notable is that the appearance and quality of 1990s HMDs were dictated not by design, but by limitations in the materials, computing components and other technologies of the day. Despite this, the foundational aesthetic of contemporary HMDs is largely unchanged, with design simply smaller and more ergonomic. Watching science fiction cinema from that period, in films such as *Back to the Future Part II* (Zemeckis 1989) and *Johnny Mnemonic* (Longo 1995)—and for a further all the more surreal example, an episode from *Murder, She Wrote* (*A Virtual Murder,* Smith 1993)—we can see that the concept designs for HMDs in this generation are strikingly similar to some of the actual systems being released at present. A comparable finding is also revealed in CAVE (Cave Automatic Virtual Environment) VR—a physical cubic space onto which between three and six of the walls are projections of a virtual environment. CAVE systems began to appear in the early 1990s (Cruz-Neira et al. 1992) and remain the subject of considerable academic and professional interest (Ritz and Buss 2016; Ronchi et al. 2016); their overall contemporary designs almost identical to 1990s precursors.

At this point, readers would be forgiven for presuming that very little difference exists between the VR generations and feel the urge to skip to the next chapter section. For the intrepid reader staying the course, however, there is one substantial difference in emergent VR between the early 1990s and more recent times. This difference is not VR itself as an isolated technological entity, but rather changes in the global state of things that have shifted what VR means to the population. What we are specifically referring to here are five discrete phenomena: advanced production, application, corporatisation, integration and acceptance.

Advanced Production

In statements made in 2016 relating to their *Daydream VR* project, Google expressed their intention not to simply create cutting-edge pieces of technology, but rather a global technological ubiquity driven by products could reduce the so-called barriers to VR. These barriers include attenuating motion-sickness, reducing HMD weight and increasing the longevity of use by the way of enhanced power efficiency, greater battery capacity and faster charging mechanisms.[2] How confident they are in their ability to realise these ambitions has largely originated from their power to implement advanced production methods. In terms of manufacturing, the VR industry, like most commercial bodies, has changed due to the impact of technology and contemporary manufacturing practices. Advances in robotics have meant that many aspects of VR hardware production can be fully automated (Brookings 2016). This not only increases productivity, enabling more devices to reach the market, but also supports adaptive manufacturing (see Nielsen et al. 2015), in which companies can respond more quickly to changes in consumer expectations and demands. Adaptive manufacturing creates a sizeable difference when comparing VR of the 1990s against modern systems. Whilst the former was unwieldy and incapable of timely response to user feedback, the latter could be regularly upgraded and refined based directly upon the opinions and demands of the consumer. This ability to quickly evolve had a substantial effect upon VR. Specifically, it enabled the industry to resist gimmickry and work towards mainstream status. First-generation VR by contrast struggled to shrug its public perception as little more than a novel prototype, a technological curiosity worth experiencing only a handful of times before exposing its core limitations; its hardware then forsaken to collecting dust atop shelves or underneath beds.

Application

Though still labouring under popular perception as a games technology, VR has numerous diverse applications beyond such recreational use. To preface this point, technology cannot exist within a vacuum (speaking metaphorically—for a literal vacuum see Samsung's endeavour to test mobile phones in outer space[3]). Its design, construction, marketing and reception are significantly affected by the associated technology of the day, essentially a form of technological ecology. Possibly, one of the most significant components of VR's technological ecology is computing power. According to a 1990s edition of the magazine InfoWorld,[4] £3400 (roughly £7000 as a 2017 equivalent value) would enable a consumer to purchase a 33MHz processor, 4MB of RAM and a 200MB hard drive. An equivalently priced system as of 2017 would afford the buyer a 10-core 3.1 GHz processor, 32GB of RAM, 12GB proprietary graphics memory and 12TB of hard disk space. Calculating Moore's Law between 1991 and 2017 puts a rough exponential power increase at 2 to the power 13 (a multiplied increase of 8192). Whilst in terms of processing power the huge increase between the generations is not quite 2 to the power 13 and the consensus is now that Moore's Law is no longer applicable for processors (see Waldrop 2016), the increase in RAM is almost spot on and the disk space increase is actually closer to 2 to the power 16 (a multiplied increase of 65,000). However, computing power is not just about the numbers. In practical terms, these advances have dramatically increased the range of functions, in both VR and computing in general, that simply could not have existed in the previous VR generation. Discussed directly in Chap. 9, it is this array of new applications that has markedly contributed to our changing perceptions towards VR. Increased graphical capabilities now enable detailed and interactive three-dimensional models of human organs for the purpose of training surgeons (see Vosburgh et al. 2013). More powerful processors enable complex data sets to be visualised in real time, creating powerful tools that can analyse the human genome (Pavlopoulos et al. 2015). These are only two of many examples that illustrate how contemporary computing technology has facilitated a substantial widening of VR's perceived value, moving away from something only good for playing games and towards acknowledgement as a multi-faceted technology with diverse function and purpose.

Of course, the technological ecology is not limited to the effects of computing power. Advances in hardware, not initially developed for the purposes of VR, have had significant impacts. Prominent examples include digital photography (the essential underlying technology that facilitates most of what we commonly describe as AR), networking technologies and the Internet (facilitating numerous VR functions that include telepresence, teleoperation and shared/multi-user VR experience), human–computer interface technology (including biometric control devices and multisensory feedback systems that yield significant accessibility advantages for VR, can contribute hugely to evoking user presence and also have wide applications for pain distraction and limb rehabilitation) and mobile/smartphone technology (building upon the advantages of networking to create a significantly different approach to VR experience). By the way of increased computing power and integration (see below), these technologies have all contributed to a huge expansion of both what VR can do and how well it can do it. Second-generation VR is no longer just a game or a novel experience, it is also a serious device for learning, building, designing, communicating and more.

Corporatisation

In a technological context, corporatisation describes another significant change to the modern meaning of VR. Whilst the term traditionally refers to transforming ownership products or services from the public state to corporations, here it refers to the way in which VR technology has transitioned from being developed and manufactured by a fledgling company built solely around that product, to but one branch of a much larger, typically multinational, corporation. From the *Stuntmaster* HMD by Future Vision Technologies to the *VFX-1* by Forte, first-generation VR systems were largely (but not entirely—see Nintendo's *Virtual Boy*) the products of smaller companies that dealt exclusively in VR technology. Comparing this against the second-generation VR and most devices, from Microsoft's *HoloLens* to Facebook's *Oculus Rift* and HTC's *VIVE,* evidences how VR technology is now firmly in the hands of corporations with exceedingly deep pockets, generating worldwide revenues of £67.5 billion, £14.2 billion and £3.3 billion, respectively.[5] This influences our perceptions of VR both directly and indirectly. The direct effect is an improvement to the profile of VR. This is due to consumers associating it with a well-known multinational corporation, thereby raising connotations of prestige and higher quality. The indirect impact relates to the earlier discussion on advanced production. With great monetary resources available for investment in VR, the hardware can benefit from the common ability large organisations have to keep costs down through mass production and bulk deals that enable the hardware to be significantly more affordable and more widely available.

Integration

With this aspect of global change, we are discussing ways in which second-generation VR has become more diverse in terms of its hardware platforms by integrating with more established technologies. Noteworthy examples of integration include smartphone-mediated VR, 3D televisions and console game controller compatibility. As we shall discuss further in Chap. 8, integration also incorporates changes for VR developers; 1990s development required those producing the software to either code everything from the ground up or navigate their way around highly complex and inaccessible development environments. Contemporary pipelines, however, integrate VR development with a wide array of toolsets and prebuilt components, from game engines for building the virtual environment to audio source development kits for quickly creating detailed soundscapes. This integration enables VR projects to be exponentially more ambitious, both in scale and in quality. For the wider applications of VR, integrations with numerous physical interfaces and multisensory feedback devices also demonstrate ways in which VR has become substantially more connected to associated technologies. The effect of this in terms of VR's perceived status is that it is no longer an isolated product, but rather a way of interacting with the world by the way of a diverse range of hardware interfaces and software environments. The consequence of this great increase in VR integration transitions us into the final global change: increased acceptance of VR technology within our everyday lives.

Acceptance

This leads us neatly into the next section of this chapter, in which we explore some of the ways in which the virtual has become the everyday. Arguably, all the preceding points (production, application, corporatisation and integration) ultimately feed into acceptance; as VR spreads in its application, the hardware becomes more affordable, the quality reflects user demands, corporate marketing and branding increases desirability, and VR features become embedded in numerous hardware products. This presents us with the most significant difference between the generations of VR. This time, we're ready for it.

How the Virtual Has Become the Everyday

Contemporary VR technology is evolving. As observed in the previous section, its development is benefitting from advances in various associated technologies and it is addressing many past mistakes to acquire worldwide acceptance and join the landmark technologies that have preceded it. We previously noted how various global changes have contributed to increasing public acceptance of the new technology, strengthening VR's bid for ubiquity as consumers are now, more than ever, well positioned to appreciate its value. This section looks to unpack the matter of acceptance a little further, by observing how, through various technological developments, the virtual as a concept has been steadily creeping into our collective consciousness. Over decades, perhaps even centuries, these technologies have been slowly managing our expectations and our values, to normalise the virtual into something we expect, take for granted and consider part of our everyday lives.

The Telephone

Whilst the history of the virtual certainly predates it, one of the first significant instances of virtual technology just so happens to be an auditory example. A story littered with debate and controversy, the origins of the telephone appear to revolve around Mr. John Philip Reis: 'the discoverer, some say, of the art of electrically transmitting speech' (Evenson 2000, p. 209). From tumultuous beginnings during which time multiple individuals laid claim to its discovery, the humble telephone presented substantial appeal by offering adopters the ability to converse with others across great distances with immediacy and clarity, as if that person was actually in the room. With the exception of some rather crude and unsubstantiated comments regarding women's addiction to the telephone (discussed in Rakow 1988), very little academic research exists concerning the societal effects of the fixed-line telephone. That said, the proportion of fixed telephone lines across the developed world peaked at 57% between 2000 and 2001.[6] Looking to their childhoods, many readers will remember their landline home telephone, usually perched upon a telephone table no less, aside a little notepad full of handwritten contacts and a bright yellow directory book. The fixed-line telephone was a massively popular technology that virtualised person-to-person communication, bypassing the requirement for physical proximity. It was utilised so extensively that it caused many to worry that the frequency and volume of person-to-person conversations were dropping because of the telephone. As such, the telephone meets the three requirements that we shall also observe in the subsequent technologies discussed within this chapter that provide further examples of the virtual becoming the everyday. Firstly, it virtualised an existing notion, object or process in a way that overcomes physical limitations. Secondly, it became ubiquitous, a commonplace entity that proliferated society and was utilised across the globe. Lastly, the technology brought with it some form of displacement (i.e. it is used regularly and to the extent that it reduces people's engagement with its 'actual' counterpart).

The Television

Although we may have begun with an auditory example, the visual modality is most certainly not without representation and possesses a powerful exemplifier in the television. An argument could be made that this discussion is failing to acknowledge the photographic camera. However, whilst it certainly meets our three requirements, the television does so to a somewhat greater extent, particularly in terms of ubiquity, and we shall be returning to virtuality and the camera at later points within this book. Beginning in the nineteenth century, the history of television is a complex blend of technological, sociological, cultural and economic narratives that cannot be done justice within a brief couple of paragraphs (for a comprehensive account see Hilmes 2003). In the UK, for the week beginning 25 July 2016, an average of 42 million people (roughly 71% of the population) watched television each day, with an average weekly viewing of 22 hours and 34 minutes watched per person[7]—almost an entire day per week spent in front of the television. Across the world, approximately 1.57 billion households currently possess a television, with projections for 2021 increasing to 1.68 billion.[8]

As human technology develops overall, one noteworthy trend is the substantial increase in the diversity of virtual functions a single new product can present consumers with. This is particularly apparent when considering television which, due to its diverse programming, can present viewers with virtual alternatives to a range of activities, from accessing current events information from television news rather than a print newspaper to satisfying their need for escapist fiction by the way of the drama series instead of the novel. As far back as the 1960s, researchers were addressing widespread concerns that the television was replacing printed media. However, analysis at the time largely reflects the current state of affairs in which people are generally eschewing print in favour of television specifically for the headlines and surface details whilst still relying on text-based outlets for deeper analysis and editorial opinion (Belson 1961). A study by Pommerehne and Kirchgässner (1986) posited that television had contributed to the decline of conventional culture, one specific form of which was a significant reduction in audience attendance at live theatre. Similarly, a review article into the impact of television in the Netherlands has argued that reading has been significantly reduced as a result of increased television watching (van der Voort 1991). As with the telephone, the television transcended the physical limitations of a physical counterpart, including the theatre and the printed newspaper. The benefits of its virtual nature were powerful enough to make the technology an item that sits in almost every living room, and it has raised concerns of displacement that we shall all become slaves to the box.

Network Technology & Social Media

Like the telephone and television before it, the Internet represents possibly the most significant landmark in wider virtual technology. Much contemporary research has been carried out examining individual applications of network technology and their displacement impact. With roughly 2.6 billion users, 205 billion messages sent daily[9] and a capacity for outright replacing letter writing in most contexts, e-mail arguably meets the criteria for virtual technology as part of our everyday. Network technology is also now an established and powerful platform for buying and selling. A study by Shim and colleagues (2000) examined the impact of e-commerce on high street retail. They concluded that Internet-based shopping was a substantial competitor but specifically in cognitive rather than experiential sense. Items that did not require direct sensory evaluation and could be readily assessed by the way of quantitative information (specifications, user reviews, etc.), such as personal computers, were where online shopping was shown to be most disruptive. More recent studies implicitly testify to the impact of Internet shopping by exploring the effect that opening a physical outlet will have on a previously web-only company (see Pauwels and Neslin 2015), reversing the focus and bringing us to a time where e-commerce has become so popular, and opening a physical location has now become something of a curiosity.

Possibly, the most striking example of how the Internet has facilitated the virtual in our everyday is in its effects upon our relationships, both in physical and in psychological terms. With regard to the former, Carl Carlson and Lenny Leonardson once told Homer: 'You're the internet's number one non-pornographic site, which makes you ten trillionth overall'.[10] The figures from this quote may not be particularly accurate, but the sentiment is something that clearly resonates with the truth. Research studies have described the exposure of young people to pornography on the internet 'as a normative experience' (Sabina et al. 2008, p. 691). A 2013 report conducted by the BBC[11] stated that there is much in the way of sensationalism, exaggeration and false claim regarding Internet pornography statistics. That said, their own findings reveal more considered ratios for pornography-related web searches, and sites delivering pornographic content were still notably high, at 14 and 4%, respectively. A review article by Short and colleagues (2012) analyses numerous papers to reveal noteworthy inconsistencies between studies in terms of how Internet pornography was defined and measured, and how conclusions were drawn. Alongside

Short and Colleagues, numerous research studies argue that the impact of Internet pornography is widespread, but it includes both positive and negative effects (see Hald and Malamuth 2008). Whilst the literature largely rejects the demonisation of Internet pornography in favour of a more balanced conclusion, it does not ignore the negative effects, some of which match our displacement criteria for normalising the virtual. Whilst predating Internet-mediated consumption, Zillman and Bryant (1988) conducted a study testing for the effects of pornography upon sexual satisfaction, their results presenting significantly lower ratings for actual sexual experience for the group regularly viewing pornography. An article in New York Magazine (Wolf 2013) voiced the displacement concern directly, arguing that the Internet was responsible for creating a scenario in which women's inability to match unrealistic 'porn-worthy' male expectation was causing men to lose sexual interest in actual women. Whilst this claim of displacement is treated with more caution within academia, its persistence in mainstream media[12, 13] supports the assertion that Internet pornography is potentially another powerful instance of the virtual becoming normalised.

Whilst the Internet presents several other avenues for the normalisation of the virtual in a psychological relationship context, from dating apps to instant messaging, it is social media that arguably presents us with the most prominent example. In the UK, 73% of adults with access to the Internet also use social networking sites.[14] According to Statistica,[15] the number of social media users globally stands at roughly 2.34 billion with the Facebook platform topping the rankings with approximately 1.59 billion 'active users' (a term that, according to Facebook, only accounts for those visiting the site directly and excludes individuals who interact with it via third-party applications[16]). Facebook is first and foremost a tool for social interaction, but has been shown to provide additional functions such as entertainment, self-status seeking and information (Park et al. 2009). Of course, such benefits are present in direct person-to-person contact but arguably not to the same extent and not without requiring greater effort. Should you wish to find out what a previous acquaintance from school was up to, you could give them a call and then meet with them in person, but then of course you might actually have to talk to them. Instead, social media offers us the opportunity to discover contacts' information, compare their network personas against our own (facilitating hierarchical positioning of our own self-status) and be entertained by their recent cat pictures, all with the minimal investment of time and effort.

Social media does of course have a fair share of detractors, and research has raised genuine concerns to try and inform the public as to best use of the technology with regard to health and well-being. Recently, the term 'Facebook depression' has emerged in research papers. In most cases, links between clinical depression and social networking activity have not yielded significant connections (Datu et al. 2012; Jelenchick et al. 2013). That said, a more recent article featured in the Review of General Psychology argues that research up to 2015 may have examined social media users too broadly and that, were we to filter users (by concentrating only on those who have a proportionately large number of friends, frequently and for long period of time read posts from these friends, and these posts are largely of a bragging nature), then hypothetically, a clearer association between social media and depression would emerge (Blease 2015). Of course, correlations with depression are not necessarily direct indicators of virtual normalisation but another issue concerning social media is its effect upon actual person-to-person interactions. As with many of the above technologies, mainstream media is the prevalent source of this issue with headlines implying social media is the 'the death of real-world interaction', 'sabotaging real communication' and 'destroying our social skills'. When read in context, these three quotes are all posed as questions rather than statements and, whilst the headlines heavily imply that the impact of social media face-to-face interaction is both negative and a genuine problem, the articles themselves rarely take a conclusive position. Academic literature does attempt to address the issue, but only a few studies appear to have been conducted and largely reach conflicting conclusions. A paper by Kujath (2011) evidences the argument that social networking provides an enhancement to in-person interaction. Respondents to this study self-reported that they utilised Facebook primarily to maintain existing relationships and form new ones. Very few felt that they had a tendency to communicate with their friends more online than in-person. Conversely, Grieve and colleagues (2013) posit that social media and in-person connectedness did not conclusively correlate and were arguably independent of one another, thereby questioning the assertion that social media either enhances or detracts from real-world socialising. Whether social media actually displaces face-to-face communication is unclear, but the belief that it does is itself very powerful with regard to normalisation of the virtual. Demonstrated by the way that it is being presented in mainstream media, there is a commonly held belief, whether truth or myth, that social media is taking over from in-person communications. It is this belief, first raising objection but slowly becoming accepted, that enables the virtual normalisation to creep in.

Mobile Technology

The mobile (cellular) telephone as a basic communications device offers relatively little in the way of product-unique access to the virtual (telephone calls, texts and e-mails are all functions that are available by other means). Mobile technology does, however, provide two relevant and highly substantial facilities. Firstly, it presents a single interface that houses (potentially) every single one of the above virtual technologies. Telephony, television, digital games, the Internet and social media are all accessible from a single device, and the accessibility and efficiency of this dramatically increase the potential for ubiquity across all sources of virtual. Secondly, the capacity to provide the virtual content without the restriction of physical location (typically the home) provides consumers access from a much wider range of locales and at more times throughout the day (such as when at work or during a commute), thereby also increasing the amount of time spent consuming the virtual content (and conversely, decreasing time spent doing anything else).

Sarwar and Soomro (2013) trace the history of the smartphone back to 1993 and its first incarnation: IBM's 'The Simon', which integrated fax, email and cellular paging functionality (plus personal digital assistant tools and a touchscreen interface) into a mobile phone. Various additions and refinements to both smartphone hardware and software have accompanied the dramatic increases in their commercial success.

The first-quarter UK statistics for 2016 reveal that 71% of adults now own a smartphone. This figure has risen consistently by 5% every year since 2014. As of 2016, roughly 2.1 billion individuals use a smartphone across the world.[17]

Whilst social media is a prominent specific example of how virtual systems have infiltrated the everyday, the smartphone represents a great leap in access to an array of virtual functions and content, arguably transforming the landscape by normalising the virtual in a much more general sense. 'There's an app for that' has become something of a catchphrase and one perfectly summing up the diversity of virtual function that the smartphone offers. Type the phrase into your search engine and articles appear that describe a seemingly infinite number of possibilities. Want to translate language in real time? There's an app for that.[18] Need to find the nearest yoga class to your location from almost anywhere on earth? There's an app for that too.[19] Want to send automated message to your significant other rather because even texting has become too taxing? The list goes on.

Holding the Door Open for VR

The recent publicity and marketing that has encompassed modern VR presents us with various taglines, each promising numerous ways in which the technology will revolutionise our lives. Following Facebook's acquisition of the Oculus Rift in March 2014, Mark Zuckerberg's strategic plan for the technology prioritises VR ubiquity as a primary ambition. This is in terms of both its availability to the masses and its consistent use throughout an individual's typical day. Whilst consumer VR headsets are being marketed predominantly as digital game peripherals, it's difficult not to notice that much of the future-facing ambitions for the technology exist outside the application of digital games and are further examples of displacement. They are not offering entirely new experiences or interactions. Instead, they propose new ways of experiencing and interacting with that which already exists (and most largely that which is commonplace), essentially a virtual means of interfacing with the everyday or alternatively, a means of interfacing with the virtual everyday. Were it presented to us fifty years ago, VR would have been met with an aggressive Luddite rejection and the technology branded as 'unnatural' and 'unreal'. This all emphasises the emergent nature of VR, which exists not in a vacuum, but in our collective consciousness and wider culture. How it is produced, why it exists, what it can do, what came before it and how we believe it will add value to our lives, all contribute to the emergent picture and our understanding of precisely what VR is.

What Is Virtual? Positioning the Virtual, the Actual and the Real

The virtual is everywhere, both in space and in across time. It permeates every facet of our daily lives. As a technology, VR has accumulated a substantial degree of trust as we engage with it under the assumption

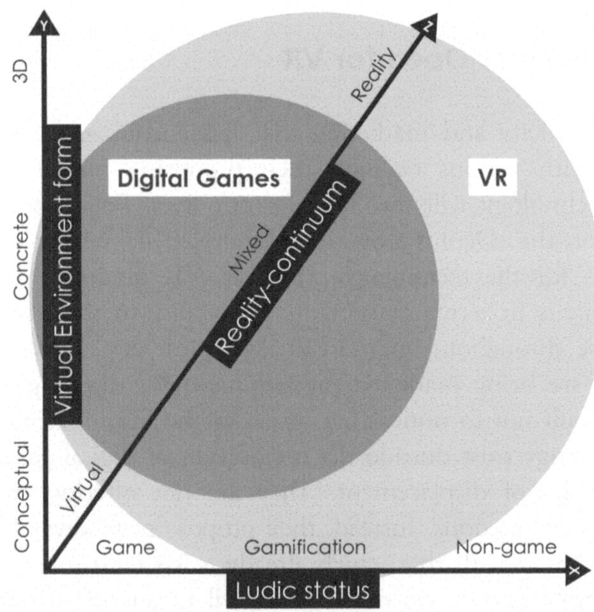

Fig. 2.1 A three-dimensional conceptual framework for VR and digital games

that the virtual is safe and of benefit to us. Despite this trust, what is meant by the virtual, virtuality and VR is something that we still struggle to define and explain. Are these terms descriptive of the same thing? If not, how are they different? Are their definitions stable or constantly changing as both our technology and understanding of the world develop at an ever-quickening pace? This section examines these terms (and a few more besides) to provide a general overview of contemporary thought, primarily describing and contrasting the virtual from technological and organisational positions. The intention is to add another vital component of the emergent framework of VR and to help us better contextualise our subsequent discussion on the role of sound in VR.

The Virtual as Technology

As a fledgling research field, the very definition of VR is a noteworthy point for ongoing debate. Existing definitions range from broad and encompassing to highly specific. They typically centre on experience, with examples including VR as a synthetic experience (Kim 2005), an immersive, interactive experience generated by a computer (Pimentel and Teixeira 1993) and an experience in which the user is immersed in a responsive virtual world (Brooks 1999). A review article by Muhanna (2015) gives us a concise overview of contemporary theorists' definitions by presenting us with five elements that they argue are essential for something to be classified as VR: (1) a virtual world—a medium presenting non-physical space by the way of graphical representations and a set of governing rules and relationships; (2) immersion—qualitative experience denoting attraction to, engagement with, the virtual world; (3) feedback—information received by the user who can interpret the input as response to their own actions within the virtual world; (4) interactivity—primarily the ability to dynamically manipulate and modify elements within the virtual world; (5) participants—the presence of an actual person experiencing the VR. Are these the essential technological characteristics of VR and how do we confidently draw the line between what is VR and what is not? Across many definitions, one feature perceived to be particularly fundamental is digital content, generated by way of a computer. To a lesser extent, three-dimensional objects/environments and user interaction also reliably feature in descriptions of VR (Dioniso and Gilbert 2013). Where we find less consistency is with regard to display type and sensory modality. Whether a true VR system has to provide visual feedback (rather than auditory, olfactory, haptic, etc.) is unclear. That said, reviewing the several definitions of VR presented above, not a single one directly mentions graphics or visual content. Equally ambiguous is whether a HMD or multi-screen display is essential to 'being VR', or whether a single flat-screen display can also be worthy of inclusion. Whilst it is difficult to get a conclusive answer regarding these questions, it is possible to make inferences by examining what technologies various literature sources identify as VR. For example, 'Audio-only VR' identifies a VR system not dependent upon any visual material (e.g. Patterson et al. 2004), whilst 'flat-screen VR' presents VR without head-mounted or multi-screen setups (e.g. Sveistrup 2004). Of course, such instances very much represent the minority and the likely reason that few people go on record to argue that graphical content delivered by the way of a HMD is fundamental to VR, is because such an assertion is quite simply taken for granted.

To complicate things a little further, the virtual as technology extends to incorporate further virtual forms. Alongside his colleague Fumio Kishino in the early 1990s, Paul Milgram (1994) put his name to the reality–virtuality continuum, distinguishing several discrete terms within. Going from left to right, reality describes the natural world, with no virtual content. Augmented reality (AR) blends natural with virtual content, overlaying the physical environment with digital content. Augmented virtuality (AV) is essentially the reverse, a virtual environment that integrates some physical content. Finally, virtuality refers to an entirely virtual environment and is closest to what a popular audience will most commonly interpret as VR. A paper by Koleva and colleagues (1999) posits that the difference between AR and AV is something of a 'whoever comes first takes precedence', meaning an environment that is founded in elements from the 'real world' and has virtual elements superimposed over it would be classified as AR. Conversely, AV begins with a virtual world and then embeds representations of physical objects within it. What counts as an embedded representation is less clearly defined, and there is a possible paradox created when we consider the presence of a human user as an essential component of any VR experience. As the user is a natural component that influences the nature of a virtual world, pure virtuality is impossible and what we traditionally think of as VR would be more correctly identified as AV.

The terms above collectively sit underneath the umbrella term 'mixed reality', with the obvious exception of reality. Differentiating various classes of mixed reality appears throughout academic literature for multiple purposes. Milgram and Kishino (1994) utilised the continuum to categorise different types of visual display in tandem with alternative approaches to interfacing, whilst Billinghurst and Kato (1999) use it to explore the effects of shared experience in network-mediated collaborative work (e.g. a virtual world becomes increasingly mixed reality when the individual interacts within it alongside more and more human players). What is consistent across most of the relevant literature is the position that the virtual is separate and distinct from the real. Any computer-generated content, even if it is overlaying an otherwise physical environment, transforms that environment overall into something that is no longer real.

Another difficulty of definition with regard to VR concerns digital games. The question is raised regarding whether digital games inherently qualify as VR by design and does the 'gamification effect' dictate that VR is inexorably tied to games? Depending on genre, and considering recent developments in motion/gesture-control games interfaces, digital games can be 3D, interactive, responsive to player movement/position and evoke feelings of immersion and presence, matching several of Muhanna's (2015) requirements for VR. Games cannot function without a player to play. They also provide multisensory feedback in both direct and indirect responses to player actions. Consequently, if we do not require HMDs in our qualification, then digital games arguably *are* instances of VR. This should not be extrapolated to assert that VR is equal to digital games, with the truth hidden in the finer details as part of a more complex idea. Figure 2.1 illustrates a broad framework for both differentiating the two terms and part-describing their relationship. The three dimensions of this model reflect

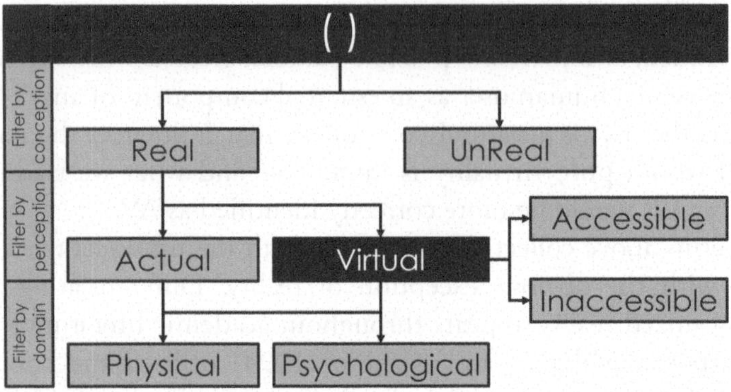

Fig. 2.2 A taxonomical perspective for the virtual based on organisational-virtual theory

the broad means by which we can consider both digital games and VR. 'Ludic status' refers to the extent to which something embodies traditional gameplay properties. 'Virtual environment form' denotes how the content of the virtual world is conceptual (an idea) and concrete (represented in some tangible digital form or a fully three-dimensional rendered environment). Finally, 'reality continuum' (which was described in more detail a few paragraphs prior) describes the degree to which various characteristics of the subject reflect either the natural world or the virtual world. In this framework, VR encapsulates digital games but also extends beyond them, in non-gaming applications and also in terms of environment form, as technical elements such as HMD-based head tracking, and stereoscopics differentiate things which are presented on semi-immersive (single flat screens) and those that are fully immersive (HMD, CAVE, multi-screen display, etc.). This framework is useful in that it helps delineate some of the key factors that contribute to our understanding of VR from a technological perspective. Whilst certainly not the whole story, these points still form a significant part of the emergent puzzle.

Overall, the virtual from a technological perspective offers us value in several notable ways. Firstly, it gives us an appreciation of how popular culture views the virtual (even if this understanding is flawed) and, by proxy, helps us to elucidate some of the consumer expectations and preconceptions, arguably an important asset for anyone wanting to understand user experience in VR or those working to develop VR applications. Secondly, these perspectives bring together the key elements of VR as an object and help us to compartmentalise some of the VR variables that pertain to VR hardware and software. Whether visual feedback is or is not essential for a system to be classified as VR may not be an important question, but it is one that indirectly highlights the notion of VR systems built around alternative sensory modalities. As a result, such a question still manages to offer us a highly interesting avenue for further investigation. To summarise, the virtual from a technological position revolves around the contemporary computer technology commonly referred to as VR. It is a system-side focus. It gives us a sense of VR as an object or a system, with empirical qualities that can be measured and manipulated. It includes the display, forms of feedback, objective aspects of immersion, controller, tracking methods and interface hardware. What it does not do is describe the phenomenological (i.e. user experience) relationship between the system and the user. The virtual is an opportunity, afforded to us by technology, to experience other worlds. These worlds may be entirely virtual or they may be mixed with elements of the physical world. Lastly, the virtual from this perspective is in opposition to reality, a key point that the following discussion presents an alternative to.

The Virtual as Organisational Theory (Virtuality)

In most books that explore VR, the perspectives under discussion tend to fit within the technological perspectives outlined above. The organisational position, however, encapsulates additional and, crucially, broader thought on the virtual that is partially, but not entirely, oppositional to the technological perspective. Some of the most prominent examples of the organisational position can be found within the works of Gilles Deleuze and his exploration of virtuality. Deleuze's virtuality is markedly unattached to technology and is instead a part of broader philosophical thought. In an article by Linstead and Thanem (2007), the virtual, as understood by Deleuze, is succinctly described as an integral piece of reality, *not* an opposition to it: '[…] the virtual *is* everything and […] is *in* everything—a principle of connectedness' (p. 1492). VR is therefore encompassed within the virtual. Although the virtual is a component of reality, it is not equal to reality and sits alongside its counterpart, 'actuality'. For something to be actual, it must be consciously attended to by a perceiver in the precise here and now, what the individual perceives to be the present moment in both space and time. The virtual is everything else. It is everything that contributes to the present actualisation and every actualisation of the past and future. This goes against two significant assertions of the technological perspective: that the virtual and the real are distinct and separate entities, and that the virtual is born from (and inherently tied to) computer technology.

The explanation for the virtual as part of reality centres around two primary arguments. Firstly, in our experience of the world, immediate sensory input is inseparable from 'projection'—'a throwing of existence ahead of itself' (Roe 2003). In simple terms, we cannot at any point or circumstance help but consider (be it conscious or subconscious) the future. This is because without such consideration, any action within the world becomes impossible. How can you reach out to grasp an object with no concept of the outcome of your action? How can a writer type any text without some concept of the completed script as it is being written? Our actions and our being are inseparable from our perception of them, and they are also rooted in space and time. The actual can only exist in association with the virtual, making both equal partners in propagating reality.

The second argument posits that virtuality is real because of its observable (and often quantifiable) impact upon our existence. As Ramiller states: '[t]he virtual […] is not simply anticipatory and prescient, but also generative' (2007, p. 355). This idea is reflected upon in popular culture, in *Imagination* Land (Parker 2007) an episode of satirical cartoon South Park, where the protagonists reflect on the powerful impact that characters and themes from fiction can have upon their daily lives. Although this is not a perfectly accurate depiction because the virtual is not limited to the imagination, a similar sentiment is presented nevertheless. An imaginary entity that inspires change in the physical environment is powerfully real. Equally so, the virtual does not merely accompany the actual, it determines its precise form and launches it into actuality. It is also, powerfully real.

Understanding the virtual from an organisational position becomes even more complex when we consider that to do so additionally requires us to make sense of nothingness and existence, alongside reality, virtuality and actuality. Embodying the emergent perspective, these additional entities all intertwine with the virtual as it is positioned within a form of existential ecology that we can use to better explain virtual itself.

If there was a grand structure for *everything* that could be exhibited in a framework (leaving aside the point that does so is a rather arrogant thing to do… see Fig. 2.2), the uppermost level of the structure would, theoretically, need to be entirely undefinable as any descriptor would inescapably generate a conceptual space outside itself. If we were to entitle it as 'existence', then it follows that something could somehow be positioned outside existence; if we were to call it 'reality'… and so on. Taking a step down the hierarchy, we can position the 'real' (existence) and the 'unreal' (nothingness), the distinction between the two being quite simply that the former can be conceived of whilst the latter escapes from physicality and contemplation (at any time and by any being, not just humans). The unreal is not comparable to the unknown (referred to in the taxonomy as 'inaccessible') because the latter refers to entities that defy conception by the individual at the present, but that nevertheless have the potential to be conceived of by someone or something at some point in time. For example, the Internet would have been inconceivable in the nineteenth century and, during its first conception circa 1960, remained inaccessible to all but a few. Today, it can be conceived of by almost every human on the planet.

Whilst the unreal is not subject to further subdivision, reality constitutes the virtual and the actual, two terms that are differentiated in relation to perception. To be classified as actual, an entity must be being perceived within the here and now—as determined by the attendance of at least one perceiver. By contrast, the virtual is that which exists outside immediate perceptual space. Both the actual and the virtual can be subdivided into physical (embodies matter or energy) and conceptual (non-physical) variants. A physical entity is not automatically actual; without attendance/perception, it is virtual. In terms of movement between these states, transition between the real and the unreal is essentially inconceivable because the unreal is itself inconceivable, and therefore, we cannot comprehend any entity transitioning between the two. We can, however, regard movement within the lower levels of the hierarchy. Something that was once virtual may become actual, subject to being consciously perceived, and then return to being virtual post-perception.

Virtuality theory, much like its technological counterpart, presents us with value when applied to VR design. Firstly, the notion of projection raises important considerations relevant to user experience in VR, specifically that users will not (and cannot) experience VR stimuli in isolation and that whatever is presented to them comes tethered to their prior experience and future expectation in addition to all connected aspects of space and time. By presenting human experience as a complex interaction of vast numbers of variables that are unique to the individual, it also highlights the inevitable limitations of homogenous design practices and assumptions that designer intent will automatically match user experience. Thirdly, the organisational perspective encourages us to consider more ecological approaches to design. It encourages us to more consciously attend to the background elements of our design. For example, the increase in body temperature we experience when wearing a HMD, or the subtle ambient soundscape in the background of a VR application, may not seem important at first thought, but addressing such points within the overall craft of VR design has great potential to dramatically improve the overall experience.

Virtuality is the underlying theory that leads to the conception of sound as an emergent perception (Grimshaw and Garner 2015), and it is now the foundation upon which this book is built. The collection of components presented in the emergent frameworks at the end of this book represents the individual pieces of the virtual as it pertains to the emergent actualisation of VR sound. When many of us consider VR sound, it is likely that the actualisation is formed of a relatively small collection of virtual components, as many others are presently inaccessible (i.e. they are simply not known to that individual). This book is itself, a means of helping the reader to enhance their actualisation of VR sound by making accessible some that which was once inaccessible.

The (hopefully not too arrogant) intention is that the reader, upon completion of this book, will actualise a renewed concept of VR sound. Possibly one that is very different to that which emerged prior to reading. Ideally one that is richer yields greater understanding and can facilitate better design in both VR and VR sound.

To summarise, the virtual can be thought of as a technological concept, built of objective and tangible elements to facilitate experiences beyond that of the physical world. From broader organisational perspective, the virtual is a fundamental underpinning of perception. It is the mechanism by which our unique experiences of the world come forth and cycle in an endless, infinitely complex, loop of sentient existence. Ultimately, both perspectives reveal significant value, both regarding our conceptual understanding of VR and as sources of practical design guidance. To acknowledge both is to position VR across the virtual, the actual, the physical and the psychological. All at once, VR encapsulates a physical component in its hardware and radiating stimuli, a psychological presence in our individual conceptualisation of VR, a virtual aspect incorporating every relatable unit of physical and psychological content that shapes experience and an actuality—our felt experience of VR within the here and now. We certainly have much to consider.

Chapter Summary and References

This chapter has begun our exploration of VR and VR sound as emergent concepts by reviewing different ways in which it can be defined and understood, but also how perspectives are rapidly changing. Two alternative perspectives are discussed, one asserting the virtual as technology and the other as an organisational theory. The central argument here is that what we understand as virtual should not be solely restricted to an experience of technology but rather a component of all experience. We can observe the broader applications of VR are revealing hundreds of new functions, whilst powerful corporatisation is enabling adaptive design and manufacturing. Better integration is contributing also, as VR reaps the benefits of advances in various technologies and our perception of VR is consequently becoming increasingly positive.

This positivity has fuelled production and implementation of VR on a trajectory that arguably is moving the technology towards ubiquity. When we consider this alongside how the virtual as both concept and an experience has permeated so many facets of our everyday lives, from the telephone to social media, the forthcoming acceptance of VR as a permanent mainstay technology is a pretty safe conclusion.

Whilst the discussion so far has admittedly focussed upon VR, this is of course relevant and leading into VR sound. Within an emergent framework of VR sound, all the above points are still of significant relevance. How we interpret and attach meaning to auditory content in a virtual environment is inexorably connected to VR and the broader issues that contribute to its nature. The organisational theory with which this chapter closed has particular relevance by the way of it covering some of the foundational concepts from which a framework of sound as an emergent perception is constructed. This framework is addressed in the following chapter as we explore the notion of Sonic Virtuality, the emergent understanding of auditory perception that was the main influence of the theoretical positions that are presented throughout this book.

Notes

1. Full definition of virtual reality. Merriam-Webster Online Dictionary: www.merriam-webster.com.
2. Barriers to VR. Variety (2016) *How Google is Tricking us All to Embrace Virtual Reality*, http://variety.com/2016/digital/news/google-daydream-vr-ecosystem-1201781195/.
3. Testing mobile phones in space. Techradar.com (2016) *Could your smartphone survive in space?* http://www.techradar.com/news/phone-and-communications/mobile-phones/could-your-smartphone-actually-survive-in-space—1220275.
4. ’90s computing power. InfoWorld (1991) *Northgate Computer Systems* (advertisement), 13(19), 62.
5. VR technology revenues. Financial Times Online (2016) *Equities*, http://markets.ft.com/data/equities.

6. International Telecommunications Union, http://www.itu.int/ITU-D/ict/statistics/ict/.
7. Weekly TV viewing statistics. BARB (2016) *Weekly* Viewing, http://www.barb.co.uk/viewing-data/weekly-viewing-summary/.
8. TV statistics. Statistica (2016) http://www.statista.com/statistics/268695/number-of-tv-households-worldwide/.
9. Email statistics. Radicati (2015) http://www.radicati.com/wp/wp-content/uploads/2015/02/Email-Statistics-Report-2015-2019-Executive-Summary.pdf.
10. Quote from *The Simpsons*.
11. The reality of internet pornography. BBC (2013) Web porn: Just how much is there? http://www.bbc.co.uk/news/technology-23030090.
12. Negative effects of the internet on physical relationships. Mail Online (2011) http://www.dailymail.co.uk/sciencetech/article-2051902.
13. Men's Health (2012) http://www.menshealth.com/sex-women/porn-debate.
14. Social networking statistics. Ofcom (2016) Facts and Figures, http://media.ofcom.org.uk/facts/.
15. Social networking statistics. Statistica (2016) http://www.statista.com/topics/1164/social-networks/.
16. Facebook users. Recode (2015) http://www.recode.net/2015/11/5/11620408/facebook-just-changed-its-definition-of-active-user.
17. Smartphone statistics. Statistica.com, http://www.statista.com/statistics/330695/number-of-smartphone-users-worldwide/.
18. Mobile app for real-time translation. Word Lens, http://word-lens-translator.en.softonic.com/android.
19. App for finding nearest Yoga class. Yoga Trail, http://www.yogatrail.com/blog/the-yogatrail-app/.

References

Belson, W. A. (1961). The effects of television on the reading and the buying of newspapers and magazines. *Public Opinion Quarterly, 25*(3), 366–381.

Billinghurst, M., & Kato, H. (1999). Collaborative mixed reality. In *Proceedings of the First International Symposium on Mixed Reality*, pp. 261–284.

Blease, C. R. (2015). Too many 'friends', too few 'likes'? Evolutionary psychology and 'Facebook depression'. *Review of General Psychology, 19*(1), 1.

Bricken, W. (1990). Learning in virtual reality. SIGGRAPH 1990.
Brookings. (2016). *How technology is changing manufacturing*. https://www.brookings.edu/blog/techtank/2016/06/02/how-technology-is-changing-manufacturing.
Brooks, F. P. (1999). What's real about virtual reality? *IEEE Computer Graphics and Applications, 19*(6), 16–27.
Cruz-Neira, C., Sandin, D. J., DeFanti, T. A., Kenyon, R. V., & Hart, J. C. (1992). The CAVE: Audio visual experience automatic virtual environment. *Communications of the ACM, 35*(6), 64–73.
Datu, J. A. D., Valdez, J. P., & Datu, N. (2012). Does Facebooking make us sad? Hunting relationship between Facebook use and depression among Filipino adolescents. *International Journal of Research Studies in Educational Technology, 1*(2), 83–91.
Dionisio, J. D. N., & Gilbert, R. (2013). 3D Virtual worlds and the metaverse: Current status and future possibilities. *ACM Computing Surveys (CSUR), 45*(3), 34.
Evenson, A. E. (2000). *The telephone patent conspiracy of 1876: The Elisha Gray-Alexander Bell controversy and its many players*. USA: McFarland & Company Inc.
Grieve, R., Indian, M., Witteveen, K., Tolan, G. A., & Marrington, J. (2013). Face-to-face or Facebook: Can social connectedness be derived online? *Computers in Human Behavior, 29*(3), 604–609.
Grimshaw, M., Garner, T., & Garner, T. A. (2015). *Sonic virtuality: Sound as emergent perception*. USA: Oxford University Press.
Hald, G. M., & Malamuth, N. M. (2008). Self-perceived effects of pornography consumption. *Archives of Sexual Behavior, 37*(4), 614–625.
Heim, M. (1994). *The metaphysics of virtual reality*. USA: Oxford University Press.
Hilmes, M. (Ed.). (2003). *The television history book*. UK: British Film Institute.
Iwata, H. (1990). Artificial reality with force-feedback: Development of desktop virtual space with compact master manipulator. *ACM SIGGRAPH Computer Graphics, 24*(4), 165–170.
Jelenchick, L. A., Eickhoff, J. C., & Moreno, M. A. (2013). "Facebook depression?" Social networking site use and depression in older adolescents. *Journal of Adolescent Health, 52*(1), 128–130.
Jerald, J. (2015). *The VR book: Human-Centered design for virtual reality*. USA: Morgan & Claypool.

Kim, G. J. (2005). A SWOT analysis of the field of virtual reality rehabilitation and therapy. *Presence, 14*(2), 119–146.

Koleva, B., Benford, S., & Greenhalgh, C. (1999). The properties of mixed reality boundaries. *ECSCW'99* (pp. 119–137). The Netherlands: Springer.

Krueger, M. W. (1995). Olfactory stimuli in virtual reality for medical applications. *Interactive technology and the new paradigm for healthcare* (pp. 180–181).

Kujath, C. L. (2011). Facebook and MySpace: Complement or substitute for face-to-face interaction? *Cyberpsychology, Behavior, and Social Networking, 14*(1–2), 75–78.

Linstead, S., & Thanem, T. (2007). Multiplicity, virtuality and organization: The contribution of Gilles Deleuze. *Organization Studies, 28*(10), 1483–1501.

Longo, R. (1995). *Johnny Mnemonic*. USA: Tristar Pictures.

Milgram, P., & Kishino, F. (1994). A taxonomy of mixed reality visual displays. *IEICE TRANSACTIONS on Information and Systems, 77*(12), 1321–1329.

Muhanna, M. A. (2015). Virtual reality and the CAVE: Taxonomy, interaction challenges and research directions. *Journal of King Saud University-Computer and Information Sciences, 27*(3), 344–361.

Nielsen, I., Dang, Q. V., Bocewicz, G., & Banaszak, Z. (2015). A methodology for implementation of mobile robot in adaptive manufacturing environments. *Journal of Intelligent Manufacturing*, 1–18.

Park, N., Kee, K. F., & Valenzuela, S. (2009). Being immersed in social networking environment: Facebook groups, uses and gratifications, and social outcomes. *CyberPsychology & Behavior, 12*(6), 729–733.

Parker, T. (2007). *Imagination land*. USA: South Park. Comedy Central.

Patterson, D. R., Tininenko, J. R., Schmidt, A. E., & Sharar, S. R. (2004). Virtual reality hypnosis: A case report. *International Journal of Clinical and Experimental Hypnosis, 52*(1), 27–38.

Pauwels, K., & Neslin, S. A. (2015). Building with bricks and mortar: The revenue impact of opening physical stores in a multichannel environment. *Journal of Retailing, 91*(2), 182–197.

Pavlopoulos, G. A., Malliarakis, D., Papanikolaou, N., Theodosiou, T., Enright, A. J., & Iliopoulos, I. (2015). Visualizing genome and systems biology: Technologies, tools, implementation techniques and trends, past, present and future. *GigaScience, 4*(1), 1.

Pimentel, K., & Teixeira, K. (1993). Virtual reality through the new looking glass. Windcrest: Intel.

Pommerehne, W. W., & Kirchgässner, G. (1986). *The decline of conventional culture: The impact of television on the demand for cinema and theatre performances*. Berlin, Finanzpolitische Forschung: Freie University.

Rakow, L. F. (1988). Women and the telephone: The gendering of a communications technology. *Technology and women's voices: Keeping in touch*, 207–229.

Ramiller, N. C. (2007). Virtualizing the virtual. *Virtuality and virtualization* (pp. 353–366). USA: Springer.

Ritz, L. T., & Buss, A. R. (2016). A framework for aligning instructional design strategies with affordances of CAVE immersive virtual reality systems. *TechTrends*, 1–8.

Robinett, W. (1994). Interactivity and individual viewpoint in shared virtual worlds: The big screen vs. networked personal displays. *ACM SIGGRAPH. Computer Graphics, 28*(2), 127–130.

Roe, P. (2003). That-which-new media studies-willbecome. *Fibreculture Journal, 2*.

Ronchi, E., Nilsson, D., Kojić, S., Eriksson, J., Lovreglio, R., Modig, H., & Walter, A. L. (2016). A virtual reality experiment on flashing lights at emergency exit portals for road tunnel evacuation. *Fire Technology, 52*(3), 623–647.

Sabina, C., Wolak, J., & Finkelhor, D. (2008). The nature and dynamics of Internet pornography exposure for youth. *CyberPsychology & Behavior, 11*(6), 691–693.

Sarwar, M., & Soomro, T. R. (2013). Impact of smartphone's on society. *European Journal of Scientific Research, 98*(2), 216–226.

Shim, S., Eastlick, M. A., & Lotz, S. (2000). Assessing the impact of Internet shopping on store shopping among mall shoppers and Internet users. *Journal of Shopping Center Research, 7*(2), 7–43.

Short, M. B., Black, L., Smith, A. H., Wetterneck, C. T., & Wells, D. E. (2012). A review of Internet pornography use research: Methodology and content from the past 10 years. *Cyberpsychology, Behavior, and Social Networking, 15*(1), 13–23.

Smith, L. (1993). *A virtual murder*. Murder, she wrote. USA: Universal Television.

Steuer, J. (1992). Defining virtual reality: Dimensions determining telepresence. *Journal of Communication, 42*(4), 73–93.

Sveistrup, H. (2004). Motor rehabilitation using virtual reality. *Journal of Neuroengineering and Rehabilitation, 1*(1), 1.

van der Voort, T. H. (1991). Television and the decline of reading. *Poetics, 20*(1), 73–89.
Vosburgh, K. G., Golby, A., & Pieper, S. D. (2013). Surgery, virtual reality, and the future. *Studies in Health Technology and Informatics, 184*, vii.
Waldrop, M. M. (2016). The chips are down for Moore's law. *Nature News, 530*(7589), 144.
Wolf, N. (2013). The porn myth. *New York Magazine*. http://croker.harpeth-hall.org/Must%20Know/Psychology/WolfPornography.pdf.
Zemeckis, R. (1989). *Back to the future: Part II*. USA: Universal Pictures.
Zillmann, D., & Bryant, J. (1988). Pornography's impact on sexual satisfaction. *Journal of Applied Social Psychology, 18*(5), 438–453.

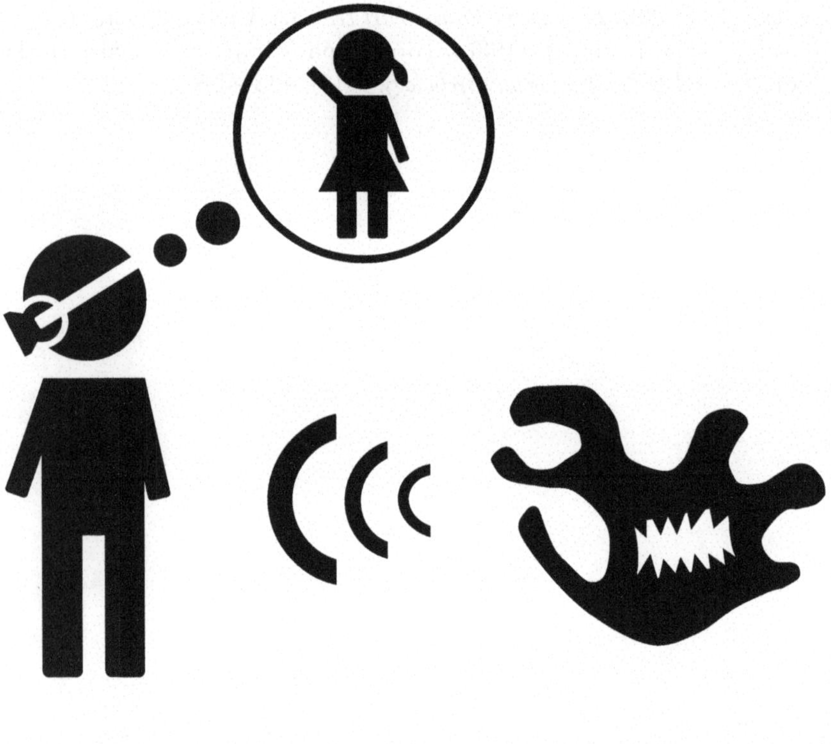

3

Sound and the Virtual

Studies examining the fundamental nature of sound have furnished us with a range of perspectives, most of which find various points of contention with one another. How sound can be defined and characterised, where sound can be located (if anywhere) and what sound means is all subject to contrasting theories, from the empiricism of acoustic perspectives to the phenomenalism of Sonic virtuality. This chapter continues our journey towards building of an emergent framework of VR sound with a critical examination of various perspectives on sound. Commencing with a review of the most prevalent positions and debates in the conceptualisation of sound, this chapter progresses to assert the importance of a holistic and contextually flexible understanding. Embodied cognition, modes of listening and other concepts highly pertinent to auditory perception are also discussed. Illusory sound and auditory imagery illustrate the 'sound without a sound wave' problem, highlighting limitations amongst the more commonplace perspectives, before we examine emergence theory and finally Sonic virtuality. The focus here is very much on the nature of sound as it pertains to the organisational aspects of the virtual. This discussion does address the technological elements to some extent, but this aspect is covered in greater detail in Chap. 4.

T.A. Garner, *Echoes of Other Worlds: Sound In Virtual Reality*,
Palgrave Studies in Sound, DOI 10.1007/978-3-319-65708-0_3

Perspectives on Sound: What (and Where) Is It?

Establishing an accurate conception of sound is somewhat trickier than one might first think. The chief reason for this difficulty originates from the complex relationship between sound as a physical form and sound as a sensual experience. As Goldsmith (2015) observes, 'many of the impressions sound makes on us are related to its physical parameters but not reducible to them' (p. 10). When, for example, a piano key is depressed, we can draw associations between the amplitude and frequency of the waveforms emitted and our sensation of hearing the sound, but this information cannot fully explain the sound as we hear it. As anyone who has spent many a tedious hour working on a piano synthesiser can testify, trying to accurately recreate the analogue sound of an acoustic upright by manipulating amplitude and frequency data is far from easy. The following sections examine various conventional perspectives on sound, as we look to how contemporary thought on the relationship between the physical stimuli and the sensuous experience of sound might be reconciled.

The Acoustic Definition

To acquire what is arguably the most standard understanding of sound, we can refer to the Oxford English Dictionary. The first definition contained therein explains the term as: '[v]ibrations that travel through the air or another medium and can be heard when they reach a person's or animal's ear'[1]. Within this definition, the 'that which can be heard' and 'the area or distance within which something can be heard' are two key elements that imply sound is equal to a sound wave and its source, with the latter suggesting that sound is (at least in part) a volume of physical space that supports the propagation of sound waves. These descriptions reflect what is typically referred to as the acoustic definition of sound (sound is a sound wave). One further point that we can take from the dictionary definition is the description of sound as 'continuous and regular vibrations', distinct from noise. This suggests that for something to be sound, it must possess a degree of organisation.

What is unclear from the above definition is the precise nature of the relationship between the receiver (listener) and that which is heard. A hearer or listener is a consistent element across dictionary definitions. The confounding implication of this being, without the presence of someone to hear it, the travelling vibrations alone are not sound. This conundrum has become an established trope of sorts across popular culture, with a noteworthy satirical example being: 'If a man speaks in the forest, and there is no woman to hear him, is he still wrong?' (O'Connell 1999). For George Berkeley (1874/1710), perception precedes reality. This is a phenomenological and existential sentiment that has been reflected in the perspectives of cultural icons including Andy Warhol and Bob Dylan. Here, sound cannot exist without a listener who is able to perceive it. From this perspective, a sound wave with an imperceptibly low amplitude (or of a frequency that is outside the listener's audition range) that cannot stimulate the hair cells that adorn the cochlea is therefore *not* a sound. Likewise, a vibrating object or multiple objects interacting to cause the emanation of sound waves cannot themselves be classified as sound in the absence of a listener.

Even if we acknowledge the requirement of a listener for the presence of sound, what the requirements of that listener are still remain ambiguous. Does the sound wave reaching the ear constitute sound? What if it traverses the inner ear, 'swims' the length of the cochlea and achieves the feat of mechanotransduction, deflecting the hair cells to generate a neural impulse; is that sound? Or is sound only that which occurs when the auditory cortex is activated? Bearing in mind the extremely complex wiring of the brain, what constitutes a legitimate activation? Does this progression of activity need to transcend from the physical domain to the psychological, that is, can sound only be that which has been consciously attended?

Returning to the dictionary definitions[1], whether a sound is organised or not is indicative of a perceptual process because the judgment of organisation is interpretive. What is music to the child can be noise to the parent, and vice versa. For some, white noise (a by-definition random audio signal with equal intensities across its frequency spectrum) is entirely devoid of organisation and quite simply noise, whilst for others it is closely reflective of waves gently crashing against the shore. So if we are to agree that sound requires the presence of a listener, does that describe simply the physical sensory reception of the sound wave or the subjective perception of sound as an experience? Finally, if these entities under scrutiny cannot be called 'sound', what do we call them? From this, we now step forward, entangled in these confounding questions, to explore the work of those foolhardy enough to go searching for the answers.

The following subsections discuss six interrelated theoretical positions on sound. The first three are identification theories that define sound in terms of what it is. The second three are location theories, representing alternate assertions of where sound is. Across the six, we can observe some perspectives that support the acoustic view of sound and others that argue against it. This section briefly outlines these perspectives and discusses some of the questions and criticisms that have been levied against them.

Sound as an Object

Approaching sound as an object is discussed by Pasnau (1999), who suggests that sound is first and foremost a spatial entity and that we as listeners inherently locate objects (and ourselves by proxy) in three-dimensional space upon receiving auditory information. The train of thought follows that we bind a received sound wave to a source object as an automatic perceptual process. Subsequently, that object is both the 'what' and the 'where' of sound, connecting the notion of sound as object to distal theory (discussed below). At the time of writing this very line, a dog can be heard barking in a garden outside. Here, the object is the dog which, according to object theory, is the most significant percept and therefore both the identification of the sound ('*the sound is a dog*') and its location ('*the sound is at the dog*').

Sound as an Event

Following a similar progression of logic, O'Callaghan (2009) identifies sound as an event and argues that it is the interaction between physical objects that best defines sound ('event theory'). In terms of localisation, event and object theory are subtly different, with the former positioning sound *about* the source object(s). Like Pasnau previously (1999), O'Callaghan supports event theory is by way of asserting its spatial aspect. For O'Callaghan then, the sound is therefore not the dog but the bark and it is located about the object as opposed to *at* it.

Sound as Object and Event

In a similar vein, Nudds (2009) suggests that any sound we attend to is inherently tied to both an object and an event. He asserts that the fundamentals of perceptual processing dictate that we inescapably formulate object and event-related information as automatic responses to attended sound wave stimuli. Nudds illustrates this point with straightforward physical interactions, such as a stone being dropped into water. In this example, the sound reveals both the event (i.e. the act of the stone dropping into water) and several properties of the object being dropped, such as its size and the speed at which it was travelling before impact. Both object and event are perceptually 'equal', and Nudds later reinforces this position by asserting that the laws of physics fundamentally insist upon a precise causal relationship between the physical properties of the object(s) and the event on one side of the equation, and the properties of the resultant sound waves on the other. Take a basic hand clap. Were all physical properties (size of objects, proximity to listener, etc.) fixed except the directional force of the objects' clapping movements, the sound wave produced would possess greater amplitude at the point of hearing. The physical causality of sound would imply that you quite simply cannot clap your hands harder to produce a softer sound. As we shall discover later in this chapter, such an assertion may not be quite as secure as one would first think.

Proximal, Medial and Distal Theory

To help us better conceptualise these different perspectives regarding where sound is, Casati and Dokic (2009) helpfully cite names for three alternative locations (that make up the latter half of our subsection heading): proximal, distal and medial. Proximal theory positions sound where the listener is (sound at the point of hearing/reception), whilst distal theory asserts that sound exists at the source (the vibrating object). Finally, medial theory places sound in the space in-between and is comparable to the definition of sound as a property of an acoustic medium. Considering the above positions of

Pasnau, O'Callaghan and the others, the distal theory appears to be the singular point that they all most agree upon. Despite this consensus, positioning sound at the medial point (sound is a sound wave) is more widely accepted amongst the broader academic community whilst arguments for distal theory are likely to raise the most eyebrows amongst academics.

Let us take a simple scenario to elucidate the previous paragraph: A clock upon a wall is steadily ticking, its smallest hand making six-degree movements each second. The ticking of the hand and underlying mechanism of cogs snapping from one position to the next creates an expulsion of energy that disturbs the position of adjacent particles, both in the materials of the clock and the surrounding air volume. One particle's displacement in turn affects another with a trajectory that we define as a sound wave. Next, provided the listener is within audible range, these sound waves reach the ear with enough amplitude to stimulate the hairs that adorn the cochlea. This prompts the conversion of the sound wave energy into electrical impulses that travel the fibres of the auditory nerve en route to the primary auditory cortex within the temporal lobe of the brain. This scenario illustrates quite a journey from origin at the source to 'final destination' at the cortex, but as we look along this path, precisely where is the sound? Distal theory would place the sound at the source and would describe the sound as the ticking of the clock (object/event). Medial theory would position the sound within the acoustic medium between the source and the listener, acoustic theory being a relative definition of sound. Finally, proximal theory locates the sound at the listener (sound is the hearing of a sound wave). Are all (or any) of these points true and if so to what extent? Is every point along the journey worthy of classification as 'sound', and how can we explain the difference between sound at the source object, sound during particle displacement and sound during transduction from sound wave energy to neural impulse?

O'Callaghan (2011) examines two positions on sound, the 'instantiation view' and the 'causation view'. The former of these describes sound as properties of the source (compatible with distal theory), whilst the latter equates sound with an effect of the source (both distal and medial). Instantiation is fully encapsulated by the source object whilst causation incorporates both object and event. O'Callaghan's primary criticism of instantiation is that it cannot explain acousmatic sound, in which the listener receives and perceptually processes the sound wave without any consideration or knowledge of the source (a commonplace design technique in horror films). A prominent example of this is Pierre Schaeffer's Musique Concréte (Schaeffer et al. 1967), a compositional style that intentionally strips source from sound to create more abstract experiences. O'Callaghan also questions the causation view, asserting that the perception of an effect does not necessarily reveal its cause: '[s]eeing smoke is not seeing fire; touching footprints is not touching a foot' (O'Callaghan 2011: p. 403).

Aspatial Sound

Whilst spatial discrimination is commonplace in everyday life, Casati and Dokic (2009) provide a compelling counter argument by referring to the act of listening underwater. As they point out, 'sound travels at about 1500 m/s in water, more than four times faster than air' (p. 102). The effect of this is that, for humans, sounds heard underwater are received with such infinitesimal latency between the two ears that the sounds defy localisation (they are aspatial). O'Callaghan (2010) has a response to this argument, however, acknowledging the underlying rationale but questioning Casati and Dokic's conclusion. For O'Callaghan then, whilst circumstances may exist in which sound cannot be perceptually positioned in space, this does not describe aspatial sound, but rather 'the experience of sound all around' (p. 136).

Elucidating the theory of proximal sound, O'Shaughnessy (2000) posits that sound cannot be heard at the source, only at the ear of the listener. In this theory, human perception creates an illusion of location as we inherently process the sound wave to determine its point of origin. Casati and Dokic (2009) argue that, in terms of localisation, there is a distinction between where a sound is and where it is coming from; it is our confusion between these two that causes our perception of sound being at source (distal). In addition to proximal, distal and medial positions, Casati and Dokic also bring a further notion to the table with aspatial theory, in which sound does not occupy any location within physical space and is effectively nowhere. Casati and Dokic are careful to differentiate the aspatial theory of sound (i.e. sound is not physically anywhere) from an aspatial theory of auditory perception (i.e. we do not hear sound to be anywhere). In a book now appreciated as a modern classic of philosophy, *Individuals* (1959), Strawson imparts arguably one of the more mind-bending thought experiments of the twentieth century: the 'No-Space world'. Taking influence from Kant and Aristotle before him, the No-Space proposition describes a world devoid of space, with time as its ontic foundation. For Strawson, only the auditory modality would be able to exist, positioning sound as an aspatial entity. The correctness of this proposition is certainly not without challenge, the most prominent concern regarding how, in a world without space, beings could distinguish themselves from sound as, by rationale, in a world where only sound can exist; such 'beings' would themselves have to be sound. As Santarcangelo and Terrone (2015) point out, Strawson's No-Space is not intended as a statement upon the empirical nature of auditory perception, but rather as a means to explore ontological frameworks of space and time. They continue, stating that this hypothetical context highlights the process-like nature of sound, as opposed to sound as an object. For Santarcangelo and Terrone, this supports the phenomenological distinction of vision from audition. Their argument is that, between two separated instances of visual perception, the individual compares image information to identify 'the same' object, whilst two separated instances of sound are perceived as *parts* that form a process.

Sound as a Phenomenon

Attempts have been made to bring a little reconciliation to the fore by bringing the various viewpoints together within a single conceptual framework. Referred to as the ecological approach to sound (Gaver 1993), this model describes sound as a phenomenon, acknowledging the distal (object/event), medial (event/medium) and proximal (listener) perspectives as pieces of a whole. This more holistic approach to sound contains several notions that we shall discuss in greater detail within the next section of this chapter. First, we can go back to the ticking clock example. This emphasises how understanding sound more holistically as a process does encapsulate the pieces of the example more fully, whilst also acknowledging the state changes from origin to reception. The vibrating second hand and cogs at the distal position *is* sound inasmuch as it is a part of the process. Likewise, the reception of the resultant sound wave at the proximal location (listener) is equally 'sound', but simply at a different stage of the process. What is unclear about equating sound with an overall process or phenomenon is that sound as a term loses precision in its descriptive power. If sound identifies all points from origin to perception, how do we differentiate one point from another? A further concern with this perspective is that it does not explicitly state whether each distinguishable point within the process is required for classification as sound. This leaves us unsure of how to confirm the presence of sound, in circumstances such as if a vibrating object and listener are separated by a vacuum, or if a listener perceives a sound when no sound wave is physically present. How can we explain illusory sound or tinnitus?

Building More Holistic Perspectives on Sound

The previous section of this chapter has reviewed various perspectives regarding the definition and relative positioning of sound: from a process of interaction between physical objects, to molecular perturbations that expand through alternative mediums, to sound waves crossing space until reaching a point of reception and, finally, a point of translation as the perturbations stimulate neural impulses that facilitate the sensual experience of sound for the listener. Although already incorporating multiple components, this process remains a relatively isolated system and one that does not explicitly account for wider situational, temporal, psychological and environmental factors. As a response, this section brings us closer to the model of Sonic virtuality with a short look at sound from the broader and more inclusive perspectives of acoustic ecology and soundscapes.

Perceptual Grouping

Citing Gestalt psychology (specifically notions from Koffka 1935), the noted neuroscientist Brian Moore (2012) outlines several general principles of perceptual organisation that raise certain questions pertinent to the 'what' and 'where' of sound. It is worth noting that Moore's concepts are not without empirical foundation and are all based upon direct observation across multiple studies. Of prominent note is the figure–ground phenomenon, possibly one of the most fundamental aspects of listening in everyday life. According to Moore, the neural wiring of the brain necessitates that auditory perception examines and interprets the complexity of everyday soundscapes by perceptually grouping available sensory information into auditory streams. Our attention subsequently prioritises an individual stream which stands out perceptually, whilst all other sound becomes less prominent. This effect is commonplace in everyday life, from our ability to follow a single voice in a room of many (also known as the Cocktail Effect, see Augoyard and Torgue 2005), to the converse inability (for many people) to recall any specific musical content from a recent popular song beyond the vocal melody. Further processes of auditory perception posited by Moore seek to explain how these streams are formed. They include similarity (sonic components [sound waves] are perceptually grouped if they are similar); good continuation (smooth changes to components shall be perceived as changes to a single sound, whilst sharp changes are indicative of a new or different sound); common fate (the perceptual grouping of two or more components within a soundscape that undergo similar changes or rates of change); disjoint allocation (an assertion that identification of source or perceptual grouping, once enacted, is difficult to reallocate to a different source or stream) and closure (an attended sound subsequently masked to inaudibility by another is perceived as continuing).

Continuing with Moore's work (2012), perceptual localisation effects (the position that we perceive a sound to be originating from) can be related to the debates surrounding the 'where' of sound. One particular example is the 'precedence effect', in which two physically separate but perceptually grouped sound components are emitted from different environmental locations. The perceptual grouping, however, causes the listener to perceive both components as originating from the location of the first. The precedence effect shows that when we experience sound, even though it is a direct reaction to a physical correlate (i.e. sound wave), it is not equal to its. Certain aspects of the sound can be completely different when we compare the properties of the sound wave to the actual perception. Our everyday hearing is awash with illusion.

Perceptual localisation effects highlight the complexity of the relationship between physical and perceptual sound. They reveal how particular arrangements of acoustic properties can dramatically alter auditory perception. However, this relationship between sound wave and perception (the physical and the perceptual) is bidirectional. Perceptual effects have a significant impact upon how we attend to and interpret acoustic sound waves. This in turn suggests that, as listeners, our ability to directly perceive the physical 'truth' of sound is very limited. Our perceptual systems group components, form auditory streams and prioritise/deprioritise these streams largely outside of our conscious control, but having a significant impact on our experience of a sound wave. In terms of categorising sound as either object, event or phenomenon, such perceptual effects arguably nominate the latter but push even further into the psychological domain. Object and event theory largely explain sound in objective and empirical terms but simply do not account for perceptual effects. Alternatively, understanding sound as a phenomenon that is inclusive of subjective experience is arguably a more comprehensive explanation.

Soundscapes and Acoustic Ecology

Sound is best described as a holistic process. In her discussion regarding soundscapes, Schulte-Fortkamp (2014) describes the term as a human perception that encapsulates, at the very least, sociocultural background, psychological state/processes and the contextualised acoustic environment. Here, sound is not simply the process of sound wave generation, transmission and reception, but rather a much more broad and complex state of affairs. The roots of acoustic ecology and soundscape studies as a discipline can be traced back to the composer Raymond Murray Schafer and the World Soundscape Project, an initiative setup with the ambition to create an ecologically balanced sonic relationship between human beings and their natural environment (Truax 1974). Simon Fraser

University published an online dictionary of acoustic ecology, within which the titular term is defined as: 'the effects of the acoustic environment […] on the physical responses or behavioural characteristics of those living within it' (Truax 1999). A good way of explaining this ecological balance of sound can be found when considering the effects that changes to the sonic environments can have upon non-human animals. Citing Krause (1993), Wrightson (2000) observes that bird and amphibian vocalisations are acoustically in keeping with the soundscape of their habitat, specifically in terms of frequency and rhythm. This acoustic relationship has crucial implications for such animals, with the industrialisation and urbanisation of many areas leading to the production of noise that masks their 'niche' frequencies, limiting the animals' capacity for communication and, in some cases, reproduction (Barot 1999). The specific sound components that represent industrial/urban soundscapes primarily include noise (i.e. sonic by-products of mechanical processes such as road traffic and construction) and what Schafer (1977) terms 'Schizophonic sound'; the artificial reproduction (through recording and broadcast) that effectively separates the sound from its original source, encompassing anything in which the physical sound waves originate from headphones or a loudspeaker. VR sound is very much a Schizophonic entity.

Wrightson's (2000) analysis of our relationship with sound across the developed world is somewhat damning, our contemporary soundscapes awash with so many competing sources that the overall intensity blurs the components of the soundscape into a single homogenised mass; what Schafer describes as 'lo-fi' or low fidelity (1977). As Wrightson states: 'the meanings sound holds for the listener in contemporary soundscapes tend to be polarised into extremes—loud and quiet, noticed or unnoticed, good […] or bad' (2000: p. 12). We identify the majority of the contemporary soundscape as noise and combat it by increasing the intensity of 'blocking' sounds. Not to be outdone, the soundscape often responds in kind by increasing its own intensity, forcing the individual to continue the cycle or submit to the noise. You will have a visceral understanding of this notion if you have ever had to share a living room with several other people, some watching television whilst others are engaging in conversation. In this particularly unfortunate series of events, the two opposing forces becoming embroiled in a battle of spiralling loudness as one party increases the volume of the television to drown out the inane chatter, before the others respond in kind by raising their voices, prompting the television volume to increase ever further—and it goes on.

The 'loudness war' (Vickers 2010) is a broader cultural example of the above concern. It describes how commercially distributed music has steadily increased in loudness for many years under the assumption (asserted by many as false) that loudness correlates to listener preference and ultimately translates to better sales. The loudness war is arguably something that can be experienced in our everyday lives and several readers may empathise with the following example, in which our unfortunate hypothetical listener receives a severe auditory shock whilst changing stations between Classic FM and BBC Radio 1. Not having the foresight to reduce the volume at least three to four clicks in advance of changing stations, they run the risk of feeling as if their eardrums have been viciously perforated.

Psychoacoustic Effects

The complex interrelating factors that influence our perceptual experience of sound are considered in detail by Augoyard and Torgue (2005), who present us with a compendium of psychoacoustic components which they group into five discrete themes: (1) elementary—objective acoustic parameters of a sound wave such as pitch, intensity and duration; (2) semantic—ways in which a sound's 'emerging signification' (p. 17) can alter a pre-existing context; (3) mnemo-perceptive—effects relating to the listener(s), inclusive of sociocultural aspects; (4) psychomotor—indicative of a bridge between auditory perception and motor action; (5) compositional—effects pertaining to complex interactions between multiple sound components. One instance of semantic effect is delocalisation, described by Augoyard and Torgue as when the listener forms an automatic impression of a sound's point of origin but simultaneously is aware that certain properties of the acoustic environment are manipulating this impression and that the origin is elsewhere. A good example of delocalisation would be when observing a ventriloquist, during which the audience perceives the voice as originating from the puppet but are concurrently aware that the actual origin is the puppeteer. Illustrating the mnemo-perceptive group, the synecdoche effect describes what is more commonly referred to as selective listening, in which individual sound components and streams are positioned into a hierarchy of attendance. The impeccable ability of the television to push the voice of one's significant other into perceptual deletion is an everyday example many will likely have experienced, though may not happily admit to. The psychomotor group encompasses effects such as phonotonie, in which a feeling of euphoria that prompts a renewed, collective or reflex gesture is the primary characterisation of the sound perception. The innate desire to dance when a particular song is heard being a familiar example of this effect.

Virtual Acoustic Ecology

Positioned centrally between an ecological understanding of sound and the Sonic virtuality framework is Grimshaw's model of acoustic ecology for first-person shooter (FPS) digital games (see Grimshaw and Schott 2008). Very briefly, the FPS acoustic ecology framework applies principles of the ecological perspective to digital games, the specific focus being upon the first-person perspective and multiplayer scenarios, presenting obvious parallels to VR. As shall be revisited later within this chapter, a holistic conceptual framework is both appropriate for our understanding of sound and of great value for VR, both in terms of theory and practical application. Ultimately, more inclusive and ecological perspectives dictate that no singular definition or spatiotemporal localisation of sound should be prioritised over any other. Instead, the individual context of the application requires careful assessment, with comparative consideration of multiple positions on sound to determine which has the greatest relevance to the particular purpose.

The ecological view of sound highlights the assertion that no sound, whether its sound wave correlate is a sine wave or a highly complex multi-stream soundscape, exists in isolation. Even if we were to manufacture an acoustic environment that could absolutely control the soundscape, it would not be possible to limit the multiple contextual effects that surround our perception. The memory of preceding sound experiences, the multimodal effects as the sonic input is perceptually combined with content from other sensory modalities plus many other variables, all impact upon the perception of sound. The ecological view offers explanatory power, such as to how two individuals listening to the same sound wave/soundscape, or a single individual experiencing repeated auditions, can describe notably different experiences. What it does not do, however, is address an important question that we shall examine now. Namely, how can we account for circumstances in which a sound is perceived but no sound wave correlate is present? The 'sound without a sound wave problem'.

The Sound Without a Sound Wave Problem

No matter where we situate sound or how we characterise or define it, the physical sound wave is, according to the above theoretical positions, inextricably tied to the concept of sound. Here, empiricism and phenomenology may prioritise the sound wave differently, but neither can fully explain how it is possible to perceive sound in the absence of a corresponding sound wave. A broad range of examples that illustrate this problem are discussed across multiple chapters within *Sonic virtuality* (Grimshaw and Garner 2015), but a very brief outline is presented here.

Tinnitus is arguably one of the most well-known instances of sound without a sound wave. Typically characterised as either a buzzing, ringing or hissing sensation, Tinnitus may be experienced intermittently or continuously. Its duration is also highly variable and intensity can range from a mild distraction to a debilitating roar (Langguth et al. 2013). Somatosensory forms of tinnitus, in which the perceptual experience can be traced to an internal physiological source, have been documented but in most cases, the sensation is entirely psychological, with no physical correlate identified. Similarly, 'exploding head syndrome' describes a rarer condition in which individuals feel an intense and sudden explosive sound with no apparent sound wave. The experience is an example of parasomnia, due to it most commonly taking place during the transitional phase between sleep and wakefulness. If you have ever been awoken by the sound of a door slamming or similar explosive event but no one else nearby has reacted even slightly and there's no trace of any possible cause, you may have experienced exploding head syndrome. Case studies conducted by Ganguly and colleagues (2013) evidence the above description and note that, to the individuals affected by the condition, the sound as it is experienced cannot be distinguished from that which has a physical correlate.

Examples of the sound without a sound wave problem extend into musical and verbal forms. The former shares a similar methodological problem with exploding head syndrome with its rarity limiting research primarily to case studies. A review article by Evers and Ellger (2004) aims to put some of the pieces together. In terms of the content of musical hallucinations, the authors note that popular music and songs from childhood, experienced bilaterally (between both ears) and with full instrumental and vocal components, best describe the most common form of the condition. The article itself gives some illumination regarding aetiology but what is of particular interest for us here are the qualitative personal reports given by the sufferers. For them, the experience was rarely emotionally neutral. Instead, reports of musical hallucination largely described intense positive (pleasantness) or negative (fear) emotional responses. This indicated that, for the individuals involved, the experiences were particularly vivid and sensually comparable to reception of a physical sound wave.

Auditory verbal hallucination (AVH) is the prominent term for what is colloquially known as 'hearing voices'. Anthony (2004) identifies the foremost features of AVH as 'hearing' mental contents such as thoughts and memories as spoken words, but 'somewhat detached from subjective ownership' (p. 108). Where AVH finds itself most correlated with psychological conditions such as schizophrenia is in circumstances in which the 'listener' establishes a perceived origin of the sound. That said, a large proportion of AVH reports have been associated with individuals classified as 'healthy/normal' (Johns and Van Os 2001). Anthony's (2004) review echoes the affective feature musical hallucinations, with individuals reporting intense emotional responses to AVH experiences. The visceral nature of the vocal sounds being perceived is often enough for the listener to identify an actual auditory source, in many cases external to the listener and perceptually located within the physical environment. This is a further indication of how 'real' such experiences are and supports the appropriateness of classifying them as sound.

In a reverse of the problem, non-cochlea sound raises the notion of 'perceived sound wave without a sound', in which the listener responds to a physical sound wave as a perceptual experience without the cochlea (which is identified as the singular point of auditory reception in the above theories on sound). An example of this is infrasonic sound (waves with frequency under 20 Hz) being detected by the skin, evoking a sensation akin to touch (Riddoch 2012).

The list of instances which raise the sound without a sound wave problem extends further, incorporating verbal self-monitoring (our functional inner voice) and auditory synaesthesia (involuntary cross-sensory activation in which [most commonly] visual stimuli is experienced as sound—see Saenz and Koch 2008). There is a philosophical point of contention here; specifically, whether hallucination is an opposition to perception, or a component of it. Another way of putting this is to question if a physical correlate is required for an experience to be a genuine perception. This again separates the empiricists from the phenomenologists. The former would argue that a physical component is what separates perception from hallucination, irrespective of how vivid the experience may be. The latter, however, would insist that it is precisely that vividness that makes the experience a genuine perception. The argument built across the last few pages arguably attests to the phenomenological. No matter which side you find to be most convincing, whether hallucinations and internalised speech/music are part of perception or an alternative to it, the question remains regarding whether such things are sound, and if not, what are they?

Building a Model of Sound as an Emergent Perception

The conceptual framework for Sonic virtuality was first formally presented as a book chapter (Garner and Grimshaw 2014) within the *Oxford Handbook of Virtuality* before being extended into a monograph the following year (Grimshaw and Garner 2015). Its origin was largely the result of the 'sound without sound wave' problem, with no existing model or definition of sound able to provide a satisfactory explanation. This section introduces the surrounding theoretical positions that helped Sonic virtuality to form (with the exception of virtuality, which was discussed in Chap. 2); reaching back to classical platonic philosophy and pushing forward to the most contemporary perspectives on embodied cognition and construal level theory. The rationale for this is not to indulge in a historical ramble, but rather to explain the crucial thinking that underpins Sonic virtuality in a way that will make the subsequent discussion on the matter more accessible.

The Cave

Of all the positions under discussion throughout this section, the one possessing the most historical gravitas is arguably Plato's Allegory of the Cave. Well known to many a student of philosophy, and rather apt for VR with CAVE being a key VR platform, Plato's Cave presents a representation of the thesis that we may only know our existence through the distorting veil of perception. Plato describes prisoners, forced to gaze at a wall upon which dance the shadows of objects positioned behind them. A great fire illuminates the objects and casts the shadows. The prisoners cannot see the objects themselves, only the shadows, which become the foundation of their understanding of the world. For Plato, it is possible for a prisoner to leave the cave by turning to face the fire, enduring a painful adjustment period as their understanding of the world is transformed. This can be interpreted to describe individuals who question their own perceptual experience and endeavour to understand the objective components of their existence by essentially becoming scientists; developing methods and tools for detecting and quantifying the objects that cast the shadows, then building conceptual models from empirical observations. This implies that whilst we can know the casting objects, we cannot experience them directly without perceptual distortion.

As the Allegory of the Cave is continues, Plato posits that prisoners who departed successfully from the cave may wish to return, infused with a feeling that their new-found understanding is superior and pitying those still confounded by the veil of perception. What they discover, however, is that, upon returning to the cave, they have lost the ability to perceive the shadows upon the wall. They have become consumed by processes, empirical observations and deductions. They can no longer simply engage with the experience. The remaining prisoners observe this and surmise leaving the cave to be harmful. As a result, the prisoners resist any force attempting to 'free' them. The prisoners and the escapees rather neatly reflect the phenomenologists and empiricists respectively, whilst in VR sound, it separates the technological aspects of VR from user-experience. Although the language used by Plato certainly suggests favour for the empiricist, neither party is comprehending both the fire and cast shadows simultaneously, indicating that both perspectives are incomplete. In this analogy, an emergent perspective seeks not to understand the shadows, the fire or the casting objects, but the cave itself, in its entirety.

Embodied Cognition and Construal Level Theory

Moving substantially forward in time, Margaret Wilson (2002) presents us with the theory of embodied cognition. Amongst other characteristics, embodied cognition asserts that all cognitive processes are contextually and environmentally situated, influenced by time-pressure, entangled with sensorimotor abilities, and can be 'off-loaded' onto the environment. Descartian dualism is rejected as embodiment asserts that the mind cannot be separated from the body, which itself cannot be separated from the world. How we think and how we feel are subject to our environment, our physiological state and also temporal and situational contexts. An example of the cognitive offloading could be our use of a calendar to structure our activities or using post-it notes to give ourselves reminders. The theory is not without challenge and in a recent critique; Goldinger and colleagues (2016) assert that embodied cognition theory incorrectly interprets empirical observations. In one particular example, they describe findings of brain imaging scans that show motor cortex activation in participants when they read the word 'kick'. According to Goldinger and colleagues, embodied cognition theory would assert that activation of the motor cortex is a mediating factor in the perception of the word (i.e. perception of the action would precede the perception of the word). They instead assert that the reverse is true; that the word is perceived first and the motor cortex activation is a subsequent response. Despite such contentions, embodied cognition has gained a good deal of traction since its beginnings to become the foundation of much of the recent research within both cognitive neuroscience and artificial intelligence studies (Matheson and Barsalou 2016). Whilst Sonic virtuality acknowledges the criticisms, that the physical body is an influencing factor in all our perceptions remains a key assertion of the framework.

In a related concept, Trope and Liberman (2010) describe construal level theory, a model of cognition in which it is asserted that our perceptions and interactions within the world are constrained by the egocentricity of psychological distance. In simpler terms, we perceive the world primarily in relation to ourselves in terms of space, time, identity and contextual relevance. Psychological distance has particular relevance to sonification systems. For example, buzzing insect in the immediate vicinity will likely evoke different feelings and actions compared to if that same buzzing was a farther away. Likewise, the auditory cue at a train station that precedes an announcement has different meaning when the train is not due for a when, compared to when it is imminent. Whilst you're at home, trying to relax in front of the television, a persistent car alarm outside is a source of irritation—until the realisation occurs that it's *your* car.

Modes of Listening

Embodied cognition and construal level theory underlie a particular proposition of Sonic virtuality, namely that sound as we experience it cannot be separated from us, our body or the wider physical environment. The shouts of your quarrelling neighbours at 3 pm are not equal to the same shouts at 3 am (provided you don't usually sleep during the day). The meaning and perceptual experience of a series of intermittent beeps indicating a countdown to zero as an individual is participating in a time-pressured game is not equal between player and observer. Some of the concepts discussed so far within this section also find distinct connections to several prominent positions on sound, specifically those that examine the relationship between sound and listener. Schafer's

(1977) seminal book on sound elucidates elements of how we attend to and make sense of sound. In a particular section regarding sound contexts, Schafer distinguishes sound into that heard by the physicist, the linguist and the composer. These listener types are matched with a perceptual prioritising of acoustics, semantics and aesthetics respectively. In Schafer's own example; for the physicist, the sound is reiterative and roughly 512 hertz at 90 decibels, whilst for the linguist the sound is an order to move. The composer, however, hears an unpleasant and aggressive sound. The sound is a car horn. What Schafer is alluding to here is that, as listeners and as individuals, we have different focuses regarding the information we wish to extrapolate during the process of hearing. These differences direct us to alternative means of attuning to a sound, in what is referred to as modes of listening.

Listening frameworks, such as that by Michel Chion (1994), help us to better understand our relationship with sound by differentiating alternative listening modes that are primarily distinguished by the intentionality of the listener. Chion's model is comprised of three modes: causal, semantic and reduced. For causal listening, the intention is to better understand the source object or event whilst semantic listening incorporates processes of informational encoding or decoding (such as language) and the intention is to interpret some sense of meaning. Lastly, reduced listening is a nod to Schafer and denotes our intention to analyse the traits or 'qualities' of the sound (such as to describe the timbre of a musical instrument). The notion of listening modes is expanded upon further in a taxonomy by Tuuri and Eerola (2012). In this revised presentation of their original model (see Tuuri et al. 2007), the very first statement asserts that: 'listening to sounds or music is not a homogeneous act of grasping meanings by hearing [and] the intentional stance of a listener is [often] overlooked' (p. 137). For Tuuri and Eerola, how we listen is affected by three dimensions: attention (the extent to which we are focussed upon the sound), intentionality ('listening mode': how we as the listener perceive the sound as a result of what we are attempting to understand about the world) and disposition ('listening style': which listening modes we are most predisposed to utilising). Intentionality is the focus of the taxonomy and the modes themselves are grouped into three categories (experiential, denotive and reflective) and positioned along a continuum of cognitive processing. Experiential modes are largely preconscious and utilise comparatively little cognitive resource. They include reflexive (autonomic physiological or behavioural responses), kinaesthetic (sense of responsive movement, e.g. dancing) and connotative (pre-established semantic connections) listening. Denotative modes describe more complex analyses with causal (establishing the source object/event), empathetic (assessing the intentionality of the source), functional (similar to empathetic but more oriented towards the action/event/purpose) and semantic (newly-established meanings with a focus on broader situational context) listening. Finally, reflective modes are fully conscious and require high-level cognitive processing; they include reduced (reflective analysis of the listening experience itself) and critical (generating new meanings through re-evaluation of those already formed) listening.

The modes of listening are also a particularly good example of the connectivity that exists between the many concepts we are exploring, specifically sound as a perceptual entity and our understanding of VR. The modes of listening can effectively be translated into modes of experience. To elucidate, each of Tuuri and Eerola's (2012) listening modes can be directly applied to reveal important elements of a user's VR experience. Whilst immersed in a VR world a user may undergo temporary changes in heart-rate, respiration of sweat secretion (reflective experience). They may engage in repeating patterns of physical movements such as poses, gestures and synchronised button presses (kinaesthetic experience). Pre-learned associations such as user-expectation or recalled prior experiences may also be felt (connotative experience). Equally so, the user may extrapolate denotative meaning from the content relative to an object or event (causal experience), an intention of an identified character (empathetic experience), or features of a character's specific actions (functional experience). As user-interaction continues, new meanings are formulated (semantic experience) and the user continually refreshes and updates their existing understanding as new information is received (critical experience). Finally, reduced experience describes moments in which the user evaluates the design quality of the VR environment. This emphasises how many of the theories discussed here can be readily reconstituted and applied to auditory processing, VR and VR sound.

The above is only a snapshot of the research that exists concerning the particulars of our relationship to sound, with further concepts including entrainment (the synchronising of internal/physiological activity with an external rhythm; e.g. attentional periodicity adjusting to match the tempo of a song—see McAuley et al. 2006) and archaeological listening to speech (a psychoanalytic technique in which the analyst seeks to interpret underlying, subconscious meaning from a patient's speech—see Josephs 1988), to name a couple. As we move ever closer to discussing Sonic virtuality directly, one further point is of particular interest for us here. When a taxonomy of listening such as that by Tuuri and Eerola (2012) is considered alongside the 'what' and 'where' perspectives on sound (see the first section of this chapter), we can observe that different listening modes both redefine and relocate sound. For example, causal listening would identify the where of sound as at the source whilst the semantic mode would place it more broadly, including the source, listener and surrounding environment. Furthermore, the critical mode is indicative of prolonged cortical activation as the sound wave is carefully evaluated and re-evaluated. Compare that to reflexive listening, which would require significantly less central processing and could potentially bypass the primary auditory cortex altogether. The processes of attributing these various forms of information to a sound wave are highly relatable to our everyday experiences as we interact with our sonic world. Empirical support of different listening modes is also available in the form of numerous studies that reveal changes to neural activity when participants attend to the same sound wave but in different ways (Brattico et al. 2010; Vannest et al. 2009). With this in mind, it seems rational to question any notion that sound can be defined in limited terms or positioned within a particular space. Instead, the meaning and position of sound is something that is in constant fluctuation, shifting with each adjustment of our perceptual processing as we constantly analyse, evaluate, reflect and repeat.

Emergence

Emergent perception is one of the several theories of emergence, the foundational premise of which asserts that the whole of any given system or process within nature is the outcome of a highly complex matrix of interrelating variables, comprehensive understanding of which requires holistic analysis. Arguably a concept with more leanings towards phenomenology, the primary assertions of emergence theory are succinctly identified by Simpson (2011) as: (1) the emergent 'whole' of a system is greater than the sum of its parts; (2) one therefore cannot establish the characteristics of the whole solely by understanding of its component parts; (3) changes to the whole of a system cannot be entirely reduced to changes in its individual components; (4) emergence occurs at multiple levels of structural organisation. In terms of sound as an emergent perception, and referring back to the perspectives of O'Callaghan (2009) and Casati and Dokic (2009), an emergent model of sound is something not definable solely as a component of the perceptual process, nor can it even be explained as the process in its entirety. Emergent sound is instead best thought of as a new entity, actualised from (but not equal to) all characteristics of the process. As a multi-layered system, it simultaneously feeds back into the process or can become component for another emergent actualisation.

To illustrate the above concept, consider the nature of a song as you listen to a live band performance. As you listen to the song, you may conceive of such a thing entirely as just that, a piece of music, but you may also unpack it to reveal multiple layers of organisation. The performance comprises multiple instruments, each contributing different melodic, harmonic or rhythmic content. Each instrument has multiple physical components and configurations. The song is formed of several different sections with every chord and melody formed of different notes, every note formed of different frequency components, and so on. Observing this reductionism in reverse reveals to us the emergent process; the note emerges from the frequencies, the chord from the notes, the progression from the chords, until we reach the organisation level of 'song'. However, if we were to analyse all of this material empirically, we still could not fully comprehend the song as a whole. The reason for this is that the song, like all things, does not exist inside a perceptual vacuum. Also relevant to the perception is the listener and the vast, multi-layered systems they bring with them. The song, listener, band and an incalculable number of other systems and variables are all present. Ultimately the closest thing we have to an actual uppermost point and a finite system would be existence in its entirety, under the assumption that existence cannot be positioned within or adjacent to anything else. Therefore, to understand the song (or anything else for that matter) with absolute comprehensiveness we require an unlimited understanding of everything. This exposes the cost-benefit compromise of an emergent framework. In embracing this perspective, it is possible to understand the actualised subject in very broad and comprehensive terms. However, in doing so, we have (conceptually) taken a big step back, at the cost of not being able to see the finer details.

You could be forgiven if the conclusion that you have reached upon reading the previous paragraph was that the ambitions of emergence theory are rather lacking in practicality. As a central tenet of Sonic virtuality, and indeed many applications of emergence, the intention is not to insist upon absolute understanding, but rather to encourage more holistic analysis as a way of reducing the risk of important factors being neglected when constructing a theoretical model and working towards a better understanding.

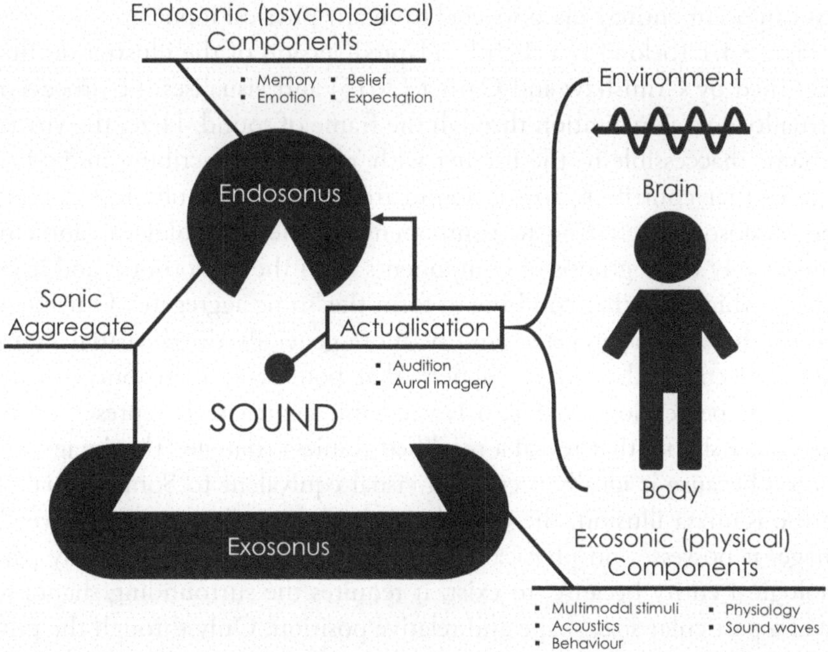

Fig. 3.1 The Sonic virtuality model

Sonic Virtuality

In Chap. 2, we deliberated some alternative perspectives regarding the virtual, differentiating the technological aspects of VR from organisational perspectives of virtuality (broader theory concerning our experience of the world). Sonic virtuality takes influence from the latter, in tandem with the main principles of emergence theory. To briefly recap, Gilles Deleuze's notion of virtuality refers to the assertion that the virtual is a component of reality and an opposition to actuality. Under fundamental restrictions of human perception, we may only directly experience the actual. The virtual is not available to us but is no less real because it is only through the virtual, that the actual is able to exist. The actual is not to be confused with the physical. Equally so the virtual is not analogous to the psychological. Both the physical and psychological are facets of the virtual and they cannot be actualised singularly. One cannot perceive a physical object without engaging some psychological process and, likewise, no psychological process can occur entirely disconnected from any physical input.

Figure 3.1 (below) is a slightly adapted version of the illustration first published by Grimshaw and Garner (2015) and visualises the process of virtuality and actualisation through the frame of sound. Here, the virtual remains inaccessible to the listener with 'exosonic' describing individual entities that contribute to the sound from within the physical domain and 'endosonic' referring to components of the psychological domain. Collectively these groups of components form the 'endosonus' and 'exosonus', which together combine to form the 'sonic aggregate'. Outside of sound, the sonic aggregate could be more generally titled 'virtual aggregate' and encapsulate every element that potentially contributes to the emergent perception. In Fig. 3.1, the sonic aggregate is represented by the darker shapes that reveal a modified Kanizsa triangle. This image was chosen because it ideally presents a visual equivalent to Sonic virtuality. In the Kanizsa illusion, the triangle is not drawn. As part of the overall image, it possesses no physical form, yet it also cannot be a purely psychological entity because, to exist, it requires the surrounding shapes to be of a particular size, shape and relative position. Only through the process of perception is the triangle revealed to us, and it is this moment of revelation that the Sonic virtuality thesis identifies as actualisation.

The Sonic virtuality model is inclusive of perceptual effects, from the ways in which we group sound waves into auditory streams in Moore's figure—ground phenomenon (2012) to how the social rank we assign to an individual during a cocktail reception can influence the clarity with which we hear their speech in Augoyard and Torgue's synecdoche effect (2005). The model also acknowledges physical causality (see Nudds 2009) but reveals problems with the assumptions commonly drawn from causal theory. That a sound event of greater intensity will consistently lead to a louder sound is an example of such assumptions. As was alluded to earlier, such a conclusion is false because it overlooks crucial endosonic and exosonic variables within the causal framework. For example, a sufficiently intense initial sound wave could affect the inner ear, creating a temporary 'blunting' effect that attenuated the subsequent sound wave so that it was perceived as quieter even though its amplitude was greater.

Sonic virtuality also shares much with emergence theory in terms of conceptual content. Referring back to Simpson's (2011) four tenets of emergence, sound as an actualisation is not simply the sum of its endosonic and exosonic components. The sum is the sonic aggregate and it is through the process of our existence and experiencing of the world that the sound emerges from that aggregate. As with emergence theory, knowing the aggregate is not equal to knowing the actualisation and the model illustrated above represents a single level of emergence as it pertains to an individual listener at one point within space and time. If two or more individuals are present within the same soundscape then multiple actualisations are occurring simultaneously, each potentially influencing the other in a continuous intertwined exchange by way of their affects upon the listeners' behaviour. The model also illustrates the perceptual feedback mechanism as the actualisation is processed through our memory, thereby becoming another component of the endosonus as a subsequent actualisation emerges. Essentially, our experience of sound in this moment becomes a potential contributor to our experience of sound in the next. Triangles begetting triangles.

Sonic virtuality resonates significantly with Plato's Allegory of the Cave. Actuality is the wall and the dancing shadows cast upon it, whilst virtuality is represented by the fire and the objects that cast the shadows. The light of the fire reflects the psychological domain of the endosonus whilst the objects, being of more concrete form, represent the exosonus. Just as the prisoners have the physical capability to turn their heads towards the fire and the objects, so to do we as listeners have the conceptual ability to understand the virtual aggregate and the components within it. The prisoners who do turn to face the fire may become accustomed to it and find that the framing of their perception shifts in its direction. Likewise, in Sonic virtuality, the listener who disregards their inherent perceptual experience of sound in order to try and understand its underlying components and mechanisms, may discover that their very experience of sound changes. With no prior musical education, our first hearing of a piano sonata is a broadly affective experience. We possess no preconceptions, expectations or relevant abstract knowledge and we 'feel' the music much more than we analyse it. Should we then embark on a course of music education, numerous systems and structures become available to us, learned through revision, practice and repetition. The same sonata is then auditioned, and where we once heard a musical piece that made us feel calm and relaxed, we now hear a key signature, a modulation, a counter-melody, and so on. It is the musical equivalent of stepping into the kitchen of your favourite restaurant and discovering how your meal was prepared. As those that escaped from the cave and later returned, once you have seen beyond the veil it can be very difficult to go back. This process is not merely an apocryphal tale and neural plasticity (referring to physiological changes to the wiring of the brain and nervous system) as a result of musical training has been evidenced substantially in recent years; revealing a physical shift in the activation patterns of the brain, with significant effects upon perception (Gaser and Schlaug 2003; Münte et al. 2002).

One further point to note is that such a shift in perception is not analogous to the virtual becoming available to us as direct sensuous experience. Instead, it is more accurate to state that our individual perceptual process has adjusted and that which was once virtual has become actual. Within the Sonic virtuality model it is perfectly plausible that a component of the virtual might be actualised, but as an actualisation, the component is now profoundly different to us. If the counter-melody of the sonata, which was once an element of the sonic aggregate, has become the actualisation, it is not correct to state that our perceptual focus has merely shifted onto a virtual entity. Such a thing is impossible. As an actualisation, an individual note (for instance) can emerg by way of a unique sonic aggregate, but we cannot experience it in isolation any more than we could the full sonata. Indeed, in this instance, the sonata itself is now a virtual entity influencing our perception of the singular note. For example, if the performance was perfect until that single actualised note, the listener's response to it as a glaring mistake would be significantly different to if the piece had been littered with errors).

Sonic Virtuality and Its Application for VR Sound

Very broadly, the function of Sonic virtuality is to question the widespread assertions that sound is a sound wave and that we can conceptualise sound as something to be defined and positioned in limited terms and specific points. The complexity of sound is heavily asserted and whilst Sonic virtuality presents us with an Everest-like challenge in terms of ascertaining a comprehensive understanding of sound, it forces us to accept the ultimate limitations of the theories that came before it and push forward upon a more progressive and holistic path.

When it comes to sound that is mediated through technology, the more established perspectives outlined within the first section of this chapter are faced with further difficulty. Schafer's Schizophonic sound (1977) breaks, or at least confounds, the connection between sound and source (and equally between physical association and perceived association). Whilst indulging in a feature film at our local cinema, all sound waves originate from the speakers positioned around us yet we perceptually locate them beyond the screen to our front. At the physical level, the sound event is the oscillation of a ceramic magnet against treated paper, irrespective of whether the audience perceives the whir of a helicopter, the bang of an explosion or the roar of a Tyrannosaurus. The latter of these complicates matters even further and has particular relevance to VR when we consider 'fantastic sound'. A 'natural' object or event that generates a sound wave may mean that the perceived source matches the physical source at the time of recording. The sound of footsteps, for example, are likely to be recorded samples of actual footsteps. However, even with natural sounds the physical and the perceived may not correspond. Foley artists are well-known for sourcing audio with comparable timbral and textural qualities from otherwise completely disconnected objects and events. Quarter inch audio tape scrunched into balls bears an uncanny resemblance to footsteps upon a grassy surface, whilst punching a telephone book that has been wrapped in electrical tape provides a less casualty-inducing alternative to punching a human in the stomach, but produces sound wave that is perceptually the same (Rodrigues 1998). Of course using un-matching sources to produce audio is essential when dealing with 'fantastic' objects that, for all intents and purposes, do not physically exist. Any VR experience in a science fiction or fantasy setting requires a substantial quantity of audio material that represents objects and events that cannot possibly be directly recorded, requiring either Foley work or synthesis to achieve the desired effect. Sonic virtuality has particular application here in explaining how our emergent perception of 'Schizophonic' and fantastic sound is formed, and how designers can more effectively craft their sound content to evoke the desired experience by understanding the wider variables that contribute to it.

Audio technology adds a further layer of conceptual convolution by way of the 'lateralisation' effect, caused by the humble headphones (Moore 2012). Lateralisation refers to the localisation of sound within the head, an effect achieved by the closeness of the headphones to the ear. This closeness effectively eliminates any potential acoustic effects between the source and the listener and also enables inter-aural difference to be determined by the headphone output, causing the source to appear to be inside the listener's head. As noted earlier, the phenomenological perspective locates sound at the source whilst the metaphysical places it (typically) at the listener. The lateralisation effect, however, causes the location of sound to align (or at least overlap) between the two perspectives, and sound is both *at* the listener and *felt at* the listener. There is also a connection raised here between lateralisation and the virtual. Headphones arguably are the auditory equivalent of the HMD (see Chap. 6) and lateralisation represents a virtual effect, in which the phenomenology of sound is extended with the creation of virtual spaces that are impossibilities outside of VR. Sonic virtuality accounts for this and VR technology is acknowledged as an important contributor to both endosonic and exosonic components within the sonic aggregate. Therefore any emergent framework of VR sound is bound to be significantly different to one of natural sound or even sound in digital games.

From the above discussion points, we can observe how conceptualising sound reveals significant problems with the more established perspectives on sound. If a sound is at the object or event, how do we account for the fact that the auditory object for many VR sounds is simultaneously the physical audio source (Foley object, synthesis, etc.), the projecting speaker and the diegetic source within the game world? If a sound is at the listener, how can we reconcile the lateralisation effect? Sonic virtuality affords us a holistic perspective, centralised upon the listener, allowing sound to exist in different forms and at multiple places.

Chapter Summary and References

Just as the virtual is delineated into organisational and technological forms, so too is VR sound. The Sonic virtuality framework addresses the organisational aspect of VR sound but has both theoretical and practical application for VR. This chapter has outlined the main problems with more established positions on the meaning of sound; the primary criticism being that they remain too limited, disregarding various important elements in favour of a clean and precise understanding. Non-cochlea sound, auditory imagery, sound synaesthesia and various other phenomena expose the restricted explanatory power of these perspectives and highlight the value of an alternative, more holistic theoretical framework. Taking a position with virtuality, embodied cognition and emergence theory (to name a few) as a foundation; Sonic virtuality presents a more inclusive, listener-centric thesis on sound.

As mentioned in the opening paragraph, Sonic virtuality primarily concerns the nature of sound from an organisational perspective. That said, by way of phenomena such as the lateralisation effect and Schizophonic sound we are beginning to see how the technological and organisational forms of the virtual are linked. Following on from here, the next chapter shall begin to bring together elements of sound and VR with a focus upon the most significant common thread that connects the two; the human receiver. In terms of sound we are referring to the 'listener', whilst in VR we would use the term 'user'. Irrespective of the terminology, emergent perception inherently positions us as receivers very much at the centre of things, bringing elements surrounding us and our experience to the fore.

Note

1. Oxford English Dictionary definition of sound. Oxford (2016). *Oxford Living Dictionaries*. https://en.oxforddictionaries.com/definition/sound.

References

Anthony, D. (2004). The cognitive neuropsychiatry of auditory verbal hallucinations: An overview. *Cognitive Neuropsychiatry, 9*(1–2), 107–123.

Augoyard, J. F., & Torgue, H. (2005). *Sonic Experience: A Guide to Everyday Sounds*. Canada: McGill-Queen's Press-MQUP.

Barot, T. (1999, January 10). Songbirds forget their tunes in cacophony of road noise, *The Sunday Times*.

Berkeley, G., & Krauth, C. P. (1874/1710). *A treatise concerning the principles of human knowledge*. Philadelphia: JB Lippincott & Company.

Brattico, E., Jacobsen, T., De Baene, W., Glerean, E., & Tervaniemi, M. (2010). Cognitive vs. affective listening modes and judgments of music–An ERP study. *Biological Psychology, 85*(3), 393–409.

Casati, R., & Dokic, J. (2009). *Some varieties of spatial hearing* (pp. 97–110). Sounds and perception: New philosophical essays.

Chion, M. (1994). *The three listening modes* (pp. 48–53). The Sound Studies Reader: Sterne.

Evers, S., & Ellger, T. (2004). The clinical spectrum of musical hallucinations. *Journal of the Neurological Sciences, 227*(1), 55–65.

Ganguly, G., Mridha, B., Khan, A., & Rison, R. A. (2013). Exploding head syndrome: A case report. *Case Reports in Neurology, 5*(1), 14–17.

Garner, T., & Grimshaw, M.N. (2014). Sonic virtuality: Understanding audio in a virtual world. *The Oxford Handbook of Virtuality*, 364.

Gaser, C., & Schlaug, G. (2003). Brain structures differ between musicians and non-musicians. *The Journal of Neuroscience, 23*(27), 9240–9245.

Gaver, W. W. (1993). What in the world do we hear?: An ecological approach to auditory event perception. *Ecological Psychology, 5*(1), 1–29.

Goldinger, S. D., Papesh, M. H., Barnhart, A. S., Hansen, W. A., & Hout, M. C. (2016). The poverty of embodied cognition. *Psychonomic Bulletin & Review, 23*(4), 959.

Goldsmith, M. (2015). *Sound: A Very Short Introduction* (Vol. 451). Oxford University Press.

Grimshaw, M., & Garner, T. A. (2015). *Sonic virtuality: Sound as Emergent Perception*. USA: Oxford University Press.
Grimshaw, M., & Schott, G. (2008). A conceptual framework for the analysis of first-person shooter audio and its potential use for game engines. *International Journal of Computer Games Technology, 2008*, 5.
Johns, L. C., & Van Os, J. (2001). The continuity of psychotic experiences in the general population. *Clinical Psychology Review, 21*(8), 1125–1141.
Josephs, L. (1988). A comparison of archaeological and empathic modes of listening. *Contemporary Psychoanalysis, 24*(2), 282–300.
Koffka, K. (1935). *Principles of Gestalt psychology*. Routledge.
Krause, B. L. (1993). The Niche Hypothesis: A hidden symphony of animal sounds, the origins of musical expression and the health of habitats. *The Explorers Journal*, 156–160.
Langguth, B., Kreuzer, P. M., Kleinjung, T., & De Ridder, D. (2013). Tinnitus: Causes and clinical management. *The Lancet Neurology, 12*(9), 920–930.
Matheson, H. E. and Barsalou, L. W. (2016). Embodied cognition. In: Wixted, J. (ed.), *The Stevens' Handbook of experimental Psychology and cognitive Neuroscience [4th edition]*. New Jercy: Wiley.
McAuley, J. D., Frater, D., Janke, K., & Miller, N. S. (2006). Detecting changes in timing: Evidence for two modes of listening. In *The Proceedings of the 9th International Conference on Music Perception and Cognition* (pp. 188–189).
Moore, B. C. (2012). *An introduction to the psychology of hearing*. The Netherlands: Brill.
Münte, T. F., Altenmüller, E., & Jäncke, L. (2002). The musician's brain as a model of neuroplasticity. *Nature Reviews Neuroscience, 3*(6), 473–478.
Nudds, M. (2009). What are Sounds? *Sounds and perception: New philosophical essays, 69*.
O'Callaghan, C. (2009). Sounds and Events. In M. Nudds & C. O'Callaghan (Eds.), *Sounds and Perception* (pp. 26–49). Oxford: Oxford University Press.
O'Callaghan, C. (2010). Perceiving the locations of sounds. *Review of Philosophy and Psychology, 1*(1), 123–140.
O'Callaghan, C. (2011, October). XIII—Hearing Properties, Effects or Parts?. In *Proceedings of the Aristotelian Society* (Vol. 111, No. 3 pt 3, pp. 375–405). The Oxford University Press.
O'Connell, M. (1999). *Spin Magazine* (stand-up quotation reference). September issue.
O'Shaughnessy, B. (2000). *Consciousness and the World*. Oxford: Clarendon Press.

Pasnau, R. (1999). What is sound? *The Philosophical Quarterly, 49*(196), 309–324.
Riddoch, M. (2012). *On the non-cochlearity of the sounds themselves*. Ann Arbor, MI: Michigan Publishing, University of Michigan Library.
Rodrigues, P. (1998). *The Art of Foley*. http://www.marblehead.net/foley/specifics.html.
Saenz, M., & Koch, C. (2008). The sound of change: Visually-induced auditory synesthesia. *Current Biology, 18*(15), R650–R651.
Santarcangelo, V., & Terrone, E. (2015). Sounds and Other Denizens of Time. *The Monist, 98*(2), 168–180.
Schaeffer, P., Bayle, F., Ferrari, L., Malec, I., Mâche, F. B., Parmegiani, B., & Philippot, M. et al. (1967). *La musique concrète*. Presses universitaires de France.
Schafer, R. M. (1977). *The soundscape: Our sonic environment and the tuning of the world*. Vancouver: Destiny Books.
Schulte-Fortkamp, B. (2014). Improving sound quality measures through the multifaceted soundscape approach. In *INTER-NOISE and NOISE-CON Congress and Conference Proceedings* (Vol. 249, No. 3, pp. 4420–4424). Institute of Noise Control Engineering.
Simpson, Z. (2011). Merleau-Ponty and emergent perception. *Journal of the British Society for Phenomenology, 42*(3), 290–304.
Strawson, P. F. (1959). *Individuals: Essay in descriptive metaphysics*. London: Methuen.
Trope, Y., & Liberman, N. (2010). Construal-level theory of psychological distance. *Psychological Review, 117*(2), 440.
Truax, B. (1974). Soundscape studies: An introduction to the World Soundscape project. *Numus West, 5*, 36–39.
Truax, B. (1999). *Handbook for acoustic Ecology* [online]. http://www.sfu.ca/sonicstudio/handbook/Alphabet_list.html.
Tuuri, K., & Eerola, T. (2012). Formulating a revised taxonomy for modes of listening. *Journal of New Music Research, 41*(2), 137–152.
Tuuri, K., Mustonen, M. S., & Pirhonen, A. (2007). Same sound–different meanings: A novel scheme for modes of listening. *Proceedings of Audio Mostly*, 13–18.

Vannest, J. J., Karunanayaka, P. R., Altaye, M., Schmithorst, V. J., Plante, E. M., Eaton, K. J., … & Holland, S. K. et al. (2009). Comparison of fMRI data from passive listening and active-response story processing tasks in children. *Journal of Magnetic Resonance Imaging, 29*(4), 971–976.

Vickers, E. (2010). The loudness war: Background, speculation, and recommendations. In *Audio Engineering Society Convention 129*. New york: Audio Engineering Society.

Wilson, M. (2002). Six views of embodied cognition. *Psychonomic Bulletin & Review, 9*(4), 625–636.

Wrightson, K. (2000). An introduction to acoustic ecology. *Soundscape: The Journal of Acoustic Ecology, 1*(1), 10–13.

4

User-Experience

Obtaining a firm understanding of the nature and fundamentals of VR is arguably a trickier undertaking than one might anticipate. In what is certainly a consistent theme within this book, the study of VR habitually pushes us toward a more holistic interpretation by repeatedly revealing concepts that connect and interrelate, the state of one invariably affecting the state of another. Such a thing is apparent throughout this chapter as we look at notions of identity and the self alongside various facets of user-experience. Whilst these concepts are discrete entities that can be distinguished from one another, they are nonetheless bound together within an ecology of VR. This chapter presents two main points for discussion. Firstly, a range of user-experience concepts (that includes immersion, presence, flow, diegesis and fun) are reviewed, as we look to address the questions of how VR uniquely contextualises such things and what the nature of sound is within the broader VR experience. The subsequent section of this chapter considers how VR has affected identity and may be fundamentally altering our interpretation of the self, with consequences for VR sound as an emergent concept.

T.A. Garner, *Echoes of Other Worlds: Sound In Virtual Reality*,
Palgrave Studies in Sound, DOI 10.1007/978-3-319-65708-0_4

User-Experience in a VR Context

One prominent difficulty with understanding a murky concept such as VR, particularly from a phenomenological perspective, is that to do so requires awareness of numerous other components, all affecting each other both directly and indirectly, backwards and forwards. For added difficulty, these components also have a penchant for the obscure. Fortunately, many have been the subject of study for many years and the literature is relatively accessible. As such, the intention here is not to reinvent the wheel with anything too general or overly detailed, as there are many good books and articles that have examined user-experience and identity pretty comprehensively. Instead, the following is a concise outline of the relevant concepts and an account of their significance to our understanding of VR and VR sound.

User-experience may be something of a human-computer interaction buzzword (Hassenzahl and Tractinsky 2006), but it has become synonymous with contemporary technology design. One can hardly bring up such things without at least a brief mention of Apple Inc., the archetypal success story of design that prioritised user-experience in both its hardware and software products. Beyond this, user-experience-led design has permeated numerous industries to become a near-ubiquitous practice. Today, there is barely a single consumer facing digital technology that doesn't heavily consider user-experience in its design, from programming a washing machine to browsing the internet. Since 2011, there has even been an annual awards body for recognising products and systems that demonstrate exceptional user-experience. This is commonly known as the *UXies*,[1] 'UX' being a commonly used acronym, particularly for user-experience in the digital games industry.

User-experience encompasses a multitude of values that are discrete from more formal evaluative elements such as usability (Hassenzahl 2003). Central to user-experience is the role of user emotion, with the affective relationship between individual (user, consumer, user, etc.) and product or service put under careful scrutiny, specifically with intent to maximise positive emotional experiences (Bernhaupt 2015). By way of gamification, user-experience concepts relevant to digital games are also finding themselves being applied to non-gaming contexts to improve consumer engagement. As Deterding and colleagues (2011) point out, industries such as digital marketing employ gamification techniques (point-based scoring, rewards, leader boards, etc.) as a means of creating more engaging experiences that entice and retain more customers.

Any framework of user-experience within VR would be expected to draw several parallels with digital games, particularly those utilising a first-person view. This is because both share comparable perspectives upon a three-dimensional virtual environment (movement within three-dimensional space, localisation of audio, direct user embodiment of avatar, etc.). However, as Benson and colleagues (2016) point out in their dimensional model of VR experience, 'reality' (i.e. the extent to which the material presented is either taken from the physical/natural world or is rendered/simulated content, relating to the continuum of mixed reality) and 'locomotion' (whether and how the user can physically move throughout the environment) present two issues quite unique to VR, positing that an updated approach to understanding user-experience in this particular context is required if we are to design systems and applications that work most effectively. HMDs enable a dramatically different viewing experience compared to a flat-screen monitor, whilst the ability to manipulate virtual space by way of more natural methods of interaction (gesture, voice, etc.) dissuades the use of traditional mouse/keyboard interfacing. Whilst this supports the assertion that user-experience theory requires nuancing for VR, the fundamental principles (primarily such as those discussed below) remain relatively constant. This is because we are still dealing with the experience of virtual environments, but with a few key characteristics that require attention.

A Note on Physiological Aspects

Whilst the majority of effects associated with user-experience are grouped within the psychological domain, it is worth briefly acknowledging several physiological effects that have been attributed to VR. The most well-known term relevant to such effects is arguably 'cybersickness'. Encapsulating multiple symptoms, cybersickness has been associated with reports of eye strain, head and neck aching, vertigo (dizziness), ataxia (a lack of coordination), disorientation and nausea (see Regan and Price 1994; LaViola 2000). Of these, disorientation is asserted to be the most prevalent symptom (Rebenitsch and Owen 2016). As LaViola (2000) observes, cybersickness pertaining to HMD-based VR is explained by way of several alternate theories, of which sensory conflict theory is noted to be the most widely accepted (Rebenitsch and Owen 2016). Sensory conflict theory posits that 'discrepancies between the senses which provide information about the body's orientation and motion cause a perceptual effect which the body does not know how to handle' (p. 50). This relates to the effect VR can have upon the vestibular system when the user moves in the virtual world but not the physical world and visual feedback mismatches proprioception. Though requiring further investigation, sensory conflict theory raises the importance of sound in terms of corresponding auditory feedback. As Roach and colleagues (2006) note, unified feedback from multiple sensory inputs can potentially mitigate a single mismatching sense, thereby implying that well-designed VR sound could reduce cybersickness if the audio and visual content were closely aligned.

Research considering physiological effects is almost entirely focussed upon mitigation of adverse effects and, with the known exception of Samsung's research into galvanic vestibular stimulation for their *Entrim 4D* HMD (see Chap. 8), there are very few studies that address the potential for positive physiological effects. Consequently, our coverage of this aspect of VR is kept to a minimum, but these issues remain a significant and ongoing concern. As the technology underpinning contemporary VR continues to develop, reported instances of cybersickness have not decreased but increased (Rebenitsch and Owen 2016). Developers therefore must maintain an awareness of these issues, and they of course remain a significant element of our overarching emergent models of VR and VR sound (see Chap.10).

Presence and Immersion

What does it mean to be immersed or present? Is immersion an objective phenomenon that can be determined by factors external to the user, irrespective of the variables of subjective experience? Does presence represent a depth value of immersion or is it a phenomenon in its own right? Immersion and presence are arguably two of the most pivotal components of VR user-experience or, at the very least, those that have acquired the most recognition. As such, this section discusses these particular concepts in greater detail before progressing onto further psychological aspects of user-experience. Sound features prominently here as we review contemporary research concerning the role of sound as a powerful component of both immersion and presence. Beyond this, we push further into the murky conceptual recesses in an attempt to elucidate the connections between immersion, presence and other user-experience components, before asking what considerations need to be made regarding immersion and presence within VR.

Many authors have explored the meaning of presence. The term has even been discussed as an allegory for God (Huston 2007). For Jean-Luc Nancy (1993), presence is an elusive phenomenon woven intricately into existence, being, knowledge, representation and nothingness. Citing Hegel, Nancy asserts that presence is fundamentally associated with 'the experience of consciousness' (p. 3) whilst, for Cuddy (2015, p. 2), 'presence stems from believing in and trusting [...] your real, honest feelings, values and abilities'. This implies that the feeling of being present may be connected to our identity and perception of self.

To differentiate presence from the notions of 'telepresence' and 'virtual presence', Sheridan (1992) describes the former as feeling present with objects (either virtual or physical) that are actually located elsewhere, whilst the latter denotes the same sensation but with objects specifically generated by a computer. Whilst a distinction between telepresence and virtual presence appears straightforward, Loomis (1992) argues that, in our everyday lives, we are inherently biased towards accepting our subjective perception of the world as absolute—to us, what we perceive is what *is*. Although commonplace, this bias is potentially quite fragile and Loomis suggests that VR can facilitate an experience that engages users to the extent that they 'question the assumption that the physical and perceptual worlds are one and the same'. An individual may approach a VR environment with the 'knowledge that it is not real' but then experience something that is so vivid and stimulating they conclude that what initially *felt* real may not actually *be* real—a notion that can, unsettlingly, then be applied to everyday experience of the physical world. A contemporary example of this can be found in the television series *Life on Mars* (Graham et al. 2007/2006) in which the protagonist, Sam Tyler, awakes from a coma, returning from what he presumes to be an imaginary dream world. After a short time, he finds himself 'numb to reality', perceiving physical stimuli as deeply muted and the dream world he left behind to be far more vivid and ultimately more real.

What is particularly interesting when considering the above is that although presence is frequently identified as a component of experience within digital games and VR, it is also a prominent element of everyday experience within the physical/natural world. This raises the question of what the difference is between presence in the virtual world and presence in the physical world. Furthermore, it encourages us to consider whether the two are fundamentally different at all, or if VR could theoretically develop to the extent that it evoked a sensation of presence equal to or even in excess of the physical world.

In simple terms, immersion can be described as the sensation of being immersed; your attention is being directed toward a particular environment or activity to the extent that you are perceptually filtering out other stimuli that are physically available to you. Despite being largely conceptual, immersion can be measured by way of quantitative data. For example, Pausch and colleagues (1997) demonstrate that search efficiency (i.e. total time taken to scan the environment and time spent re-scanning the same space) within a virtual room correlates with self-report of immersion significantly enough for this objective measure to be valid. Theoretical models detailing immersion within a digital game context typically localise immersion within the psychological domain, acknowledging that physical/objective factors can have an effect and so form part of the overall ecology, but arguing that immersion ultimately should be understood as a subjective experience. For example, Jennett and colleagues' (2008) 'SCI-model' describes immersion in three discrete but overlapping forms: sensory (immersion in the environment attributed to sensory stimuli), challenge based (a balance of task challenge and user ability) and imaginative (affective engagement with narrative or characters).

The concept of place is one that is regularly raised when considering the meaning of immersion and presence. In a description by Slater (2009), presence is analogous to what he terms the place illusion: 'the qualia of having a sensation of being in a real place' (p. 1). Subsequent research has posited that immersion directly enhances the place illusion, connecting both immersion and presence to place. Here, to be immersed or to feel present is within a world or an environment, be it virtual or physical. 'Immersion in place' has also been applied to learning non-native languages by surrounding the individual in the language over a prolonged period, known as 'immersion education' (Cummins 2000). Here, audition of the speaker's native language is suppressed, and they are exposed to a continuous and surrounding stream of non-native content. What is interesting here is that, whilst language is certainly different to a geographical place, the nature of immersion in language reveals some consistencies, primarily by the way of the medium of sound. Both immersion education and immersion in VR involve some form of physical surrounding. The user experiences large quantities of attractors, over an extended duration and in a three-dimensional, physically surrounding manner. In addition, both involve attenuation (or removal) of distractors. In VR that could be sound waves emanating from outside the virtual world (noise) whilst in immersion education it could be hearing native language. In the latter context, the immersed learner may be in a familiar classroom with no change in geographical place, but the sound all around them is of people speaking in a non-native language. This soundscape is not familiar, and the learner is, in a sense, transported to a different (some might say virtual) place.

Immersion and presence are two terms quite frequently employed interchangeably. From the points raised above, it is difficult to see if they are different at all. Their definitions and relationship to one another remain a prominent point of contention in research, which we briefly review here. Jerald (2015) cites the work of Slater (2003) to give one particular differentiation between immersion and presence. For Jerald, immersion is system relevant whilst presence is user relevant. For example, specifications of a HMD (e.g. the extent to which the visual input surrounds the user or the resolution quality of that input) would determine the level of immersion as a feature of the system, whilst the perceptual experience of the user (e.g. feeling embodiment with a game's avatar or a sense of physical interaction) is indicative of presence. Examples of contemporary VR research often reinforce this perspective by referring to visual hardware that physically surrounds the user (e.g. HMDs as opposed to flat-screen monitors) as 'immersive VR' (see Slater et al. 2010).

Slater and Wilbur (1997) provide a framework that accounts the influence of immersion on presence, identifying five key mechanisms: (1) inclusiveness: the extent to which technology/hardware itself is 'invisible' or 'free from signals that indicate [its] existence' (p. 606); (2) vividness: akin to 'realism' and largely describing quantitative indicators of quality such as image resolution; (3) proprioceptive matching: the degree to which the VR content reflects the user's perception of movement and being (e.g. synchronisation of head movement and rotation of the virtual camera); (4) extensiveness: analogous to multi-modality (e.g. audiovisual with haptic feedback is more *extensive* than audiovisual alone); and (5) plot: the presence/absence and depth of narrative. For Slater and Wilbur, these five elements are characterised at the system-side to determine objective immersion. Here, a user cannot feel immersion whilst, equally, a system cannot build presence.

For an alternative perspective, Brown and Cairns (2004) use grounded theory (a branch of qualitative research that processes data first, looking for emergent themes as a means to formulate a hypothesis) to gain an understanding of immersion from the user's perspective. From their data, they observed an emerging hierarchical structure of immersion consisting of engagement (the user is drawn to and commits a degree of time and energy to interacting with the immersive object), engrossment (user is emotionally committed to the object and less susceptible to attending non-object-related stimuli, now perceiving such things as distractions) and total immersion (complete cognitive and emotional commitment to the object and entirely detached from non-relevant stimuli). For Brown and Cairns, these terms represent sequential steps of increasing immersion, and the final step, total immersion, is the phenomenon they describe as presence.

Ultimately which perspective you adopt is likely to result from how you wish to utilise the two terms to theoretically appraise your subject, be it a digital game, VR or other entity. Positioning immersion as an objective, system-side property and presence as a measure of user-experience help us to compartmentalise characteristics of the system. It puts designers in a seemingly more powerful position by marking out a range of strategies with which they can reliably 'build' immersion into their product. The difficulty here is the assumption immersion cannot be *felt* when, with regards to everyday discourse, this is obviously incorrect as 'to feel immersed' is a perfectly acceptable statement. Another problem is that this model of immersion is heavily biased towards the system, accounting for multiple variables in the technology that can affect immersion, without providing any differentiation or nuance for experience of presence beyond 'feeling more present' and 'feeling less present'. This can potentially lead to pretty short-sighted design. Presuming that a VR system that demonstrates (for example) Slater and Wilbur's (1997) five pillars of immersion will automatically generate an equal effect in terms of user-experience, is to make a serious false assumption of homogeneity. Rarely is it suggested within academic research that designers should make this assumption, but it has become a common way of thinking that has filtered through, particularly with regards to computing hardware. For example, representatives of a particularly well-known manufacturer of computing components have made claims that 'total immersion' will be achieved by way of the singular achievement of a 16K (15,360 by 8640 pixels). This problem with this suggestion is that it is essentially prioritising vividness above all else and disregarding any other factors.

As the most established alternative to the system-side/user-side perspective, viewing both immersion and presence as integrated aspects of user-experience, with presence as immersion's deepest iteration, provides us with a richer conception of VR from the position of the user. From a design perspective, it rallies against the widely held assumption that VR can be improved upon with only consideration of the system. It does this by asserting that users, as a heterogeneous population, are an equally importance aspect of immersion and presence. However, this perspective is also not without issue. If we are to accept presence solely as the deepest level of immersion, we are consequently suggesting that it is a finite sensation and that one cannot experience varying degrees or types of presence. This may be underestimating its complexity. A further limitation of this perspective is that, whilst it accounts for the nature of immersive experience, it offers no insight regarding the psychological (namely cognitive and affective) or physiological distinctions of the user that may affect the experience of immersion. In an interesting paper by Psotka and Davison (1993), this issue is investigated, and a range of potential variables regarding susceptibility to immersion are considered, broadly categorised into: (1) imagination: an individual's capacity for imaginative thought; (2) vividness of imagery: how 'real'/substantial imagined content appears to the individual; (3) concentration and attention: the duration and extent to which an individual is cognitively capable of focussing their attention upon VR content and shut out any distractors; (4) self-control: the ability and willingness of the individual to engage with the VR; (5) proneness to simulator sickness physiological conditions that affect the likelihood and individual will have an adverse reaction during a VR experience; and (6) social preferences: whether the individual is more likely to feel immersion when alone or with others. Subsequent research has progressed with these ideas to investigate further underlying 'human factors' determining immersion and presence. These include empathy (the degree to which the user experiences emotional states corresponding to those exhibited within the virtual world, see Sas and O'Hare 2003) and absorption (a tendency directly towards an experience itself rather than its underlying meaning, see Lombard and Ditton 1997). At present, there is still a long way to go in terms of comprehensively establishing and evidencing all of these human factors but there is, however, more than enough for us to acknowledge that individual user differences can have a substantial impact on the depth and form of immersion and presence.

There is most certainly no shortage of research positing the value of sound as a means of evoking and enhancing sensory immersion and presence (Ermi and Mäyrä 2005; Paterson et al. 2010; Ponder et al. 2002). In digital games, sound is argued to be a key method of both drawing the user into the virtual world and maintaining that sensory immersion for the duration of play (Sweetser and Wyeth 2005). In a VR context, the comparable characteristics of sound waves and haptic feedback (both essentially being forms of vibration), when used together, has been touted as a powerful approach to immersion by way of delivering corresponding multimodal feedback to a user (Chu 2007). In the same vein, although VR may continue to be seen as a primarily visual medium, the fundamental immersive quality of such a system is audiovisual, requiring both graphics and sound to not only be present, but deeply interconnected to present a unified experience.

As Grimshaw (2008) notes, the physical nature of a sound wave emanating from a loudspeaker is such that it interacts with the three-dimensional environment, reflecting around the material landscape and filling the space to surround the listener. Grimshaw also observes that this effect is focused upon the acoustic sound wave and becomes one of many effects when the perceptual aspects of sound are considered. Auditory perception reveals a far greater range of immersive potentials by way of factors that include conventions and listener expectation. A study by Garnier and colleagues (2010) adds a social component to sound immersion by way of an experiment that observed the Lombard effect (increasing loudness of speech) in participants when a cocktail party recording was played through their headphones. A control in which participants heard white noise of the same decibel level revealed no such effect, suggesting that it was the social nature of the perceived sound that raised the loudness of participants' speech.

Sound has been shown to support various practical applications by way of increasing presence. For example, Ice and colleagues (2007) reveal that auditory feedback in an online education course enhanced subjective ratings for user presence and information retention scores. As we shall discuss in Chap. 9, the practical application of audio content in VR, by way of its immersive nature and its capacity to enhance user presence extends to significant benefits in applications such as pain distraction, telecommunications, eLearning and embodiment for rehabilitation exercises to name a few.

So, how do we put these pieces together? In Chap. 3, the theory of Sonic virtuality positions sound as an emergent perception actualised from a fluctuating virtual aggregate; a matrix of physical and psychological components whose properties and positions ultimately determines the nature of a sound as a complete perceptual experience. In something of a recurring theme within this book, virtuality and emergent perception have conceptual application beyond sound, with one of these applications being immersion. Based on the above, we can merge the technological aspects of immersion with the human factors to begin to map out an emergent model of immersion. Here, the distinction is not explicitly between system-side and user-side factors but rather between those that are of the physical domain and those that are psychological. This positions Slater and Wilbur's (1997) 'objective immersion' (inclusiveness, vividness, narrative, extensiveness and proprioceptive matching) all within the physical domain whilst human factors are split, with most in the psychological domain (including imagination, attention, social factors, etc.) except those associated with physiological traits (susceptibility to VR sickness for example). Collectively, these factors form the virtual aggregate within which any combination of elements can interact with one another. Brown and Cairns' (2004) work can also be acknowledged, as their degrees of immersion (engagement, engrossment and total immersion) describe the characteristics of immersion as an experience that are then fed back into the virtual aggregate to become potential influencing factors of future immersion actualisations (i.e. how immersed you feel in this moment will influence how immersed you feel in the next). The emergent perception of immersion should also acknowledge the various forms of presence, including ludic (present in the gameplay/interactions), diegetic/narrative (present in the story), social (present in the characters), geographical/spatial (present in the place) and temporal (present in the time/period). Finally, an emergent model would also acknowledge wider aspects of user-experience to draw connections between our immersion in VR and our sensations of flow, fun and diegesis; three key components of user-experience we now turn our attentions towards.

Further Psychological Aspects

Immersion and presence describe only a part of the overall framework for user-experience and the following discusses several further psychological aspects that are worthy of our attention. Individually, these aspects give insight into how we as users respond to VR, and how the experience can actually feel. One common thread revealed throughout this section, however, is that all user-experience aspects are inherently tied together, the characteristics of one influencing those of another in a complex aggregate of variables, from which emerges the overarching perception; our experience in VR.

Flow

Not exclusive to the domain of VR or even digital games, the concept of flow arose in the late 1980s from the work of Csikszentmihalyi (1990) whose model of flow describes the term as an optimum balance of high challenge and equally high skill. To experience flow, an individual must experience intense arousal that they perceptually attribute to a challenging context whilst simultaneously feeling that they are in control. This sensation of control is itself ascribed to a firm belief in their skill as relevant to the task. Should the balance tip towards overwhelming challenge, the individual would experience anxiety, whilst a skew in the opposing direction would evoke relaxation or even boredom. Flow incorporates several cognitive and affective characteristics that include feelings of contentment and pleasure, attenuated awareness of stimuli external to the flow situation and a loss of 'time consciousness' (Faiola et al. 2013).

With traits such as challenge, skill and pleasure, it is unsurprising that flow was embraced by digital game studies as an approach to user-experience testing, specifically for evaluating enjoyment in games. 'GameFlow' (Sweetser and Wyeth 2005) contextualises Csikszentmihalyi's concept, retaining the centralisation of a challenge/skill balance but with further emphasis on the gameplay and user-experience in relation to social interaction. By way of this social aspect, GameFlow accounts for the multi-user-experience in many digital games, with the assertion being that flow in this context evokes a sense of competition, cooperation and community.

Reid (2004) presents us with a model of flow within VR and associates the term alongside notions of playfulness, cognitive ability, competence, creativity, satisfaction and self-efficacy by way of volitional control. For Reid, playfulness refers to elements within a virtual environment that encourages 'active involvement', specifically circumstances in which the user sets their own goals and delineates the meaning of these goals and their subsequent actions. For example, in an open world game such as *Fallout 4* (Howard 2015), it is the difference between following the prescribed storyline missions and wandering the wasteland looking for your own adventure. To possess self-efficacy is to believe that you have the abilities required to perform appropriately in a given situation (Bandura 1994). Volitional control is a facilitator of self-efficacy; 'a pattern of thoughts and feelings that predisposes and enables persons to anticipate, choose, experience and interpret behaviour' (Reid 2004, p. 455). Completing this model of flow are three outputs that can be utilised as dependent measures when evaluating flow: competence, creativity and user satisfaction. To elicit flow, the experience must stimulate feelings of self-efficacy by accurately matching the difficulty (or challenge) to the user's (typically cognitive) ability, thereby evoking volitional control and eliciting sensations of competence (as the user receives positive feedback for their actions) and creativity (as they formulate unique personal gameplay narratives—less playing the game and more playing *with* the game) that together contribute towards increasing user satisfaction.

With regards to flow, an example of the above framework would be the increased naturalistic design of VR's control interface within a first-person shooter game context. Contemporary HMDs map the audio-visual display to orientation (and in some cases, position) of the user's head, rather than by way of mouse or gamepad input. Consequently, as the user can now respond to positional sound cues (e.g. footsteps indicative of approaching enemies) with physical movement to orientate their avatar for a response action. The challenge is whether the user can successfully respond to the sound cue, shooting the enemy before they are themselves shot. The skill is in their ability to meet the challenge by way of action, and it is here that the interface reveals its effect. Being able to manoeuvre the avatar using physical movement may facilitate more accurate movement, thereby increasing user ability and decreasing challenge. Alternatively, the user may find the naturalistic interface unstable, negatively affecting ability and increasing challenge. How quickly users can adapt to such a HMD interface could also differ from adaptation to a mouse or gamepad controller. Here the underlying framework for flow remains constant but it is how the balance of challenge and ability is established and retained that distinguishes flow for VR. The interface of VR also presents an association to the work of Roger Caillois (2001 [translation from 1958]), who identifies three individual forms of flow: Agôn flow (a feeling of continuous engagement within a competition), ilinx (an unbroken sensation of presence in a non-physical space) and mimicry flow (actions in virtual space accurately represent physical-world counterparts). It is the latter of these that can be related to naturalistic interfaces as the means by which we 'mimic' physical action in contemporary VR is quite different to digital games.

Caillois' alternative forms of flow bring us neatly to a brief look at the interconnectivity of user-experience elements, specifically in his notion of ilinx, which is analogous to deep immersion and presence. During an experience of ilinx, individuals with a motivational priority towards being immersed (escapism, fantasy, etc.), experience flow more as an unbroken continuation of that immersion and a feeling of presence within the virtual world. Chen (2007) also likens flow with immersion, asserting that: 'Descriptions of the [f]low experience are identical to what users experience when immersed in games' (p. 32). Immersion is a central pillar of the GameFlow model (Sweetser and Wyeth 2005) and, for Cowley and colleagues (2008), the connection between flow and immersion positions the latter as one determining factor of the former. Another connection can be found within a conceptual framework by Chou and Ting (2003) that positions presence as a central aspect of flow experience, asserting a circular relationship in which feeling present (via ludic immersion) within a virtual environment draws all user focus towards play, in turn evoking sensations of control that contributes towards flow. This then feeds back as the experience of flow enhances presence, taking the user further down the rabbit hole, so to speak.

Diegesis

Traceable back to Ancient Greece and Plato's *Republic*, the term *diegesis*, broadly speaking, distinguishes narration from representation (which Plato referred to as 'mimesis'). A simple example of this would be, in the telling of a personal anecdote, the difference between describing the objects and events of the story (diegesis) and re-enacting some speech or a behaviour that occurred at the time (mimesis). This example resonates with the more contemporary understanding of diegesis as 'the fictional world in which the situations and events narrated occur' (Prince 2003, p. 20). For Loponen and Montola (2004), diegesis is analogous to imaginary frameworks, specifically, the construction of fictitious worlds by a receiving individual (i.e. the audience at a cinema screening or the reader of a novel) in response to the information presented to them by the media; and emphasising diegesis as an aspect of user-experience by highlighting the role of the individual. Bunia (2010) likens diegesis to 'the spatiotemporal universe of the story' (p. 679), but then continues that this understanding is flawed and that diegesis describes a more elusive and complex phenomenon. This is certainly an appropriate sentiment when we consider the many variations that exist upon the theme of diegesis.

A resurgence of interest in the subject of diegesis arose in late 1960s with the work of Gerard Génette, who positioned diegesis within more contemporary narratology and described the term roughly as the 'narrated world' (1970). A relatively recent review article on the subject by Bunia (2010) provides a detailed account of diegesis, and how perspectives on the term have developed in recent years. Here, Génette's diegetic forms are outlined that include: hetero-diegetic (the narrator does not feature in the narrative), homo-diegetic (a single individual is both narrator and protagonist), extra-diegetic (narration is performed externally to the narrative) and intra-diegetic (a protagonist performs narration within the narrative itself).

Sound has a particularly substantial presence in academic literature concerning diegesis. For instance, Percheron and Butzel (1980) consolidate theoretical perspectives relevant to diegesis in sound for motion pictures. Here cinematic sound is diegetically classified by way of synchronisation to the image, with *sound on* being that which can be clearly attributed, at that moment, to a visual representation upon the screen (diegetic sound) and *sound off* referring to sound with no synchronised image (non/extra-diegetic sound). This, for example, could differentiate two sonically identical instances of a violin playing. In the first instance, the sound is presented alongside the moving image of a woman standing motionless, whilst in the second, the sound accompanies the woman playing a violin, her movements matching the duration and articulation of the notes. Percheron explains that such distinctions are often difficult to ascertain in practice, exemplifying with a hypothetical scenario in which the sound of chirping birds overlays the image of a rolling hills landscape in which the birds cannot be seen (what is commonly referred to as 'acousmatic sound', Kane 2014). Here, the diegetic properties of the sound are not determined objectively but are instead resultant from the perception of the listener. Whilst one individual may notice the asynchronisation between sound and image, interpreting the chirping as extra-diegetic, another may semantically link the sound to the synchronous images that are present (the hills and trees) and perceive the sound to be within the diegetic world despite no birds being present on screen.

Prominent in the study of both literature and cinema, it is unsurprising that explorations of diegesis have found their way into the realm of digital games. Within this context, definitions of the term are largely comparable to those outlined above as it describes the internal world of the digital game (Lindley 2002). That said, the diegetic properties of something like a digital game reveal some of the significant complexities of the medium. All within the same moment, a character within a game may embody three distinct diegetic profiles. At once they may be a 'fictional being' (a protagonist/antagonist within the narrative), a 'ludic game piece' (elements of the gameplay or mechanical underpinnings of the game) and an avatar within a social space (relevant to multi-user games in which characters are controlled by other human users). Together, these profiles form what Schröter and Thon (2014) call 'intersubjective constructs', a singular character that is also a plurality, revealing different constructs depending on perspective.

With both sound and digital game studies describing diegesis as an important theoretical underpinning of design, it is to be expected that an amalgamation of the two provokes significant research attention. However, consideration of sound within games reveals a much greater conceptual complexity when compared to cinema or literature. As Collins (2007) explains, the interactive nature of virtual environments presents more factors to consider regarding the variables generated by the human user in that, during interaction, 'the audience is engaging directly in the sound-making process' (p. 263). Ekman (2005) delineates four forms of game sound in terms of their diegetic character by way of two factors, diegetic/non-diegetic (non-diegetic being comparable to extra-diegetic) and signal/referent ('signal' referring to the broader characteristics of the sound and 'referent' meaning the specific object or event that the signal represents). 'Signal' and 'referent' can be thought of as respectively analogous to proximal (perceived at point of sensation, e.g. the listener's ears) and distal (perceived at the point of source, i.e. object or event) theories of sound perception. Within this model a sound may be diegetic in both signal and referent (i.e. the sound is perceived to originate within the game world and be indicative of an object/event also in that world), non-diegetic in both (such as in the musical score of the game) or a combination of the two. The first two instances are simply referred to as 'diegetic' and 'non-diegetic', respectively, but in the other two circumstances, where there is a mixture, Ekman identifies diegetic signal/non-diegetic referent as 'masking sounds' and non-diegetic signal/diegetic referent as 'symbolic sounds'. To elucidate: as you begin the game a brief musical excerpt plays to establish the scene (non-diegetic). As you take your first tentative steps forward, the sound of your avatar's footsteps matches the pace of your movement and reflects the surface upon which you are walking (diegetic). Progressing further, the briefest of leitmotifs upon a double bass immediately informs you that a particular enemy is nearby (symbolic) before the sound of a glass smashing behind you, not actually tied to any game object, makes you spin around in terrified anticipation of what might have made that sound (masking). Whilst this model of game sound diegesis is somewhat problematic and overlaps with sound concepts such as the auditory earcon and acousmatic sound, it does help illustrate the increasing complexity that the modality of sound brings to understanding diegesis within virtual environments.

From analysis of the survival horror genre of digital games, Kromand (2008) presents the concept of 'diegetic collapse' positing that well-crafted sound within a game, moving between diegetic and non-diegetic sound characteristics (both in signal and in referent) can undermine user confidence as they feel less able to extract gameplay-relevant meaning from the sounds in the game; an ideal scenario when your intention as a designer is to evoke a fear response. This diegetic/non-diegetic fluctuation may mean that the listener is presented with otherwise comparable sounds that are sometimes tied to a diegetic object/event and, at other times, are not. This relates to Jørgensen's (2011) notion of 'trans-diegetic sound'. Derived from Gerard Génette's literary theory of 'metalepsis', trans-diegetic essentially describes sounds that transform from one diegetic property to another during the course of audition. A common digital game example of this would be the use of high-pass and reverberation filters upon a musical sample to give the impression that the source of the music is within the diegetic world (say, emanating from a radio—although no such object is required to be visible in the scene). Gradually, fading out the filters then creates an impression of transition between diegetic states, as the return of the lower frequencies and dryer timbres in the music reflects listener expectations for incidental (background/score) music. Here, the perceived source of the music has, without interruption to the sample, changed from being internal to the diegetic world, to being external.

Within virtual environments (digital games specifically), Jørgensen (2011) suggests that trans-diegetic sound does not necessarily require any change to the acoustic sound wave. Instead, a sound (typically speech) of diegetic origin can become non-diegetic simply by being directed at the user, as opposed to the user's avatar. In this instance, a game character converses with the user as if they are aware of the diegetic divide itself and that you (the user) are of a different world to them—what cinema commonly refers to as the 'fourth wall break' trope. Such a technique is demonstrated to tremendous effect in *The Stanley Parable* (Wrenden 2013), in which the game's narrator, frustrated at the user's reluctance to exit a broom closet, starts addressing the user in person:

> I've come to a very definite conclusion about what's going on right now. You're dead. You got to this broom closet [...] when a physical melody of some sort shut down your central nervous system and you collapsed on the keyboard.

This form of this trans-diegesis is also a regular feature of real-time strategy games, in which in-game characters address the user directly, but from within the diegetic world. This effect is also attempted in various first-person perspective games in which your avatar does not possess a name (and is often a 'silent protagonist') and in-game dialogue is contrived to evoke the illusion that the characters are conversing directly with the user. Here, immersion is the endgame intention, as the user experiences a bridging between worlds; the virtual world extending into the physical by contextualising the user sat at their computer as part of the diegetic discourse ('virtual-to-physical trans-diegesis') and the virtual world containing (or responding to) more physical-world content, such as demonstrating non-diegetic awareness ('physical-to-virtual trans-diegesis').

The nature of trans-diegetic content also has specific implications for VR user-experience. Returning to our dimensional model (see Chap. 2), the 'reality-continuum' axis concerns diegesis as we move further towards the 'reality' apex. For example, whilst a digital game largely consists of a virtual world into which the 'reality' element is limited to the user, VR is increasingly integrating physical-world and virtual-world content together (henceforth 'mixed reality'). Here, a virtual character could respond to the characteristics of a physical object. For example, biometric sensors could integrate the user's heart rate into the VR world. This information could then be sent to the character's intelligence framework, enabling them to respond to that information as if it were drawn from the virtual world. In this scenario, the user is sharing something that would otherwise be extra-diegetic with what would otherwise be a diegetic character. At present, approaches to integrating augmented and mixed content are continuing to develop, and the result is a significant further step towards diegetic collapse, far beyond what was originally described by Kromand (2008).

Mentioned briefly in Chap. 3, the acoustic ecology framework for first-person shooters (see Grimshaw and Schott 2008) presents several further diegetic forms that are unique to our interactions within virtual worlds. Whilst their focus is upon how game sounds construct narrative, these forms could arguably distinguish game content of any modality. Ideodiegesis refers to a single user's narrative (specifically *yours*) within which kinediegesis describes ideodiegetic narrative pertaining to their actions and physiology. Lastly, tele-diegesis is primarily relevant to multi-user games and shares characteristics with telepresence (described in the next subsection of this chapter). It refers to narrative formed by a second user in response to ideodiegetic content of the first user. For example, in a multi-user FPS such as the *Counter Strike* series (Le et al. 2000), sub-machine gun fire is generated from the actions of User One (kinediegetic) and relates to that user's ideodiegetic narrative: *'I am trying to kill User Two'*. This is not the narrative of User Two of course and for them the gunfire is telediegetic: *'User One is trying to kill me'*. These diegetic forms are not necessarily limited to VR or digital games and could also be applied to physical board games or shared imaginative play.

The broad intention of this Chapter is to demonstrate the interconnectivity of presence, immersion, flow, identity and various other concepts to support the assertion that VR and VR sound are holistic phenomena, best understood as ecologies and not as sets of independent factors. With this in mind, diegesis reveals close connection to immersion, specifically by presenting us with the means to more accurately describe immersive experience. Within a digital game context, McMahan (2003) differentiates 'diegetic immersion' from 'non-diegetic immersion', defining the former as being immersed within the narrative of the virtual world. The latter refers to being immersed in play and is also known as 'ludic immersion' (Ryan 2009). Connections can also be established between diegesis and flow. For example, two of Schröter and Thon's (2014) diegetic forms overlap very neatly with two particular forms of flow. Their 'avatars within social space' diegetic character partners with Sweetser and Wyeth's (2005) 'GameFlow' concept (a social-based experience of flow in multi-user games) whilst character as 'ludic game piece' is highly reflective of Caillois' (2001/1958) 'Agôn' flow.

Fun

The capacity to evoke a sensation of fun within digital games and VR is most certainly a principle evaluative measure, not only for recreational activities but also more serious applications. Significant correlations have been drawn between a digital game's capacity to elicit fun and its effectiveness as an educational tool (Rosas et al. 2003). Equally so, reviews of therapeutic digital games have posited increased efficacy correlates with experiencing fun (Kato 2010). A programme that elicits fun can give an impression of greater usability and even influence consumers' buying decisions (Carroll and Thomas 1988). Fun as a concept is most certainly one of much significance to almost all people. Martin (2016) notes that, when comparing the number of results found on Google, the search term 'fun' returned more responses than did 'sex' (update: sorry to report that at the time of writing this book, this is no longer the case, but it is still pretty close).

McKee (2016) notes some difficulty in understanding fun, stating that the most prominent specialist dictionaries in communications studies and philosophy have yet to include an approved definition for the term. Despite this, our conceptual understanding of fun is developing, in part due to the work being published in the field of human-computer interaction. For Koster (2013), fun relates to the process of learning, asserting that when we develop in some way, for a purpose that is meaningful to us, we shall experience fun. This reflects an earlier assertion by Monk and colleagues (2002) that, particularly within software designs, systems should aspire to both utilitarian and hedonic ideals: 'people are neither interested in a dull but very useful tool nor a fancy but useless toy' (p. 925). In a study by Hsu and colleagues (2005), contrasting versions of the retro arcade classic *Pacman* (Iwatani et al. 1980) were played by participants who then evaluated the designs. Factor analysis of the results revealed six measures relevant to fun: novelty and powerfulness, appealing aesthetics, interactivity, challenge, sense of control and reward. Of these measures, a sense of control, powerfulness and challenge all resonate heavily with the concept of integrated user-experience by way of their relevance to flow. For Koster (2013), fun is positioned at the upper-most intensity levels of flow, asserting that when challenge is high but the user feels able to control the situation and powerful enough to cope with that challenge, then a sensation of fun will surely emerge.

As alluded to earlier, fun also draws association with immersion. Federoff (2002) ties immersion to fun by way of usability, suggesting (relating to Caroll and Thomas 1988) that a game capable of eliciting fun increases perceived usability and that, in turn, facilitates greater immersion. Here, immersion also feeds back into fun as Federoff cites Aycock (1992), suggesting that feeling immersed within a 'separate reality' is itself inherently fun. More recent qualitative research by Poels and colleagues (2007) reasserts these connections between immersion and fun, positing that processes of 'imaginative immersion' (e.g. creative behaviour within the game, free-roaming exploration of the virtual environment) are consistently paired with sensations of fun, as described by the users themselves.

Towards the beginning of this chapter, the notion of gamification as a means of enhancing user-experience in non-game applications was outlined. This relates to the connection between immersion and fun, as can be elucidated with the concept of 'ludic language play'. Broner and Tarone (2001) explain this term as part of second language acquisition, in which gameplay and pedagogic strategies are merged to create a learning experience that is fun. Evoking this response in students not only increases engagement in the initial stages but also has increased power to retain their attention by immersing them in the language.

With a focus more specifically upon VR, we can extrapolate an established connection between fun and novelty (Hsu et al. 2005) when we consider that the inherent novelty of HMDs may be one of the leading causes of VR's capacity to evoke fun. As the hype-machine of contemporary VR is quick to point out, the new generation of HMDs, and particularly phone-based/wireless systems, is presenting users with interactive experiences quite unlike anything else. Whilst rarely the direct focus of any particular study, VR research frequently reflects this assertion in its observations of fun and enjoyment being mentioned repeatedly in qualitative feedback from participants. Reporting of this is particularly prominent in comparative evaluations of a VR therapeutic application against its 'analogue' counterpart (physical, in-vivo, etc.), and this is largely attributed to two fun-generating factors; novelty and gamification (Bryanton et al. 2006; Halton 2008; Schmitt et al. 2011). Of course, as the technology becomes more ubiquitous and normalised as part of our everyday lives, this effect is likely to fade. Consequently, designers need to be aware that novelty as a means of evoking fun within VR is a well that may soon run dry.

Whilst academic literature has not addressed the nature of fun in terms of auditory experience, some instances can be found that assert audio-only games are perfectly capable of evoking a sense of fun (Targett and Fernström 2003). Similarly, 'satisfying sound' is a term presented by Schneiderman (2004) to describe an aesthetic auditory ideal balanced between sounds that are annoying and those that go completely unnoticed. These satisfying sounds are what Schniederman describes as one of the essential features of interactive fun.

Extending beyond what we have discussed above, several further connections can be inferred to position fun as a key element within an overarching model of VR user-experience. For example, presence and diegesis can both be linked to fun by way of immersion. As the sensation of fun increases our engagement with the VR system, our attention becomes ever more fixed upon it and filters out any would-be distractors from the physical environment, causing us to feel present within the virtual world. Working the other way, it seems equally plausible that a sensation of presence would impact upon fun, potentially both as an attenuator and amplifier (or even a negative effect), this time by way of flow. Should the user feel present within the virtual environment, it would follow that this would increase their capacity for control. This effect is also discussed earlier within this chapter regarding ludic immersion where presence leads to control, then the sensation of flow, the more intense experience of which we have already connected to fun. Interestingly, the precise opposite could also be true. Whilst feeling deep presence within VR would facilitate increased user control, should the challenge presented to the user still exceed their ability, the balance of flow would be upset. The resultant failure for the individual, who is deeply present within the environment and emotionally committed to the task, would likely evoke frustration, the arch nemesis of fun.

To Summarise

User-experience unquestioningly extends beyond immersion, presence, flow, diegesis and fun. That said, these elements collectively go a long way towards illustrating the nature of VR experience as something that is highly complex and also unique to the individual. No single element of user-experience can be manipulated in isolation. Much like the frustrated software developer who fixes one system bug to discover two new bugs spring forth as a direct result, designing for fun, immersion or any user-experience aspect without consideration of wider effects, limits design control. In addition to these elements being inexorably tied, equally so is the VR system entangled with its user(s). Therefore, whether designing for VR or attempting to better our understanding, it is important that we do not fall afoul of giving undue prominence to either the system or the user. Both are crucial components of the emergent model of VR.

Reframing the Self and Identity in the VR Age

Earlier within this chapter, we examined the role of diegesis with regards to the experience of VR, in which some issues relating to identity were raised. Specifically, diegesis concerned identity in the context of an individual's position in relation to the conceptual walls that separate worlds. This final section widens the net somewhat to explore further the nature of the self and its relationship with digital games, sound, VR and beyond. As with previous terminological difficulties that have emerged so far throughout this book, the self and identity are confounded and often interchangeable terms. To help clarify the following discussion, we shall identify the two as discrete perspectives upon a wider phenomenon, with 'the self' positioned specifically as our conceptualisation of our personal identity (from within the psychological domain) and 'identity' the broader term encapsulating the self, alternative forms of identity (e.g. social, collective) and more objective and quantitative aspects.

What Is the Self?

Issues pertaining to the self and identity permeate much of our everyday lives. Burke (1997) traces the origins and development of the self and individuality across European culture, noting that identification and deliberation upon the self have occurred relatively consistently throughout history. For Carl Jung, the self can be represented as a circle, at the epicentre of which is positioned the ego (i.e. consciousness, 'self-knowledge'), a significantly smaller dot that consumed a very small proportion of the larger circle's surface area. Here the self incorporates self-knowledge but extends far beyond it, alluding to a large proportion of our identity existing outside both our understanding and control (Eisendrath and Hall 1991). From a sociological perspective, identity can be roughly reduced to fundamentals of similarity and difference that construct a framework of classifications that, in turn, form groupings that we may position ourselves within or external to (Lawler 2015). Our gender, ethnicity, sexuality, values, political leanings and so on can all be characterised by this framework.

More broadly, Cote and Levine (2015) differentiate three levels of identity: social (concerning our position within society), personal (our outward behaviours) and ego (the sense of personal agency and continuity of experience). A further form, 'collective identity' is presented by Karolewski (2009) to describe phenomena such as nationalism and other identification of individuals within large-scale groupings. Identity is also descriptive of the relationship between the individual and the place and time of their environment. In more recent times, neuroscience has been muscling in on the question of identity, utilising brain imaging techniques in attempts to reveal neural correlates. This research is beginning to match both specific regions within the brain and particular patterns of neural activity with qualitative markers of identity. For example, a study by Platek and colleagues (2004) observed heightened activity across the middle and superior frontal gyri section of the right brain hemisphere corresponding to participants' recognising themselves in photographs. More recently, research has also begun to uncover ways in which the self is represented not just by localisation of brain activity, but by characteristics of the neural impulses themselves. Tacikowski and colleagues' (2014) study examined participants' electroencephalographic responses to audition of familiar against unfamiliar names (including instances of participants' own names), revealing distinctions between groups in the amplitudes across numerous waveform components of event-related potentials (ERP: immediate neuroelectrical activity in response to a stimulus). The conclusion drawn from this study being that greater ERP amplitudes are indicative of a sense of greater subject importance being attributed to a stimulus (i.e. what we perceive to be relevant to our identity is neurologically prioritised over that which is non-relevant). Sticking with the psychodynamic theme, this can be left to your superego as to whether you feel we a biologically absolved of being selfish, it simply being something that is hardwired within our brains.

Sound and the Self

Sound possesses numerous associations with identity, not least of which is the substantial role identity has in our connection to music which, if we relate to the forms of identity above, reveals a deep and multi-layered relationship. The music we listen to may influence our friendships and, mostly in our more formative years, part-dictate our positioning within social hierarchies. How we dress, our mannerisms and gestures, even our language and vocal articulations can be heavily resultant from affiliation with a particular genre of music. Furthermore, music enjoys a powerful role within wider popular culture and is often a chief signifier of different periods within cultural history. Being simultaneously a central aspect of the sonic background to everyday life, music can differentiate between discrete periods within our own personal histories and characterise our personal continuity of experience. Finally, the power of music to reveal similarities between individuals on a massive scale has led to the formation of several collectives large enough to support sociocultural change, from Punk and anti-authoritarianism to Hip hop and Black civil rights.

Sound other than that which is language or music also discloses matters of identity. Ego identity and sound can certainly be linked when we consider both the affective potential of sound as a driver of agency and the various means by which sound facilitates continuity of experience. Sound can direct our attention and encourage particular actions whilst an enduring soundscape can effectively unify an otherwise fragmented visual scene. Sound has the potential to effectively characterise social identity: a cacophony of industry, muffled shouting and wailing sirens indicative of a very different position in a social hierarchy when compared to the sounds of pouring Champagne, the fine strike of crystal and the gentle lapping of waves upon the shore.

Sound is a powerful means of characterising the here and now, the presence/absence, positioning and relative loudness of various sonic elements of the soundscape revealing our location in both place and time, alongside further description of the environment. Situational identity and sound are of significant relevance to virtual environments, in which designers can transcend physical limitations to transport users into a seemingly limitless variety of situations. Referring again to the acoustic ecology framework of first-person shooters (Grimshaw and Schott 2008), digital game sound is classified by the alternate ways in which it situates the user; described by the authors as 'functions' of sound. A sound may, for example, differentiate between any two discernible locations ('topoplast'), facilitate the perception of in-game or real-world passages of time ('chronoplast'), or identify the historical period ('aionoplast').

The Virtual Self

Identity, or rather the distinct lack of it, is often a central component of many a digital game narrative; a point well made by Ben 'Yahtzee'

Croshaw (2010), who argues that: 'If adventure games were a medical condition, the first symptom would be amnesia and the second would be kleptomania'. The design decision to remove any identity from a game's protagonist primarily serves to enhance immersion by way of a shared cognitive circumstance. As the user, you are newly introduced to the protagonist the game wishes you to embody and the world in which they are positioned. Removal of the avatar's identity by way of narrative-amnesia mitigates the their prior historical and situational knowledge (i.e. the avatar does not 'know' who they are, where they are, what they are meant to do or the underlying meaning of their situation), thereby placing the user and avatar within a more shared cognitive space and, ultimately, increasing immersive potential. The nature of the virtual self is directly addressed in the survival horror game *SOMA* (Grip et al. 2015). Be forewarned, the following contains a major plot spoiler. The prelude of the game sees protagonist Simon Jarrett visiting a brain imaging specialist who claims to be able to digitally replicate Simon's mind. The purpose of which is to then run simulations upon the digital mind in order to determine optimal treatment for Simon's brain injury. During the scan, Simon passes out and awakes in what he later discovers is a deep sea research facility in the year 2104. At the mid-point of the game, both the protagonist/avatar and the player discover that Simon's consciousness is in fact the digital copy and the physical 'Simon' has been dead for many years. For the player, this is a fascinating transition of self, as the continuity of their experience during play essentially tricks them into misunderstanding the identity of their avatar, and as many users have become immersed in the game and embody the character, they have also misunderstood themselves.

Interaction within virtual worlds opens up a new dimension in terms of understanding identity and a great deal of research explores the concept of the self within the context of digital games. Back in the mid-1990s, Sherry Turkle (1994) explored the reconstruction of the self when playing in the *MUD* (Multi-User Dungeons). MUDs are analogous to contemporary Massively Multiplayer Online (MMO) worlds, describing networked virtual environments in which large numbers of users interact with each other by way of their bespoke/customisable avatars. In MUD/MMOs, individuals are interacting socially within a virtual environment in ways highly comparable to the physical equivalent but what is most interesting is that the differences between virtual and physical interactions are primarily ways in which the virtual worlds can transcend the limitations of the physical world. VR and digital games can offer users options for designing and editing their identities that would be significantly difficult or even impossible to attain in the physical world.

Turkle (1994) notes that MUD users have the freedom to construct an identity that reflects a personal ideal or, equally, one that closely reflects their own perception (i.e. an identity that matches the self). This freedom to construct an identity has also been utilised by users to explore particular characteristics of themselves that they dislike, feel ashamed of or fear by imbuing their avatar with such traits. Users are able to experiment with gender swapping, non-human forms and multiplicity of identity. Perhaps most intriguing is the capacity afforded by technology for 'invisibility', in which an individual's identity can be disassociated from its progenitor, essentially generating a non-identity and giving them the opportunity to act with anonymity. This invisibility within shared virtual worlds raises the notion that, whilst the self is intrinsically analogous to the individual, it nevertheless resonates beyond this to impact more widely upon communities. In 1993, the first known case of 'cyber-rape' was committed in the MUD game *LambdaMOO* (still active at time of writing). MacKinnon (1997) discusses this as an event that encouraged us to reappraise the social construction of rape and appreciate that a physical interaction is not required to induce severely damaging psychological effects, specifically in circumstances such as this in which the virtual object being assaulted is, at the time of attack, being deeply embodied by a human user.

As with our discussion regarding immersion and presence, there is evidence to suggest that identity, both in a perceptual sense and as something that guides outward behaviours, is highly susceptible to environmental stimuli, including that of a virtual environment. The 'Proteus effect' (Yee et al. 2009) refers to observations of changes in users' behaviour resulting from visual characteristics of their character-avatars. In their study, Yee and colleagues exemplify this principle, presenting a significant link between the height of a user's avatar and the aggressiveness of their play style, positing that a virtual identity has the power to influence actual behaviour.

Our exploration of identity in virtual environments finds itself entangled with some of the arguments within Chap. 2, in which it was discussed how the virtual was becoming 'everyday'. In a paper entitled 'The Cyborg's Dilemma', Biocca (1997) describes an emerging circumstance in which continued interactions with artificial interfaces in virtual technology become perceptually more 'natural' over time. The effect of this is unnatural adaptations in the user's thinking and behaviour. As the machines become more human, we at the same time become more mechanical. More specifically, Biocca explores the relationship between avatar representation on the distortion of body image and body schema (an individual's conceptual model of their self). Here, immersive designs such as accurate mapping of movement between user and avatar enable the virtual body to compete with the physical body over what we perceive as our 'phenomenal body'. In a related experiment, Fox and Bailenson (2009) noted that individuals regularly exposed to a virtual avatar (representing themselves), that 'gained/lost weight' in response to physical activity, would voluntarily exercise significantly more than the control group. The conclusion being that identification with the avatar evoked greater motivation for exercise. Observing the avatar lose weight felt to the user, like progress. These are examples of embodiment, which is succinctly described by Csordas (1990) as when '[…] the body as a methodological figure [is] nondualistic, i.e., not distinct from or in interaction with an opposed principle of mind' (p. 8). This reiterates the effects observed above; the deeper that the associations are between our mind and a body (not necessarily 'ours' or even of material form), the more the former is susceptible to influence from changes to the latter.

Csordas' 'methodological body' is revealed in more recent research to yield substantial explanatory power. An experimental study by Slater and colleagues (2010) investigated the perceptual phenomenon of body transfer, within a VR context. Participants to this study were presented with a virtual environment by way of a HMD. The environment was an unremarkable room using relatively low-fidelity design. Participants first observed two female virtual characters (one standing and one seated) on the other side of the room, the one standing stroking the other's shoulder. During the experience, the test group participants' perspectives would then change to that of the seated woman, at which point the researcher would stroke the participant's shoulder synchronously to the movements of the virtual character. Following several minutes of this experience, the standing virtual character would stop the stroking action and suddenly strike the other avatar. Slater and colleagues collected qualitative feedback and also heart rate deceleration (in response to the sudden strike) to determine the degree of body transfer. Their results found that the participants largely experienced significant stress and physiological change in response to the virtual avatar being hit and reported high levels of ownership towards to avatar, experiencing what was essentially a basic graphical representation of a body to be their own.

The capacity of VR to induce ownership of a body other than their own has been described as akin to an out-of-body experience or 'disembodiment' (Lenggenhager et al. 2007). Feeling as though you are leaving your physical form arguably overlaps with our understanding of presence, specifically telepresence (see Chap. 9). Telepresence refers to the experience of being present in a place other than one's physical location. Once again, this ties back to the virtual becoming the everyday as telepresence is certainly not limited to what we would typically describe as VR technology or even digital systems. As we shall discuss further in later chapters, a good novel or film can certainly induce the feeling of being present elsewhere, as the reader or viewer engages cognitively and emotionally with the content to the extent that the physical domain (extending to both their material body and surrounding environment) falls out of their attentional focus and the other world draws them in.

De Ridder and colleagues (2007) differentiate disembodiment, from 'depersonalisation', referring to a break not from the body, but from the self. Aardema and colleagues (2010) explore depersonalisation from within a VR context. Their study asserts that people can be positioned along a continuum of depersonalisation. Along this continuum, daydreaming constitutes a mild iteration of the condition and depersonalisation disorder one that is more severe. For Aardema and colleagues, this position can shift in response to several factors. VR is argued to be one such factor, as their study found that numerous qualitative measures of depersonalisation (incorporating aspects such as a sense of presence, emotional connectivity to the physical world and observed disturbances in memory and cognition) were shown to increase following VR exposure.

Whether VR applications are extending our methodological bodies into the virtual domain or simply disassociating us from our physical form, both effects have substantial value within the field of rehabilitation. Addressed fully in Chap. 9, VR-mediated rehabilitation procedures are a rapidly advancing area of research, within which both bodily extension and disassociation feature prominently. For example, mirror therapy of Phantom Limb Pain (see Ramachandran and Rogers-Ramachandran 1996) has been built upon by way of VR. Here, patients observe a virtual limb, through a HMD, that is displays position and movement matching patient's perception of their phantom. As they no longer possess the particular physical limb, the virtual analogue becomes an extension of their physical body, 'able to evoke vivid sensations of movement originating from the muscles and joints of amputees' phantom limbs' (Murray et al. 2009). Equally, disassociation facilitates notable therapeutic value in the form of VR applications that attenuate physical presence and connection with one's body for analgesic purposes. A well-established example of this is *Snow World* (Hoffman et al. 2003), a VR game in which patients typically suffering from severe burns throw barrages of snowballs at penguins and snowmen in an arctic environment.

Connecting Identity to User-Experience

Alongside notable connections to VR, digital games and sound identity and the self do not buck the trend of integration with other concepts of user-experience. Returning to Csikszentmihalyi and flow (1990), disappearing concern for the self is positioned as a central tenet of the conceptual model. This reveals that the experience of flow during gameplay correlates with a disassociation from identity (when we experience flow our focus is so much upon the task that our identity merges with it). Jennett and colleagues (2008) connect identity, immersion and flow in their SCI-model, in which challenge-based immersion (balancing skill with challenge) is analogous to most descriptions of flow whilst being immersed imaginatively (i.e. forming an identity-mediated bond with game characters or narrative) reflects aspects of the self. Gerard Génette's diegesis pays much regard to the identity of the narrator, fundamentally linking diegesis with identity. Within cinema, as Koutsourakis (2012) notes, the diegetic identity of the bodies projected upon the screen can often be obscure. For example, if an actor presents a particularly convincing affective display, this can make it difficult for the audience to discern if what they are witnessing is an act (identity as character) or a genuine expression (identity as actor). For Koutsourakis, even the camera (our viewing perspective) possesses identity that is routed in diegesis. When considered, the camera is typically non-diegetic, a portal of the physical world enabling the viewer to experience the fictitious. By way of its movements, however, it can acquire the identity of a diegetic character. For example, by simulating head movements appropriate to the narrative, the camera can simulate a diegetic character's viewpoint, creating the illusion for the audience that they are observing events thorough a particular character's eyes.

Chapter Summary and References

In addition to providing a more detailed account of VR user-experience theory, this chapter has sought to illustrate the importance of recognising VR as a unique medium for studying user-experience and identity, whilst also highlighting the relevance of sound across all points of discussion. Throughout this chapter, a holistic perspective has been championed by observing the numerous ways in which the individual components of user-experience and elements pertaining to identity are all interconnected. The intention is to emphasise the importance of not identifying each user-experience component as an isolated entity, as we add yet more pieces to the emergent framework of VR sound.

Throughout the next few chapters, this book traces the historical progression of VR and audio technologies, seeking to address the question of how the relationship between VR and sound has developed over the years. Throughout these subsequent chapters, a bi-directional relationship begins to reveal itself, including both how our changing understanding of sound is reflected in VR (conceptualisation, technological developments and software designs) and also how VR and associated technologies (such as biometrics, binaural systems and head-related transfer function) are influencing our understanding of sound. The purpose of these discussions is to help elucidate the meaning of VR sound and also to provide an account of how an emergent framework can ultimately lead to better design and implementation.

Note

1. The UX Awards (UXies): https://userexperienceawards.com.

References

Aardema, F., O'Connor, K., Cote, S., & Taillon, A. (2010). Virtual reality induces dissociation and lowers sense of presence in objective reality. *Cyberpsychology, Behaviour and Social Networking, 13*(0), 1–8.

Aycock, H. (1992). Principles of good game design. *Business Source Premier, 14*(1), 94–98.

Bandura, A. (1994). *Self-efficacy*. Wiley.

Bernhaupt, R. (2015). User-experience evaluation methods in the games development life cycle. In *Game User-experience Evaluation* (pp. 1–8). Springer International Publishing.

Benson, J., Olewiler, K., Daniels, J. W., Knoop, V. & Wirjadi, R. (2016). *An Experience Framework for Virtual Reality*, medium.com: https://medium.com/@Punchcut/an-experience-framework-for-virtual-reality-f8b3e16856f7#.ii9mv9hbq.

Biocca, F. (1997). The cyborg's dilemma: Progressive embodiment in virtual environments [1]. *Journal of Computer-Mediated Communication, 3*(2), 0–0.

Broner, M. A., & Tarone, E. E. (2001). Is it fun? Language play in a fifth-grade Spanish immersion classroom. *The Modern Language Journal, 85*(3), 363–379.

Brown, E., & Cairns, P. (2004). A grounded investigation of game immersion. In *CHI'04 Extended Abstracts on Human factors in Computing Systems* (pp. 1297–1300). ACM.

Bryanton, C., Bosse, J., Brien, M., Mclean, J., McCormick, A., & Sveistrup, H. (2006). Feasibility, motivation, and selective motor control: Virtual reality compared to conventional home exercise in children with cerebral palsy. *Cyberpsychology & behavior, 9*(2), 123–128.

Bunia, R. (2010). Diegesis and representation: Beyond the fictional world, on the margins of story and narrative. *Poetics Today, 31*(4), 679–720.

Burke, P. (1997). Representations of the self from Petrarch to Descartes. In: Porter, R. (Ed.) *Rewriting the self: Histories from the Renaissance to the Present.* USA: Routledge. 17–28.

Caillois, R. (2001). *Man, play and games.* 1958. *Trans.* Meyer Barash. Urbana and Chicago: University of Illinois Press.

Carroll, J. M., & Thomas, J. C. (1988). Fun. *ACM SIGCHI Bulletin, 19*(3), 21–24.

Chen, J. (2007). Flow in games (and everything else). *Communications of the ACM, 50*(4), 31–34.

Chou, T. J., & Ting, C. C. (2003). The role of flow experience in cyber-game addiction. *CyberPsychology & Behavior, 6*(6), 663–675.

Chu, L. L. (2007). *U.S. Patent No. 7,208,671.* Washington, DC: U.S. Patent and Trademark Office.

Collins, K. (2007). An introduction to the participatory and non-linear aspects of digital games audio. *Essays on sound and vision*, 263–298.

Cote, J. E., & Levine, C. (2015). *Identity formation, youth, and development: A simplified approach.* Psychology Press.

Cowley, B., Charles, D., Black, M., & Hickey, R. (2008). Toward an understanding of flow in digital games. *Computers in Entertainment (CIE), 6*(2), 20.

Croshaw, B. (2010). Amnesia: The Dark Descent [internet video review]. *The Escapist.* http://www.escapistmagazine.com/videos/view/zero-punctuation/2092-Amnesia-The-Dark-Descent.

Csikszentmihalyi, M. (1990). *Flow: The psychology of optimal experience* (1st ed.). New York: Harper & Row.

Csordas, T. J. (1990). Embodiment as a paradigm for anthropology. *Ethos, 18*(1), 5–47.

Cuddy, A. (2015). *Presence: Bringing Your Boldest Self to Your Biggest Challenges.* Hachette UK.

Cummins, J. (2000). *Immersion education for the millennium: What we have learned from 30 years of research on second language immersion*. Retrieved April, 16, 2006.

De Ridder, D., Van Laere, K., Dupont, P., Menovsky, T., & Van de Heyning, P. (2007). Visualizing out-of-body experience in the brain. *New England Journal of Medicine, 357*(18), 1829–1833.

Deterding, S., Sicart, M., Nacke, L., O'Hara, K., & Dixon, D. (2011). Gamification. using game-design elements in non-gaming contexts. In *CHI'11 Extended Abstracts on Human Factors in Computing Systems* (pp. 2425–2428). ACM.

Eisendrath, P. Y., & Hall, J. J. A. (1991). *Jung's self psychology: A constructivist perspective*. Guilford Press.

Ekman, I. (2005). Meaningful noise: Understanding sound effects in computer games. *Proc. Digital Arts and Cultures*.

Ermi, L., & Mäyrä, F. (2005). Fundamental components of the gameplay experience: Analysing immersion. *Worlds in play: International perspectives on digital games research, 37*(2), 37–53.

Faiola, A., Newlon, C., Pfaff, M., & Smyslova, O. (2013). Correlating the effects of flow and telepresence in virtual worlds: Enhancing our understanding of user behavior in game-based learning. *Computers in Human Behavior, 29*(3), 1113–1121.

Federoff, M. A. (2002). *Heuristics and usability guidelines for the creation and evaluation of fun in digital game*. Doctoral dissertation, Indiana University.

Fox, J., & Bailenson, J. N. (2009). Virtual self-modeling: The effects of vicarious reinforcement and identification on exercise behaviors. *Media Psychology, 12*(1), 1–25.

Garnier, M., Henrich, N., & Dubois, D. (2010). Influence of sound immersion and communicative interaction on the Lombard effect. *Journal of Speech, Language, and Hearing Research, 53*(3), 588–608.

Genette, G. (1970). Fronteras del relato. *Barthes, R. y otros, Análisis estructural del relato. Buenos Aires: Tiempo Contemporáneo.*

Graham, M., Jordan, T., Pharoah, A. et al. (2006-2007). *Life on Mars*, Kudos Film and Television, BBC Wales.

Grimshaw, M. (2008). Sound and immersion in the first-person shooter. *International Journal of Intelligent Games and Simulation, 5*(1), 119–124.

Grimshaw, M., & Schott, G. (2008). A conceptual framework for the analysis of first-person shooter audio and its potential use for game engines. *International Journal of Computer Games Technology, 2008*, 5.

Grip, T., Nilsson, J., Thomas, I., et al. (2015). *SOMA*. Sweden: Frictional Games.
Halton, J. (2008). Virtual rehabilitation with digital games: A new frontier for occupational therapy. *Occupational Therapy Now, 9*(6), 12–14.
Hassenzahl, M. (2003). The thing and I: Understanding the relationship between user and product. *Funology*, (pp. 31–42). Springer Netherlands.
Hassenzahl, M., & Tractinsky, N. (2006). User-experience-a research agenda. *Behaviour & information technology, 25*(2), 91–97.
Hoffman, H. G., Richards, T., Coda, B., Richards, A., & Sharar, S. R. (2003). The illusion of presence in immersive virtual reality during an fMRI brain scan. *CyberPsychology & Behavior, 6*(2), 127–131.
Howard, T. (2015). *Fallout 4*. USA: Bethesda Game Studios.
Hsu, S. H., Lee, F. L., & Wu, M. C. (2005). Designing action games for appealing to buyers. *CyberPsychology & Behavior, 8*(6), 585–591.
Huston, P. (2007). *Martin Buber's journey to presence* (No. 7). Fordham Univ Press.
Ice, P., Curtis, R., Phillips, P., & Wells, J. (2007). Using asynchronous audio feedback to enhance teaching presence and students' sense of community. *Journal of Asynchronous Learning Networks, 11*(2), 3–25.
Iwatani, T., Funaki, S., Kai, T., et al. (1980). *Pacman*. Namco: Japan.
Jennett, C., Cox, A. L., Cairns, P., Dhoparee, S., Epps, A., Tijs, T. et al. (2008). Measuring and defining the experience of immersion in games. *International journal of human-computer studies, 66*(9), 641–661.
Jerald, J. (2015). *The VR Book: Human-Centered Design for Virtual Reality*. Morgan & Claypool.
Jørgensen, K. (2011). Time for New Terminology?: Diegetic and Non-Diegetic Sounds in Computer Games Revisited. In: Grimshaw, M. N. (Ed.) *Game Sound Technology and Player Interaction: Concepts and Developments* (pp. 78–97). IGI Global.
Kane, B. (2014). *Sound unseen: Acousmatic sound in theory and practice*. USA: Oxford University Press.
Karolewski, I. P. (2009). *Citizenship and collective identity in Europe*. Routledge.
Kato, P. M. (2010). Digital games in health care: Closing the gap. *Review of General Psychology, 14*(2), 113.
Koster, R. (2013). *Theory of fun for game design*. O'Reilly Media, Inc.
Koutsourakis, A. (2012). Cinema of the Body: The Politics of Performativity in Lars von Trier's Dogville and Yorgos Lanthimos' Dogtooth. *Cinema: Journal of Philosophy and the Moving Image, 3*, 84–108.

Kromand, D. (2008). Sound and the diegesis in survival-horror games. *Audio Mostly 2008.*

LaViola, J. J., Jr. (2000). A discussion of cybersickness in virtual environments. *ACM SIGCHI Bulletin, 32*(1), 47–56.

Lawler, S. (2015). *Identity: Sociological perspectives*. Wiley.

Le, M., Cliffe, J., et al. (2000). *Counter-Strike*. USA: Sierra Studios.

Lenggenhager, B., Tadi, T., Metzinger, T., & Blanke, O. (2007). Video ergo sum: Manipulating bodily self-consciousness. *Science, 317*(5841), 1096–1099.

Lindley, C. A. (2002). The gameplay Gestalt, narrative and interactive storytelling. In *Proceedings of computer games and digital cultures conference*, Finland: Tampere, 6–8 June.

Loomis, J. M. (1992). Distal attribution and presence. *Presence: Teleoperators & Virtual Environments, 1*(1), 113–119.

Lombard, M., & Ditton, T. (1997). At the heart of it all: The concept of presence. *Journal of Computer-Mediated Communication, 3*(2), 0–0.

Loponen, M., & Montola, M. (2004). A semiotic view on diegesis construction. *Beyond role and play*, 39–51.

MacKinnon, R. (1997). Virtual rape. *Journal of Computer-Mediated Communication, 2*(4), 0–0.

Martin, P. (2016). *Play Hard—Have Fun: A Philosophy for Life*. Lulu.com.

McKee, A. (2016). *FUN!: What Entertainment Tells Us About Living a Good Life*, Palgrave Entertainment Industries.

McMahan, A. (2003). Immersion, engagement and presence. *The video game theory reader, 67*, 86.

Monk, A., Hassenzahl, M., Blythe, M., & Reed, D. (2002). Funology: Designing enjoyment. *CHI'02 Extended Abstracts on Human Factors in Computing Systems* (pp. 924–925). ACM.

Murray, C. D., Pettifer, S., Howard, T., Patchick, E. L., Caillette, F., Kulkarni, J. et al. (2009). The treatment of phantom limb pain using immersive virtual reality: Three case studies. *Disability and rehabilitation.*

Nancy, J. L. (1993). *The birth to presence*. Stanford University Press.

Paterson, N., Naliuka, K., Jensen, S. K., Carrigy, T., Haahr, M., & Conway, F. (2010, September). Design, implementation and evaluation of audio for a location aware augmented reality game. In *Proceedings of the 3rd International Conference on Fun and Games* (pp. 149–156). ACM.

Pausch, R., Proffitt, D., & Williams, G. (1997). Quantifying immersion in virtual reality. In *Proceedings of the 24th annual conference on Computer*

graphics and interactive techniques (pp. 13–18). ACM Press/Addison-Wesley Publishing Co.
Percheron, D., & Butzel, M. (1980). Sound in cinema and its relationship to image and diegesis. *Yale French Studies, 60*, 16–23.
Poels, K., De Kort, Y., & Ijsselsteijn, W. (2007). It is always a lot of fun!: Exploring dimensions of digital game experience using focus group methodology. In *Proceedings of the 2007 conference on Future Play* (pp. 83–89). ACM.
Ponder, M., Herbelin, B., Molet, T., Scherteneib, S., Ulicny, B., Papagiannakis, G. et al. (2002, November). Interactive scenario immersion: Health emergency decision training in JUST project. In *VRMHR2002 Conference Proceedings*.
Platek, S. M., Keenan, J. P., Gallup, G. G., & Mohamed, F. B. (2004). Where am I? The neurological correlates of self and other. *Cognitive Brain Research, 19*(2), 114–122.
Prince, G. (2003). *A dictionary of narratology*. University of Nebraska Press.
Psotka, J., & Davison, S. (1993). Cognitive factors associated with immersion in virtual environments.
Ramachandran, V. S., & Rogers-Ramachandran, D. (1996). Synaesthesia in phantom limbs induced with mirrors. *Proceedings of the Royal Society of London B: Biological Sciences, 263*(1369), 377–386.
Rebenitsch, L., & Owen, C. (2016). Review on cybersickness in applications and visual displays. *Virtual Reality, 20*(2), 101–125.
Regan, E. C., & Price, K. R. (1994). The frequency of occurrence and severity of side-effects of immersion virtual reality. *Aviation, Space and Environmental Medicine, 65*(6), 527–530.
Reid, D. (2004). A model of playfulness and flow in virtual reality interactions. *Presence, 13*(4), 451–462.
Roach, N. W., Heron, J., & McGraw, P. V. (2006). Resolving multisensory conflict: A strategy for balancing the costs and benefits of audiovisual integration. *Proceedings of the Royal Society of London B: Biological Sciences, 273*(1598), 2159–2168.
Rosas, R., Nussbaum, M., Cumsille, P., Marianov, V., Correa, M., Flores, P. et al. (2003). Beyond Nintendo: design and assessment of educational digital games for first and second grade students. *Computers & Education, 40*(1), 71–94.

Ryan, M. L. (2009). From narrative games to playable stories: Toward a poetics of interactive narrative. *Storyworlds: A Journal of Narrative Studies, 1*(1), 43–59.

Sas, C., & O'Hare, G. M. (2003). Presence equation: An investigation into cognitive factors underlying presence. *Presence, 12*(5), 523–537.

Schmitt, Y. S., Hoffman, H. G., Blough, D. K., Patterson, D. R., Jensen, M. P., Soltani, M. et al. (2011). A randomized, controlled trial of immersive virtual reality analgesia, during physical therapy for pediatric burns. *Burns, 37*(1), 61–68.

Shneiderman, B. (2004). Designing for fun: how can we design user interfaces to be more fun? *Interactions, 11*(5), 48–50.

Schröter, F., & Thon, J. N. (2014). Video game characters. Theory and analysis. *Diegesis, 3*(1).

Sheridan, T. B. (1992). Musings on telepresence and virtual presence. *Presence: Teleoperators & Virtual Environments, 1*(1), 120–126.

Slater, M. (2003). A note on presence terminology. *Presence Connect, 3*(3), 1–5.

Slater, M. (2009). Place illusion and plausibility can lead to realistic behaviour in immersive virtual environments. *Philosophical Transactions of the Royal Society B: Biological Sciences, 364*(1535), 3549–3557.

Slater, M., & Wilbur, S. (1997). A framework for immersive virtual environments (FIVE): Speculations on the role of presence in virtual environments. *Presence: Teleoperators and virtual environments, 6*(6), 603–616.

Sweetser, P., & Wyeth, P. (2005). GameFlow: A model for evaluating player enjoyment in games. *Computers in Entertainment (CIE), 3*(3), 3–3.

Tacikowski, P., Cygan, H. B., & Nowicka, A. (2014). Neural correlates of own and close-other's name recognition: ERP evidence. *Frontiers in human neuroscience, 8*.

Targett, S., & Fernstrom, M. (2003). Audio games: Fun for all? All for fun!. In *Proceedings of the 2003 International Conference on Auditory Display*, Boston. MA, USA. 6–9 July 2003.

Turkle, S. (1994). Constructions and reconstructions of self in virtual reality: Playing in the MUDs. *Mind, Culture, and Activity, 1*(3), 158–167.

Wrenden, D., Pugh, W., Higueras, J., et al. (2013). *The Stanley Parable*. USA: Galctic Café.

Yee, N., Bailenson, J. N., & Ducheneaut, N. (2009). The Proteus effect: Implications of transformed digital self-representation on online and offline behavior. *Communication Research, 36*(2), 285–312.

5
Representations

Worlds beyond the physical have existed conceptually, long before they reached the digital platform of a computer. The influential power of fictional representations in literature, cinema and television is such that it has significantly affected our present perceptions and expectations for VR. This chapter reveals how numerous fictional depictions of VR have turned out to be highly prophetic and are indicative of a technological vision to which we humans have remained doggedly loyal. Throughout this chapter, we trace the precursors of VR back through fiction, from the utopias of ancient classicism, to early twentieth-century science fiction and fantasy, to contemporary works published during the first and second generations of commercial VR. We consider how underlying notions of virtuality, alternate reality and simulacra can inform us about the meaning of VR. We also observe the circularity of influence between actual and fictive VR technology, noting how fiction inspires reality which in turn inspires fiction. Sound remains an important aspect throughout as we examine the nature of virtual soundscapes as depicted in fiction, be they virtual, simulated or alternate realities. How sound is treated in fictional visions of VR hardware is also addressed and the overarching assertion is made that sound has been a consistently underrated

T.A. Garner, *Echoes of Other Worlds: Sound In Virtual Reality*,
Palgrave Studies in Sound, DOI 10.1007/978-3-319-65708-0_5

component of VR, both before and after its technological realisation in the computing age. More contemporary visions of VR and their underlying desires are also discussed and it is revealed that, despite setting impossibly high expectations for actual VR technology, these desires have been ongoing aspects of the human condition, powerful enough to survive the 1990s crash and return even stronger in present day. Before we begin, a friendly warning: because this chapter discusses thematic and narrative content, it unavoidably contains a handful of rather significant spoilers of several novels and digital games. Effort is made throughout to identify the particular work before a spoiler is exposed, so that a specific revelation can be safely skipped. Proceed with caution.

Fictional Worlds

Whilst the philosophers of classical antiquity may not be able to lay claim to being the progenitors of other worlds as conceptual forms, the third book of Plato's *Laws* presents us with a prominent example from that era. In the dialogue, three speakers have been discussing governance and law before progressing to the origins and developmental processes of political systems. Towards the end of the book, it is revealed that one of the speakers has intentions to build a new city, Magnesia. The following fourth and fifth books of Laws proceed to conceptually construct Magnesia as a utopian society by way of reasoning upon governance, possessions, virtue and happiness (Bobonich 2011). From Euhemerus' island paradise of *Panchaea* to Tao Yumming's *Peach Blossom Spring*, the notion of utopia is one that has developed in both ancient Eastern and Western cultures. In Renaissance literature, Thomas More's *Utopia* (1516, trans. 1975) and Tommaso Campanella's *City of the Sun* (1602/2012) both, like Plato before them, utilise fictional locales as the foundation for discussions on political philosophy. The utopia is, by definition, an ideal against which we can compare actual society but is itself decidedly virtual, with the term's literal meaning 'nowhere' or 'not-place'.

To be clear, the term 'fictional world' (or 'literary world') is used to denote the setting of a book and the entirety of its characters, locales and narrative. As we shall discuss below, it need not be limited to a planet but may extend out to encompass a universe or even multiverse, megaverse or omniverse ('a universe that is spatiotemporally four dimensional',[1] for those of you who are interested). Equally, it could be limited to an environment smaller than a single room. This section of this chapter is built from the assertion that VR in its contemporary form is, in many ways, a digital descendant of literature's fictional worlds. This is observed not only in terms of objective qualities that include structure, mechanics and aesthetics, but also with regards to subjective user-experience values such as function and received meaning.

The composition of literary worlds is of course, not limited to utopias. Equally so, their function extends beyond comparing an ideal with our flawed, physical universe for the purpose of political discourse. Jonathan Swift for instance, conjured multiple worlds (his 'remote nations') that were neither utopian in character nor for the purpose of political discussion. Instead, *Gulliver's Travels* (Swift 1726) describes worlds with a variety of unique characteristics, some good and others not. Centring the narrative upon an everyman protagonist with which the reader can readily identify, Swift's remote nations offer an experience of the fantastic; from conversing with 72ft. tall giants in Brobdingnag to encountering the gravity-defying city of Laputa. Swift is famously quoted as stating his intention for the work was 'to vex the world rather than divert it' (Winton 1960). The function of his fictional universe is one of satire and amusement.

Before we look at how literature addresses VR in terms of its hardware design and functionality, this chapter first explores the nature of virtual worlds in times that largely predate VR technology; the aim being to uncover some of what the worlds of VR can learn (and indeed have learned) from those of literary fiction. Specifically, points of scale (i.e. the geographical size of the world), realism (the degree to which the world reflects our observable, physical reality) and dimensional portals (how the world is separated into multiple dimensions that may or may not include a 'real world') are discussed in addition to considering effects of genre and aesthetics. We start with scale.

Fictional Scale

The different scales of literary worlds present a prominent feature for differentiating them in terms of form. As noted earlier, the term 'worlds' is akin to that which is diegetic and refers to all content within the piece of fiction. However, a world may exist, implicitly or explicitly, within a delineated space. As with Campanella's City of the Sun and More's Utopia, town and city-level worlds are relatively prominent within fiction. A more contemporary literary example is that of Derry, the setting for many a Stephen King novel, which positions a fictional town within a non-fictional state (Maine) and country (USA). A fictional locale within one that is non-fictional can be reduced to even smaller scales, such as St. Scholastica's College of Philip Pullman's *Northern Lights* (1995), which is located with the very non-fictional University of Oxford. Anthony Trollope's *The Warden* (1855) extends the boundaries somewhat and the book is geographically encapsulated in the fictional English county of Barsetshire, beyond which is the implied presence of England, Europe, Earth and so on. In *Far from the Madding Crowd* (1874), Thomas Hardy's 'merely realistic dream country' of Wessex encompasses multiple fictive reflections of actual counties whilst Aldous Huxley's *Brave New World* (1932) presents a fictive iteration of Earth. Global-level fictional worlds may also be planets other than our own, such as the Discworld, first featured in Terry Pratchett's *The Colour of Magic* (1983).

One fictional world that encapsulates an entire fictional multiverse is the creation of schoolmaster and satirist, Edwin Abbot, with his abstract novella *Flatland: A Romance of Many Dimensions* (1884). Consisting predominantly of two-dimensional geometric shapes, as its name implies, the Flatland world, spins the yarn of a square named 'A Square', who through his dreams and visitations, experiences the alternate realities of Lineland (a universe of only a single dimension), Spaceland (three dimensions) and Pointland (no dimensions). We shall return to scale later in this chapter, when the influence these qualities of fictional worlds have had upon VR is discussed. Before that however, we move on to discuss the matter of realism.

Realism

Abbot's tale of Flatland (1884) highlights another facet of fictional worlds separate from scale, that of degree and form of realism. Indeed, literary analysis can classify works by their realism characteristics. Surrealism, for example, describes worlds that exist within the psychological domain and which are an expression of dreams or of the imagination (Marcus and Nicholls 2004). In contrast, magic realism refers to worlds that are 'realistic' in most aspects but include small but significant fantastic elements; 'the realism of the real is permeated by magic just as the world of the magical is underpinned by the real' (Hart and Ouyang 2005, p. 4). A well-known instance of this form is Joanne Harris' *Chocolat* (1999) where, the 'domestic magic' of protagonist Vianne Rocher is subtly placed within the other-wise realistic setting of Lansquenet-sous-Tannes.

A fictional world may aspire to emulate, or recreate, the non-fiction world almost entirely, creating a work in which the fictitious aspect is very subtle. In Lucy Montgomery's *Anne of Green Gables* (1908), much of the fiction bears close connection to the non-fictive elements that inspired it. Montgomery modelled the appearance (and to an extent, personality) of book's protagonist, Anne Shirley, from an image of early twentieth-century American actress Florence Evelyn Nesbit. Many of the other characters were also based heavily upon people known to the author, whilst fictional locales such as the community of Bolingbroke Redmond College are delicately fictionalised versions of non-fictive places. When both the setting of the fiction and the narrative and characterisation therein closely reflect the author's perception of our 'real' world, this is what we would typically refer to as a 'realistic' form.

Abbot's Flatland (1884) represents the opposite size of the scale in terms of the physical nature of its universe, yet it is not devoid of realism. The narrative (characterisation, plot, dialogue, affective motivations, conflict etc.) features deep connections to our non-fictive world, without which the satirical commentary that serves as the book's function would not work. Such literary conventions arguably concern the humanness of the text and are noticeably consistent across the many fictional worlds depicted within these stories. To exemplify this point with an instance of the 'literary nonsense' genre, Lewis Carroll's *Alice's Adventures in Wonderland* (1865) may have absurdity woven into most aspects of the text, from the characters to the physics. However, the disconnection from actuality is far from total. Much like Lemuel Gulliver before her, our protagonist Alice is highly relatable and reflective of the actual. Furthermore, whilst many of the creatures that populate Wonderland are strange and incredible, their interactions with Alice reveal them to be largely anthropomorphic which, by definition, presents them as composites of non-fiction and the fantastical.

Portals

A final point to note in this brief look at the nature of fictional worlds concerns portals. Whilst many literary worlds describe a single realm, Tolkien's Middle Earth (*The Hobbit*, 1937), for example, many others also incorporate multiple, clearly differentiated realms. In many instances, one of these worlds is far more comparable to our non-fictive world. It may be identified as 'Earth', 'reality', 'the real world', 'our world' and so on. This kind of realm is identified by Tolkien himself (1997) as a 'primary world', with fantastic realms alternatively known as 'secondary worlds'. A common function of the primary world is as a narrative device, analogous to that of many protagonists in such literature (who often originate from this primary world—see Alice and Gulliver); a method of grounding the fantasy within a foundation that is relatable for the reader.

In this multi-realm fiction, the term 'portals' describes any object or process that connects the primary and the secondary worlds together. The nature of the portal has a bearing upon how we conceptualise the story. For Alice, the material properties of the rabbit hole that bridges her world (the primary world) with Wonderland (the secondary world) gives the impression that the latter physically exists within same universe as the former. Of course, Carroll confounds this interpretation at the novel's conclusion when Alice is returned to her world in a moment of awakening, seemingly from a dream. Works such as J.M. Barrie's

Peter and Wendy (1911) and Roald Dahl's *The BFG* (1982) also present their fantastical realms (Neverland and Giant Country respectively) as adjacent to a primary world. In these stories, the portals take the form of obscure pathways through the sky that give no clear indication of precisely where these secondary worlds are in relation to Earth. Methods of obfuscation are commonplace in fantasy literature, with some works documenting multiple portals to fantastic realms, none of which clearly demarcate their exact route. In J.K. Rowling's *Harry Potter* series (1997-2007), Hogwarts may be reached (to name a few) by way of a portal in London St. Pancras station, by teleportation (with the use of Floo Powder, Portkeys or a Vanishing Cabinet), or by way of undisclosed land, air or sea pathways. Despite the range of options, none of the details regarding these portals give any indication of precisely where Hogwarts is; whether it is a region located within the primary world, or somewhere that exists upon a secondary dimensional plane is never explicitly made clear.

When it comes to the nature of fantastic realms and their relative position to the primary world, science fiction literature commonly takes a somewhat different approach to fantasy. The genre is particularly well known for presenting 'alternate reality' Earths, drawn from specific hypothetical projections, much like the utopias of Magnesia and Panchaea. Science fiction often differs from such visions of paradise however, typically by projecting their vision further into the future, integrating organisational and political elements with speculative advances in science and technology. More often than not, they will present the reader with a less utopian and more dystopian vision. The World State of Huxley's *Brave New World* (1932), for example, is formed from the specific consideration of how mind-altering drugs and state-controlled artificial reproduction methods would shape a civilisation, Huxley's answer being a world that is deeply oppressed by a totalitarian authority. In this example, the portal is not a specific object or pathway that connects the primary and secondary world, but rather the 'what if' proposition that the author has identified as the point of divergence that separates the timeline of the story from that of the non-fictional world. This counterfactual history approach to building literary worlds became prominent during the twentieth century and has been utilised to explore questions across several academic disciplines that include cognitive science, geography and political theory (see Singles 2011). Prominent literary examples include Phillip K. Dick's *The Man in the High Castle* (1962) and Winston Churchill's *If Lee had not Won the Battle of New Gettysburg* (1931). The divergence point of the former is the assassination of Franklin Roosevelt in 1933, leading to a world in which the Axis Powers have overcome the Allies and the United States has become a totalitarian state. Churchill's counterfactual narrative diverges from a point during the American Civil War, in which Confederate General Robert E. Lee wins the Battle of Gettysburg, starting a chain of events that ultimately leads to the circumventing of the Great War. With regards to literary worlds, these divergence points provide the figurative portals that separate their fiction from non-fiction.

In addition to the figurative portal of counterfactual history, Science fiction literature does also employ objects and space-time pathways that function as physical portals between the fantastic and more realistic realms. Where science fiction is again differing from the fantasy genre, is in terms of the explicit detail it typically provides regarding the nature of its portal(s). For example, the setting of Philip K. Dick's *Do Androids Dream of Electric Sheep?* (1968) is our Earth 24 years in the future, relative to the book's year of publication. The major events between publication year and that of the books setting are explicitly detailed, giving the reader a clear sense of the temporal pathway that has led from their primary world to the secondary world of the novel. Within the main narrative cannon of the series that began with Frank Herbert's *Dune* (1965), the position of Earth in relation to the planet Arrakis (the setting for much of the series) is given, both in terms of space and time.

Points of scale, realism and portals provide means of analysing and differentiating the worlds of fiction. They also reveal to us the some of the audience's expectations relative to the genre of the work. Following on from these observations, the next section inquires as to how these points have relevance to virtual worlds and VR.

How Literary Worlds Have Inspired Virtual Worlds

The relevance of literature to VR in terms of how their worlds are formed is highly significant in a number of ways. Firstly, the alternative forms of fictive worlds have provided a framework for developing virtual ones. This can be observed across the three points of world characterisation discussed earlier in this chapter, namely; scale of the world, degree or form of realism, and implementation of portals.

In terms of scale, fictional worlds can vary dramatically. Similarly, virtual worlds also utilise specific scales and boundaries to define themselves and subsequently influence our conception of them. To elucidate this by way of contemporary digital game titles, we can contrast the parochial setting of *Everybody's Gone to the Rapture* (Curry and Pinchbeck 2015) against the (almost) universe-spanning world of *No Man's Sky* (Murray et al. 2016). The former transports the player to the 1980s and the fictional village of Yaughton, which is explicitly stated to sit within the non-fictional county and country of Shropshire, England. The latter, by contrast, utilises procedural generation to, theoretically, enable the player to explore an infinite number of galaxies.[2] This difference in scale has a striking effect upon the overall nature of the games and ties into multiple aspects of the overall experiences. One particular example worth noting is the effect scale has upon gameplay mechanics, specifically player movement. Within Rapture, the player can only travel on foot and, whilst there is a means of running, this mechanic is deliberately unintuitive to encourage players to explore slowly and carefully. By contrast, No Man's Sky does facilitate a walking mechanic but also includes a jetpack for terrestrial exploration and largely encourages movement by way of your spacecraft that facilitates travel between planets in minutes and near-instant warping between entire star systems.

Considering the two virtual worlds of Rapture and No Man's Sky, there is also a consistency in terms of how they can be differentiated by way of genre and within that; realism, scale, portals and aesthetic. In terms of genre, Rapture presents us with what is arguably a world of magic-realism. In addition to its smaller-scale fictionalised setting that is positioned inside larger, non-fictive locales, it also utilises a single and specific 'magical' element within a virtual world that otherwise, every effort has been made to make it highly comparable to a typical rural village in the west of England. This in turn drives the aesthetic. Yaughton is rendered in sumptuous detail to emulate the look and feel of an actual environment; from the weathered welcome sign that reads 'please drive carefully' to the Tambour mantel clock atop a fireplace which was, in our primary world, an exceptionally common adornment in rural households during the 1980s. Within this relatable setting, unexplained patterns of light and the vanishing of all the village's inhabitants provide the magic in magic-realism. Finally, the portal connecting the fiction with the actual is also matching of the genre in that there is no explicit or physical portal present as there is only a single reality within the game. Consequently, the portal is that which connects the player to the game. The real world is the player's world and the portal is effectively the screen. This reflects typical magic-realism literature, where the portal between worlds is the written page. No Man's Sky by contrast embodies the space opera subgenre of science fiction. This also resonates equally with literary expectations in terms of scale (a vast expansive universe), realism (a mechanical/pseudoscientific set of processes 'explaining' fantastical elements), portals (a universe that either implicitly or explicitly contains our actual Earth that can be reached via travel through space) and aesthetic (clean, polished and minimalist), all of which reveals many consistencies with the genre as it is presented in literature.

This is not to say that the nature of either fictional (whether in literature or other media) or virtual worlds is restricted to a single genre's 'vanilla' template. Examples across both forms can be found in which, even in the specific terms of world-building, genre is bended, hybridised, or has one or more of its components (scale, realism, portal and aesthetic) altered. Just as David Mitchell's (2004) *Cloud Atlas* brings us a swashbuckling adventure within a mystery/thriller within a dystopia (and that doesn't exactly do the genre-complexity of the novel any real justice), so does the forthcoming (at time of writing) title *Fabular: Once Upon a Spacetime* attempt to blend in equal parts the conventions of science fiction and fantasy. Whilst there are undoubtedly more examples of this kind of experimentation within literature and cinema than there are in virtual world-building, there certainly appears to be an emerging trend in that direction. The numerous successes and deeply interesting fictional worlds that have been crafted through experimental designs arguably lead us to conclude that future virtual worlds will continue to explore avenues of experimentation, with their literary and cinematic precursors as guides.

In addition to providing a partial framework for defining and expressing their nature, virtual worlds also owe much of their aesthetic to literary precursors. The look and feel of many virtual worlds can be clearly attributed to specific subgenres of literature, from the neo-Victorian Steampunk world of *Bioshock Infinite* (Levine 2013) that owes a debt to works such as those of H.G. Wells and Jules Verne, to the galaxy-spanning virtual world of *Mass Effect* (Hudson 2007), its aesthetic built upon literary worlds of the space opera subgenre. Gothic horror as an artistic direction in digital games takes a substantial influence from literature of the same genre, particularly examples of the late eighteenth and early nineteenth century. Let us take the case of Edgar Allan Poe's *The Fall of the House of Usher* (1839) wherein he describes the titular house:

> I looked upon the scene before me - upon the mere house, and the simple landscape features of the domain - upon the bleak walls - upon the vacant eye-like windows - upon a few rank sedges - and upon a few white trunks of decayed trees

More than a century later, the Usher homestead has been directly attributed to the design of Derceto, the haunted mansion of the original *Alone in the Dark* (Raynal 1992) survival horror game. As the game's protagonist, Emily Hartwood, ascends the driveway on approach to the Derceto mansion, we can observe in a single shot the ghostly teal walls of the exterior, the twisted black trees in the grounds, and the single lit attic window that is seemingly gazing down at Emily as she nears the entrance. An eerie reflection of Usher.

Similarly, the post-apocalyptic aesthetic in virtual worlds also reveals distinct literary cues. For instance, the barren and dilapidated landscape of *STALKER: Shadow of Chernobyl* (Bolshakov et al. 2007) is heavily influenced by the description of Harmot, a fictional Canadian town in *Roadside Picnic* (Strugatsky and Strugatsky 1972). Characters and objects that populate virtual worlds and contribute to its overall aesthetic can also reveal inspirations from literary sources. For example, the design of the 'air screamer' creature from *Silent Hill* (Toyama et al. 1999) directly descends from an illustration and description within Arthur Conan Doyle's *The Lost World* (1912). Clear comparisons can also be drawn between Huxley's *Brave New World* (1932) and the recent digital game title, *We Happy Few* (Sears and Clayton 2016), both of which prominently feature a widely distributed happiness drug, 'Soma' and 'Joy' respectively, that suppresses independence and encourages conformity.

In addition to direct routes by which virtual worlds take influence from literary ones, many indirect examples can also be found. In such instances, cinema often provides the more direct source of inspiration but the literary precursor remains important as the motion picture that inspired the virtual world is typically influenced, in turn, by the novel. Such a trajectory can be observed as we trace the cyberpunk aesthetics of the digital game *Deus Ex* (Spector et al. 2000) to the motion picture *Blade Runner* (Scott 1982), itself an offspring of *Do Androids Dream of Electric Sheep?* (Dick 1968). Rather than reinventing the wheel here by listing further instances, this indirect connection can be observed in a great many examples online.[3] The site features a database of films, books and digital games, and outlines numerous sources of inspiration for each, typically citing interviews with the creators themselves. From the examples that can be found on this site, it is made very clear that, whilst many of the games attribute much of their aesthetic and design to film, this line of inspiration links inexorably to a literary source.

From the above, it is clear that the virtual worlds of digital games and VR have been greatly influenced by literary precursors with regards to various aspects of their design. But what about sound? The next section looks specifically at the relationship between fictive worlds and sound. Specifically, we examine how literature utilised descriptive language to present sound within fictional worlds before seeking to uncover how VR sound in particular is reflected in fiction.

The Sound of Fictive Worlds

There is an argument to be made that literature is, in fact, one of the earliest forms of sound recording. Predating the advent of acoustic recording technology, literature provided (and continues to provide) a means of capturing and communicating auditory information. In much the same way as notation captures music, a text (be it fiction or non-fiction) can utilise descriptive language to document the tone, flow and character of a sound. As Calanchi asserts, literature is 'the most powerful reservoir of sounds from our past' (2015, p. 11). This section briefly explores methods of expressing sound and soundscapes by way of text (described as 'sound-language' from here on in) for the purpose of world-building.

Using *The Fall of the House of Usher* (1839) to exemplify her point, Calanchi (2015) notes that Poe's poem uses sound-language to describe its world. Poe's method for characterising the soundscape is not by way of descriptive language, but rather by phonological effect; selecting words that, when spoken, reveal a timbre that helps characterise the world. As St. Clair (2013) points out, contemporary soundscape studies are primarily focussed upon media that produce physical sound waves and literature is largely overlooked as a vehicle for soundscapes. Indeed, it is rather difficult to uncover much in the way of research exploring world-building language that prioritises sound over image. Adams (1989) however, does discuss a particular text which is described in its own foreword as: 'the noisiest novel ever written'. Henry Roth's *Call it Sleep* (1934) earns this accolade arguably within its first few pages: 'All day the guttural, the high-pitched voices, the astonished cries, the gasps of wonder, reiterations of gladness had risen from the decks in a motley billow of sound. But now her decks were empty, quiet…' (pp. 10–11). Roth's attention to the soundscape of his fictional world reveals the significant descriptive power that sound can afford to an author, particularly as a means of characterising difference and change over time. Comparing descriptive soundscapes in literature to VR, expressing an initial soundscape to better characterise a subsequent one has certainly been applied to the design of many virtual worlds. It is particularly prominent in survival horror games, where the oppressive droning of the background soundscape suddenly disappears, leaving a gaping hole of silence that grabs the player's attention and pushes sound (ironically in absentia) to the foreground. A sonic juxtaposition is created, that can effectively undermine player-expectation and evoke a sense of fear.

Revisiting some of the literary worlds referenced earlier in this chapter, we can observe that sound-language presents several further functions with regards to world-building. For example, in Conan-Doyle's *The Lost World* (1912), sound-language is utilised to create a sense of physical space: '…a frightful outburst of sound, the uproar of the carnivora cage when the step of the bucket-bearing keeper is heard in the distance' (p. 69). A similar effect is achieved in Philip K. Dick's *Do Androids Dream of Electric Sheep* (1968), in which acoustic descriptors are used to create an impression of the fictional space: 'Deckard found an echoing, noisy, slightly miscontrived rehearsal taking place' (p. 44). Huxley's *Brave New World* (1932) demonstrates how sound-language can very efficiently populate a literary world whilst also evoking a sense of dynamics: 'This hive of industry, as the Director was fond of calling it, was in the full buzz of work' (p. 99). In equal measure, the opposite effect is realised in Roald Dahl's *The BFG* (1982), in which silence evokes a static or still world:'The house was absolutely silent. No voices came from downstairs. There were no footsteps on the floor above either' (p. 1).

The above techniques reveal noteworthy comparison to virtual worlds, with sound design being a highly effective way of characterising the world in a way that the visuals alone would not be able to do with the same level of efficiency. A contemporary example of this appears in the not-particularly concisely titled (but nevertheless, excellent) *Dr. Langeskov, the tiger and the terribly cursed emerald: A whirlwind heist* (Pugh 2015). Throughout the game, the majority of graphical objects are static with almost all dynamic elements being presented by the game audio. When the titular tiger itself makes an appearance it is only revealed as sound. At various in-game locations, when the player gets close enough to some of the walls and doors, they can hear numerous muffled voices and sounds on the other side, indicating that much of the action is occurring somewhere else. This effectively evokes a virtual world that is both larger and more populated than the visuals would suggest. *Everybody's Gone to the Rapture* (Curry and Pinchbeck 2015) demonstrates how sound in virtual worlds can evoke both a sense of movement within a relatively static visual scene and also a feeling of stillness and emptiness by way of an attenuated or silenced soundscape. Winds whispering through the trees, birds chirping in their branches, and the rustling of long grass and bushes as the player pushes through the vegetation. These events and interactions fill the soundscape of Rapture's virtual world but are, in most cases, not rendered graphically. The opposite effect is observed as the player enters The Stars at Night public house and approaches the bar. At this point, the gentle ticking of the clock upon the wall appears almost invasively loud, highlighting by contrast the deafening silence of the overall soundscape.

Science fiction, particularly the space opera subgenre, often exploits sound-language to indicate the dynamic motions and physicality of objects central to the narrative. A good example is in Isaac Asimov's *Foundation* (1951): 'The ship landed in a medley of noises. There was the far-off hiss of the atmosphere cutting and sliding past the metal of the ship. There was the steady drone of the conditioners […] and the slower rumble of the engines enforcing deceleration' (p. 17). Likewise, virtual worlds that belong within the space opera subgenre, such as *Mass Effect* (Hudson 2007) will regularly use a variety of sounds to represent the physical properties and dynamics of a spacecraft, from the heavy metallic clunks as the ship docks, to the phase-sweeping wind sound as the craft slows from a jump through hyperspace.

The points above specify a number of connections between how sound is represented through text within literary worlds and by way of crafted audio soundscapes in virtual worlds. However, possibly the most prominent realisation that emerges when considering sound within these two mediums, is just how consistently sound is deprioritised against visuals. If you were to read through most of the literary examples presented above, amongst many others, you would observe the reliability with which the author first describes the setting in terms of what can be seen, and then by what can be heard (if the soundscape is mentioned at all). In terms of virtual world design, this ranking of the sensory modalities is equally stark, with digital game research repeatedly acknowledging this issue. As we shall address in more detail in later chapters, this does not necessarily present a problem for VR, but rather an opportunity. As a relatively fledgling technology that affords us new ways of creating and interacting with sound (see Chap. 8), VR need not follow the same path as its forerunners. It could instead champion the benefits of a more balanced multimodal approach to virtual world design.

Literary Visions of VR

So far, this chapter has examined the nature and construction of fictional worlds in literature as a means of better understanding the virtual worlds of digital games and VR. We continue this chapter with a look at how VR is, and has been, explicitly depicted in literature, with a view to uncovering how literary visions of VR may have been, and may continue to be influential. A quick reminder: this section contains spoilers.

Simulated Reality

In much the same way as the utopias that facilitated Plato's philosophical explorations presented the reader with a highly idealised subject, so too has much of fiction approached VR from an idealistic perspective. The term 'simulated reality' is, as a diegetic entity within fiction, arguably more commonplace than VR. Broadly defined (most prominently in science fiction literature), simulated reality is the theoretical apex of VR, a fully immersive, multisensory, artificial world that is perceptually indistinguishable from the 'genuine' or 'real' world and it is therefore no longer 'virtual' (in terms of no longer being almost or approximately real).

One common feature of the simulated reality trope involves the author exploiting the indistinguishability of the simulation from reality as a means to confound characters within the narrative, making them believe that the world is not a simulation before revealing the truth at a later point. This technique is also regularly used upon the reader, providing the twist in the tale as everything up until the revelation has to be re-evaluated in this shocking new context. Tolkien might describe such a thing as 'a secondary world masquerading as a primary world'. A second frequent characteristic of simulated reality is that it has been manufactured by some form of creator, whose existence is often positioned beyond the space-time boundaries of the simulation. These points are well illustrated in an early instance of literary simulated reality, *Simulacron-3* by Daniel Galouye (1964). In this story, protagonist Douglas Hall is one of a team of scientists tasked with constructing a 'total environment simulator' within which, digital humans will possess consciousness but be unaware of their true nature—all for the purposes of conducting mass-market research. Following a series of events, including the vanishing of his colleague, Hall begins to conclude that these occurrences are evidence that he is himself a computer-generated being who exists within a world that is as much a simulation as the one he created. In cinema, the truth behind a simulated reality is often withheld from the protagonist but not typically the audience; such as in *The Matrix*[4] (Wachowski and Wachowski 1999). Simulated reality as a key narrative point is not a regular occurrence in digital games, particularly as a means of deceiving either the protagonist or the player. *Assassin's Creed* (Désilets et al. 2007) is one instance, depicting the Animus, a simulated reality system that generates its environments from 'genetic memories'.

Simulated reality as a concept extends to philosophical thought and is the subject of a fascinating discussion concerning our (primary) universe. Barrow (2007) presents an assertion that resonates with the story of Simulcron-3 in stating that, were our universe a simulation, unavoidable imperfections in its code (or influences from the creator-universe) would inevitably cause glitches and drifts in the physical constants of nature to appear over time. In his discussion, Barrow characterises the discussion, citing work by Bostrom (2003) and Hanson (2001) who assert respectively that the theoretical capacity to generate simulated universes is such that our universe being simulated is in actual fact a statistical likelihood, and that consequently, we should all strive to participate in more significant events and interact with more influential people as this will raise our status within the narrative of the simulation and thereby increase our chances of continued existence. Stand out or else risk early deletion.

Whilst there is no intention for this chapter to engage in any kind of theological debate, there are obvious resonances between the philosophical concept of simulated reality and the universe as depicted in the book of Genesis within the Old Testament of the Bible. Whilst it is ambiguous as to whether the creator is positioned outside a spatial, temporal or dimensional perimeter, Genesis implies the existence of some form of reality beyond our known universe and nominates an entity from that reality as 'constructor'. Whether the creationist interpretation of the universe's conception is to be considered fiction is not an assertion being made here, only that the concepts being described within the opening pages of Genesis arguably reflect subsequent literary depictions of simulated reality. The Bible may be a somewhat tenuous example of the simulated reality concept, but it is one of the more well-known instances that predates the genre of science fiction.

Regarding twentieth century literature, certain works reveal how the simulated reality narrative arose not from a technological concept, but from broader ontological questions concerning the nature of being. To elucidate, Robert Heinlein's *They* (1941) is one of the earliest fictions to explore the concept of simulated reality, doing so from a solipsist perspective that has itself also become a well-established narrative device. In They, our unnamed protagonist is confined within a mental institution due to his insistence that everything he experiences in the world is a false creation, propagated as part of a conspiracy by 'real' entities that exist in another world, outside of the one in which the protagonist is situated. The narrative proposition of They has resounding similarity to that of *The Darkness 2* (Carter et al. 2012) which transforms the scenario into an interactive dilemma. Throughout the game the setting alternates between two alternate worlds. The first, a supernatural 'noir' underworld in which protagonist Jackie Estacado is endowed with dark, fantastical abilities. The second is an asylum in which he is a patient. The question of which is the true reality culminates towards the end of the game when Jackie finds himself teetering atop the edge of the asylum roof and the player is given a choice; accept the asylum as real and climb down from the ledge, or insist that the underworld is real and throw Jackie from the roof.

Moving forward a decade to the mid-twentieth century, and simulated reality in a technological guise begins to appear, as science fiction attempts to make explicit the mechanisms by which such a thing could potentially exist. *The Veldt* (Bradbury 1950) prophetically depicts a technological concern that would continue to have impacted more than 60 years later. Bradbury's tale is of two children, Peter and Wendy, who become deeply addicted to the spoils and efficiencies of their nursery simulation. Their parents slowly come to realise (albeit, far too late) that they have no connection with, and are of no value to their children, who have now all but rejected the physical reality in favour of their virtual one. A variation upon this theme can be found in *Fallout 3* (Howard et al. 2008) in the depths of Vault 112. In this place, the antagonist Dr. Stanislaus Braun has lured the vault's inhabitants to enter the faux-utopian simulation of Tranquillity Lane by promising them a superior virtual alternative to their physical reality. Once the inhabitants have become suitably reliant upon their simulated world, Braun exploits their dependency to use them as test subjects in an endless series of macabre experiments.

As the times and the terminology progress, more contemporary novels such as Jonathan Lethem's *Chronic City* (2009), update their labelling to 'VR'. They are however, continuing to describe that which is more akin to simulated reality, in that the world that is initially presented to us as 'primary' is later revealed or implied to be 'secondary', but that the protagonist (and often also the reader) remain unaware of this until later in the narrative. However, literary depictions of that which is closer to VR can be found, and we take a look at examples of this in the next section.

Literary Depictions of VR

Although it may not be possible to know the exact first instance in literature that something unmistakably 'VR' appeared, Stanley Weinbaum's *Pygmalion's Spectacles* (1935) is widely accepted to be the beginning for

VR as a literary form (Sadler and Dooly 2014). The term 'virtual' does not appear within the short story, but Weinbaum's description of the technology is revealing. In the opening, 'Professor' Albert Ludwig is engaging in a rather convoluted sales pitch before responding to a question from the deeply sceptical Dan Burke:

> How? How? But simply! First my liquid positive, then my magic spectacles. I photograph the story in a liquid with light-sensitive chromates. I build up a complex solution–do you see? I add taste chemically and sound electrically. And when the story is recorded, then I put the solution in my spectacle--my movie projector. I electrolyze the solution, break it down; the older chromates go first, and out comes the story, sight, sound, smell, taste—all! (p. 2)

Ludwig goes on to explain that the spectacles do not directly simulate touch, but that 'if your interest is taken, your mind supplies that' (p. 2). Weinbaum's description of the spectacles likens them to a gasmask, with eyepieces that are filled with liquid and that present the virtual world when electrically stimulated. The glasses themselves are even tethered to a separate device situated upon a table that Ludwig proclaims to be 'a rectifier […] for the electrolysis' (p. 2). Bearing in mind that Weinbaum characterised his magic spectacles in 1935, what he describes is something distinctly reminiscent of the HMDs commercially released by HTC and Oculus more than eighty years later. The VIVE technical specifications describe the headset's Organic Light-Emitting Diode (OLED) display which is, by definition, a layer of organic compound that emits light patterns in response to an electrical current—a mechanism that reveals uncanny similarity in when compared to Weinbaum's 'electrolyzed chromates'.

The resemblance between the Pygmalion's Spectacles and contemporary VR does not stop at hardware specifications. Rather it extends to concepts of user experience, with Professor Ludwig's impassioned speech asserting that tactile sensation may be evoked without direct stimulation, provided the user is receptive to immersion across the other senses. This cross-modality effect, in which one sense (be it taste, touch, smell, etc.) can be evoked or enhanced by stimuli of the other senses, is something that has been investigated in actual scientific research. Interestingly, sound appears to be the more common sensory modality associated with evocation of illusory or enhanced tactile sensation. Dhruv and colleagues (2002), for example, utilise noise stimulation to enhance tactile sensation in older adults whilst Jousmäki and Hari (1998) assert that adjusting the frequency spectra (timbral qualities) of synchronised sound feedback during hand rubbing can modify the perceived roughness of the touch sensation. It's worth noting however, that the Pygmalion's Spectacles may not have been quite so prophetic, had Weinbaum read von Schiller's (1932) paper, published three years earlier, that asserted repeated noise bursts could potentially generate a tactile perception of touching a rough surface.

In the years that followed Pygmalion's Spectacles, fictional literature continued to explore concepts and characteristics pertinent to VR. These works would similarly prove to be prophetic in anticipation of the technology that would follow. Science fiction has arguably strived to continue this trend, serving as oracle, right up to the present day. The relationship between science fiction literature and non-fictive technology has even extended to the latter making use of terminology from the former; examples include Norman Spinrad's *Songs from the Stars* (1980) and Neal Stephenson's *Snow Crash* (1992) which popularised the use of the term 'avatar' within a computing context.

Evidence that science fiction literature is continuing to be prophetic of advances in VR technology, even today, can be observed in William Gibson's *Neuromancer* (1984). Gibson's story envisions a global information network in which users engage by way of a virtual environment known as the 'matrix': '…bright lattices of logic unfolding across that colourless void' (p. 2). Facebook purchasing Oculus is indicative of a forthcoming merger between social media and VR (see Chap. 9) and in recent months, applications have indeed started to become available with which the Internet can be experienced as a virtual representation of physical space. Such interfaces are poised to transform Web-based content into virtual objects and events that can be interacted with in new, more deeply immersive ways. It is also worth noting that literary depictions often go beyond VR, extending across the virtual continuum to conceptualise augmented and mixed-reality technologies. As Fenlon (2012) notes, William Gibson's work is again worthy of mention with his book *Spook Country* (2007), which describes augmented reality (AR) visors integrated with global satellite positioning technology to facilitate location-tagging, object tracking and overlaying physical space with virtual art. Throw in a Pokémon and life really is imitating art.

Looking at novels published closer to the present day, conceptions of VR continue to evolve yet arguably retain a great deal of conceptual content from earlier works. In *The Leveller* (Durango 2015), for example, the protagonist is one of many professional hired by parents to retrieve their children, who refuse to return from the virtual world of MEEP; a resounding adaptation of Bradbury's classic apocryphal tale, the Veldt (1950). In the same vein, Ernest Cline's *Ready Player One* (2011), presents the OASIS, a massively multiplayer game that began with a two-dimensional screen display and traditional desktop computer interface, but has transformed through VR technology into a shared digital universe that most of the Earth's population enter daily. The broader aspects of the book resonate with characteristics previously raised throughout this chapter, such as an oppressive capitalist dystopia and a virtual world that offers some form of escape. Developments are also apparent however, with OASIS not describing a simulated reality and instead retaining the 'virtual-ness' of VR by not being indiscernible from reality (at first). New angles on fictive VR also include the incorporation of more contemporary technology issues into the narrative, such as cybercrime and hacking.

In terms of the VR hardware that facilitates the fictional VR system of OASIS, Cline (2011) describes a kit comprising of a visor and a pair of haptic gloves. The visor is lightweight, wireless and employs lasers that project visuals directly against a user's retinas. The gloves meanwhile provide dual functionality, firstly as a means of moving the player's avatar, and secondly to enable the virtual objects to stimulate actual tactile sensation. During the 1990s, haptic feedback hardware similar to that in Cline's novel appears in several literary depictions of VR, and their usage strongly resonates with some of the actual human-computer interfaces currently being manufactured for both the experimental and commercial markets. Greg Egan's *Permutation City* (1994), for example, portrays a 'force feedback' glove that is employed simultaneously as a means of computer interfacing and to facilitate a sense of touch. Whilst not necessarily originating from literature, this concept has arguably sustained public and commercial interest when considering the numerous devices currently being manufactured which boast comparable functionality—such as the *Manus VR*,[5] a pair of gloves that enable hand and finger tracking for interfacing with VR environments, plus programmable vibration motors for haptic feedback.

Durango's MEEP (2015) by comparison draws from a vastly older concept, the realisation of which (if even possible) is arguably much further in the future. The MEEP system is essentially a brain-computer interface with all sensory stimuli bypassing the traditional bodily inputs by way of direct neural stimulation. This conceptual design can be traced back to as far as a thought experiment derived from Cartesian scepticism (1641), that raises the possibility of our existence not being a complete human form, but rather a brain in a vat, with all aspects of our physical experience an illusion resulting from complex patterns of electrical stimulation. Connecting fictive VR to actual science, the brain-in-a-vat vision of VR does have a non-fictive correlate that is worth mentioning briefly. Transcranial Magnetic Stimulation (TMS) describes a process of electromagnetic induction, in which a magnetic field generator (also known as a 'coil') is positioned close to a person's head for the purpose of stimulating a particular brain region (O'Shea and Walsh 2007). TMS is admittedly in its infancy, however the technology already shows very exciting potential in applications such as pain reduction (Lefaucheur et al. 2001) and mood regulation (George et al. 1997). Whilst there remains a substantial gap between the brain-computer VR interfaces of literature and the capabilities of TMS, work such as that by Pascual-Leone and Walsh (2001) begins to scratch the surface, with their findings that stimulation of certain regions of the brain can significantly raise an individual's visual awareness. The findings published in such research studies suggest that whilst bypassing the sensory organs to project a total-immersion experience is far from a present-day reality, there is reason to believe it might just be possible in the future.

Sound in Literary VR

Whether the reality generated by the fictional hardware devices of literary science fiction is simulated or virtual, one common element of these narratives is the presence of some form of technology evoking sensory stimulation and/or simulation. As one might expect, the visual modality enjoys centre stage across almost all literary depictions of VR. But to what extent (if at all) do these devices deal with the auditory modality? How is sound represented?

Returning to our earliest depiction of VR on record, Weinbaum (1935) prioritises the visual display of the Pygmalion's Spectacles but does also refer to sound, if briefly. His description distinguishes the mechanism for evoking sound as electrical stimulation, as opposed to the gustatory sense, which is induced chemically. Taste stimuli are delivered through a rubber mouthpiece and goggles provide the image, but there is no specific mention of any physical hardware from which sound is received.

Variations upon the brain-in-a-vat theme proliferated 1990s science fiction literature, at a time when VR was in its first commercial generation. These works were also less likely to be explicit in designating the mechanism by which sound stimuli is received. Matthew Stover's *Heroes Die* (1998) for instance presents the 'Simichair', a helmet coupled with breathing apparatus and a large needle inserted in the back of the neck. Here, sound content is presumed to be delivered alongside all other sensory modalities by way of electrical impulses sent directly to the brain. Being the decade that gave us (amongst others) the Virtual Boy, 1990s science fiction literature did also present us with headset-based VR depictions, but again, predominantly outlined vision-centric devices, such as the headset visors in Tad Williams' *City of Golden Shadow* (1996). That said, acknowledgements to sound have notably been made in a few instances. Returning to *Permutation City* (Egan 1994), this tale partially outlines a fictive functionality of VR sound, describing headsets that operate within a multi-person shared virtual space that users can customise to their liking. This extends to the soundscape as it is illustrated in a scene in which the virtual space is occupied by multiple musical performers and groups, with the audience able to select which performance they would like to see and hear by 'tuning in' to their preference by way of controls on their headsets.

Continuing forward in time to the years leading up to the second generation of commercial VR, the OASIS system in Cline's *Ready Player One* (2011) initially makes no mention of any sound element, the visor and glove being the primary interface hardware. Later in the story, an audio aspect of the system is described as what essentially amounts to a high-resolution surround sound setup, with speakers affixed at all angles across walls, floors and ceilings. Even the more recent literary illustrations continue to prioritise visuals whilst relegating sound. Daniel D.Warwick's *Hive Propolis* (2015) for example features an AR system coupled with direct brain stimulation, in which illusory holographic heads-up displays are projected into physical space by a user's mind. Again, whilst sound content is referred to in particular sections of the book, the central aspect of the technology is inescapably visual. Admittedly, it would be unfair to lay the blame for the deprioritisation of sound in literary depictions of VR squarely at the feet of the works themselves. It would also be difficult to argue against the assertion that most literature, indeed a substantial proportion of communication is visual. As we shall observe later in this chapter, visions of VR in other media reveal precisely the same sensory imbalance and there are numerous other factors that have pushed sound down the ranks in VR design. As has been stated already in this book, we should not be lamenting this state of affairs, but rather seeing opportunities for improvement.

So far, this chapter has primarily looked at fictional worlds as they exist in literary forms. We have observed the deep connections between fictive representations and non-fictive VR, both in terms of the hardware, the virtual worlds and its experiential qualities and issues. The next section proceeds to examine cinematic representations, particularly those of the 1990s, to reveal the dramatic and disastrous effect that such fictive instances had upon our perception of actual VR technology of the time.

Cinematic Representations and Their Effect upon Expectation

Films depicting VR in the 1990s covered a range of technologies, from HMDs and glove controllers to direct brain interfaces. Examining such films gives an insight into the design cues of the hardware and also reveals the value propositions (namely functionality and quality) that would contribute to the expectations consumers had for the actual VR systems of the 1990s.

A retrospective on 1990s cinema by Palmer (2009) centres upon a rather scathing observation of the prevalence of 'spin', both in film and Western culture in general. For Palmer, spin refers to: 'not just a technique of diversion, a smoke screen or con game [but rather] a reconfiguration of reality that can generate a positive change out of the most negative situation' (p. 220). This suggests that, whilst all cinema is inherently fictional, films from the 1990s were more of a departure from reality than others; drawn from a culture that 'reconfigured reality' to highlight the positives of the subject matter, even if those positives were just conceptual ideals of the time.

Prior to 1990, cinematic representations of VR technologies were relatively scarce but instances were already proving popular, in films such as *Tron* (Lisberger 1982). By 1990, VR had thematically infiltrated cinema and showed great commercial promise in examples that included *Total Recall* (Verhoeven 1990), a film adaptation of Philip K. Dick's *We can remember it for you wholesale* (1966). The film depicts a VR-like enclosure that implants memories by way of 'brain stimulation', enabling the user to experience a fantasy that is perceptually indistinguishable from their actual life. This representation of user experience is one that features prominently across both cinema and literature. In literary fiction, as discussed earlier in this chapter, a computer-generated environment that cannot be perceptually distinguished from the actual world is characteristically referred to as 'simulated reality'. In cinema and television however, another term with essentially the same meaning often features more prominently; 'artificial reality'. The use of artificial reality as a representation of VR experience is commonplace, arguably due to it being the most cost-effective form to put on screen. An environment indistinguishable from the 'real world' requires little change in production whilst a more abstract 'other-world' would likely demand expensive visual effects and filming techniques. However, not all fictional VR utilises the artificial reality trope, with a small number of examples bucking the trend.

Whilst films such as Total Recall dabbled with some of the underpinning philosophical aspects of VR, fictional representations comparable to the actual technology were not presented to a mainstream audience until a couple of years later, in arguably one of the most well-known cinematic representations of VR in the 1990s, *The Lawnmower Man* (Leonard 1992). The film highlights the gaming functionality of VR but also extends to use of the technology for education, training and cybersex. In terms of technology, the film depicts HMDs in tandem with mechanical platforms to simulate motion and gloves that track hand/arm movement to enable control of virtual limbs. Multiplayer gaming and interfaces using joystick controllers are also represented throughout the film and the aesthetic of both the hardware and computer graphics appear significantly dated when watched today. Despite the aesthetics being more representative of actual 1990s VR technology, user experience in The Lawnmower Man is most certainly an idealised vision. Use of the hardware is portrayed as highly dynamic, responsive and integrated, with characters in the film easily becoming fully immersed in the VR and reacting to it as if it were the physical world.

Examining VR technology from a different perspective, *Strange Days* (Bigelow 1995) depicts VR within a sci-fi noir aesthetic. The film's setting is in a dystopian Los Angeles in which the consumerism of VR is comparable to that of narcotics, with experiences traded on the black market. Central to the narrative of Strange Days is the value proposition of its VR system known as the 'SQUID (Superconducting Quantum Interference Device) Receptor Rig'. The function of the device is the acquisition of memories by recording electrical patterns across the cerebral cortex and storing them upon a MINI Disc (to really evoke that 1990s aesthetic). Memories could then be 'replayed', providing the user with a full, multisensory re-experiencing of the events that took place at the time of recording. In line with its functionality, the SQUID bears more semblance to a wireless magnetoencephalogram than a HMD. In fact, the virtual device featured in Strange Days is based heavily upon an non-fictional equivalent of the same name; a highly sensitive magnetometer, capable of detecting minute magnetic fields that include those produced by brain activity (see Barone 1992). As such, SQUID technology underpins various physiological measuring techniques, including magneto-encephalography and magnetic resonance imaging but its function in Strange Days is most comparable to transcranial magnetic stimulation coupled with electro-encephalography (TMS-EEG). TMS refers to the stimulation of cortical activity by way of artificial magnetic fields whilst EEG describes the measuring of that activity, detecting electrical impulses in the brain (see Rogasch and Fitzgerald 2013). However, whilst the SQUID in Strange Days is capable of generating a virtual experience akin to artificial reality, TMS-EEG is, at present, only beginning to draw connections between brain activity and broad psychological processes such as attention (Ahrens et al. 2017) and relaxation (Bonnard et al. 2017).

Released in the same year as Strange Days, *Johnny Mnemonic* (Longo 1995) features representations of more well-known VR technologies. Its roster of fictional devices include the 'Sino-logic 16' (a holographic user interface with which user's navigate the Internet), 'Thompson Eyephones' (a HMD) and 'Sogo-7 data gloves' (a hand tracker/gestural interface). Although the aesthetic of these devices is repeatedly derided in online retrospectives (see Dyess-Nugent 2012), both their functionality and operational quality far exceeded any non-fictional equivalent of the time. Through the Sino-logic 16, users could easily interact with an immersive representation of the Internet, whilst the data gloves tracked user movement with pinpoint accuracy and the eyephones presented fluid and reliable visual feedback in a device that was lightweight and easy to operate. Examples like the above begin reveal the impossibly high standards that fictional representations were setting for non-fictive VR technology, particularly with regards to matters of user-experience.

Fischer (2011) notes that movie executives of the 1990s proved rather susceptible to the hype of VR, but observes that the majority of films built upon a VR theme did not prove particularly successful at the box office. Fischer suggests that the reason for this was primarily a disconnection with the audience, many of which had little to no regular interaction with computer technology. In terms of box office performance, films depicting VR largely support Fischer's observations. The Lawnmower Man grossed just over $32 million worldwide (ranked 42nd for the year 1992, making roughly $12 million more than the box office average[6]), whilst Johnny Mnemonic took $19 million (ranked 84th) and Strange Days garnered $8 million (ranked 126th). The same is evident in terms of critical reception, with such films revealing a tendency towards poor to mediocre ratings (though Strange Days fares a little better in contemporary reviews[7]). Examining further VR-themed films across the 1990s reveals a notable downward trajectory towards the end of the decade, suggesting that the disillusionment, caused in part by cinema, had an adverse effect on cinema itself. That said, it is also worth noting that whilst Strange Days and The Lawnmower Man had modest impacts at the box office, they both attained 'cult-classic' status several years later (Graham 2017). This fact, combined with the box office figures, suggests that VR-themed films were something of a niche, ignored by many but establishing a strong cult fan base. Whilst it would be incorrect to assert that the whole world was swept up in VR during the 1990s, those that were arguably took the concept to heart.

The theme of VR was by no means not limited to cinemas, with representations of the technology and user experience gaining direct access to our homes through the television. Early depictions of VR include *Doctor Who* and 'The Matrix' (first featured in *The Deadly Assassin* [Maloney 1976]), a VR system built from memories of the deceased for the purpose of predicting the future. A few years later, the British science fiction comedy series *Red Dwarf* (Bye 1988) depicted VR at several points across the various series', during which the user experience remains relatively constant whilst the hardware undergoes several changes. For instance, in the second series episode *Better than life*, the VR system is a direct brain interface that fits over the head and is similar in design to the SQUID devices of Strange Days. Several years on and the show's representations of VR began to more closely reflect actual VR technology. The sixth series episode *Gunmen of the apocalypse* depicts VR with a HMD and haptic glove combination whilst *Stoke me a clipper* in the seventh series presents an enclosure system featuring a retractable visor and various haptic accessories. The latter in particular bears close resemblance to the actual Virtuality 1000SD systems[8] (minus of course Red Dwarf's nefarious 'groinal attachment'). As noted above, the experience for the users does not change and, similarly to The Matrix in Doctor Who, the virtual environment is indistinguishable from actuality and akin to 'artificial reality' (throughout the series, the technology itself is regularly referred to as such).

The 'Holodeck' is perhaps the most well-known representation of VR technology on television, that first appeared on screens during the early seventies in *The Practical Joker* (Reed 1974), a second season episode of *Star Trek: The Animated Series*. The Holodeck is a fictional VR system that conceptually resembles a number of actual VR technologies. The Holodeck is an enclosed room, within which the structural foundation of all content within the virtual world is composed of actual matter. Onto this matter, visual details are superimposed by way of holographic projection. Use of the Holodeck does not require the user to don any hardware and whilst the size of the projection room is limited, physical movement is mediated by 'force fields' that generate haptic feedback for the user whilst also manipulating the holographic projections so that the virtual environment accurately moves in response to user action. For us as viewers of the show, the user experience is once again an example of artificial reality. However, the visual aspect of the Holodeck bears close connection with CAVE projection systems whilst the principle of using force fields to enable movement without at all changing the user's position is reflected significantly in omnidirectional treadmills. Lastly, the ability of the Holodeck to facilitate haptic and tactile sensation without the user wearing any associated hardware, also possesses a non-fictive equivalent in 'ultrahaptics',[9] a recent VR technology that creates fields of ultrasound to project sensations onto the hand for tactile feedback.

Both before and during the 1990s, fictional representations of VR in cinema and television mostly drew inspiration from non-fictive technology, as can be seen in their design, functionality and implementation. It is therefore not surprising that general consumers of 1990s VR technology drew parallels between VR as depicted on screen and that which was available to purchase and experience. Marketing was only too happy to reinforce this, with adverts for systems such as the Virtual Boy exploiting a distinctly cinematic feel.[10] With such close ties between the fictional and the non-fictional in terms of how the hardware looked and functioned, it follows that consumers would expect comparable levels of functionality and quality.

Unfortunately, as we shall see in Chap. 7, the 'reality of virtual reality' had little chance of meeting such grand expectations. Before we address this however, the current chapter closes with a look at how the impact of fictional representations extends far beyond establishing our expectations, to be highly influential in shaping the foundational desires that we associate with VR experience.

Representations and the Desires of VR

Earlier in this chapter, we have seen how fictional representations in cinema, literature and television have shaped how we understand and define VR. They have contributed substantially towards consumer expectations for the technology's function, quality and aesthetic. They have also had a visible impact upon developers, for whom such fictional representations are an ongoing source of inspiration. This section closes the chapter with a look at one further aspect of this relationship between fictional and non-fictional VR, examining how representations in film, television and digital games are responding to recent developments in contemporary VR. Incorporating both conceptual and technological aspects, we review this relationship between fiction and non-fiction as it pertains to the underlying desires that draw us to (and away from) VR. This relationship reveals both art imitating life and life imitating art, but just as the representations of the past can be seen in technology of the present, those of the present may provide us with a glimpse into the possible future.

Across the countless fictional representations in film, television and digital games, consistent themes emerge regarding foundational human desires that VR technology of the future may one day be able to fulfil. From being able to indulge in any conceivable fantasy to achieving immortality, the vision of VR certainly reveals some grandiose ambitions. Such desires can be more broadly categorised into four related groups that we discuss below; escapism, nihilism, assimilation and conservatism.

Escapism

Beginning with that which is arguably the most commonly attributed desire of VR, escapism can refer to many things, largely depending upon the object or subject from which the individual wishes to escape. Escapism has been previously described in research as a means of avoiding undesirable psychological processes and has been connected to everyday activities that include watching television (Henning and Vorderer 2001) and consuming alcohol (Sadava et al. 1978). Digital games (and by extension, VR) are commonly labelled as escapist, based on what Calleja (2010) posits as a false assumption that the act of gameplay is in opposition to the seriousness of everyday life; an 'avoidance of the real' (p. 335). Calleja takes issue with the term 'real' as its implication is that the corresponding 'unreal' of a digital game is inferior, irrespective of context.

For Tuan (2000), escapism is a fundamental and ever-present human desire that encapsulates various forms. We may wish to escape from the Earth, from our physiological/biological being, from our baser animalistic nature, and also from what Tuan would describe as our 'intolerable uniqueness'. Each of these forms of escapism has a connection to VR. In terms of escape from the Earth, one of the fundamental aspects of VR is not only the presentation of an alternate world but, critically, the shutting out of the actual world. Our escape from our physiological/biological being is apparent across identity issues of VR experience; as we assume the role of the avatar to become a warrior, an alien, a robot or even a God. As discussed in Chap. 4, the virtual identity is greatly malleable, enabling us to change various aspects of our appearance and extend our abilities far beyond those that are largely fixed in the physical world. We may switch our ethnicity, gender and even species. Virtual systems also provide us with means of subverting our animalistic nature. This is particularly apparent in AR, with which users are encouraged to engage with mechanistically. Finally, escape from our intolerable uniqueness is also granted by VR, which can connect users to millions of others, effectively 'submerging the self in a group, thus escaping from one's singularity, frailty, loneliness and vulnerability' (Tuan 2000, p. xi).

The theme of escapism has featured in many representations of VR discussed earlier in this chapter. The fictive relationship between siblings Peter and Wendy and their VR nursery in *The Veldt* (Bradbury 1950) is arguably one of escapism, their increased dependency upon the virtual world tied to their disdain for the physical. Cinematic representations in the 1990s, that include *Total Recall* (Verhoeven 1990) also depict VR technology as a means of escape from the humdrum of everyday life. This theme remains equally prominent in contemporary representations. In literature, with works such as *The Leveller* (Durango 2015), the escapism offered by the virtual world is enough to make it an addiction, drawing in children and requiring specialist professionals to be sent in as a way of extracting them. In the science fiction anthology television series *Black Mirror* (Brooker et al. 2011), the first series episode *Fifteen Million Merits* features a gesture-controlled CAVE interface that surrounds each inhabitant's miniscule living quarters, providing an expansive virtual space that helps them escape the fact that they are living inside little more than a box. Modern cinematic representations also highlight a continuation of the desire for escapism as a theme of fictional VR. One such example is the film *Gamer* (Neveldine and Taylor 2009), in which users can sense the environment (across all sensory modalities) from the body of another person by way of a VR interface. Whilst the technology in Gamer is primarily exploited to escape boredom, a system capable of identifying and erasing specific strands of memory is presented in *Eternal Sunshine of the Spotless Mind* (Gondry 2004). Here, the technology depicted in the film enables users to escape painful past experiences, whilst all other aspects of the self are retained, enabling the user to live in what is effectively an AR.

Returning to Tuan's (2000) escapism forms, to escape from the limitations of the physical body describes a frequently recurrent theme in fictional representations of VR. This desire ties primarily to aspirations for immortality but also extends to invulnerability, omnipresence and other superhuman abilities. This form highlights connections between the desires by incorporating both escapist (the wish is to leave the body behind) and assimilation (the wish to integrate the disembodied Self within a new physical form) desires. That said, a difference is apparent between the two forms, which is illustrated when comparing films such as *Surrogates* (Mostow 2009) and *Self/Less* (Singh 2015). Both films feature a similar transhumanist fantasy in which an individual's consciousness is transferred into an alternative 'host' body. The key difference relates to the underlying motivation of the protagonist. Surrogates describe a form of telepresence that fulfils the assimilation desire to merge their identity with an ideal physical form. Users' biological bodies are retained in this instance and the robotic 'surrogates' are portrayed as more of an extension than a replacement. Self/Less by comparison describes an escapist fantasy of consciousness transference, in which the user abandons their physical body to escape from severe illness or injury.

Escapism is primarily relevant to augmented virtuality. The reason for this is that the escapist desire focusses upon dissociating from a particular component of the self, whilst the overall sense of identity is retained. An analogy would be to discard a bruised section of an apple. The bruise is a piece of the apple, but when discarded the apple remains an apple, but one that is more desirable. Augmented virtuality describes the integration of actual entities within a virtual environment, specifically entities that are relevant to the user. This would describe digital games in which the player is provided with an absent or limited avatar characterisation, encouraging them to perceive the avatar as themselves. It also relates to techniques that attempt to draw elements of the player's personality into the game, such as moral choice mechanics. In contemporary VR, augmented virtuality techniques include body tracking ('I am reaching out for the virtual object with *my* arms'), integrated biometrics (in which the player's heart rate, breathing, etc. can affect the virtual environment) and substitutional reality (the digitising of physical objects that are then placed into the virtual world—see Simeone et al. 2015). The singular purpose of these techniques is to encourage retention of the self, but place it somewhere altogether more fantastic. You're still you, but a you who is saving the planet from total annihilation, which is better.

Substitution

Whereas escapism as a fantasy of VR refers to circumstances in which the individual discards a particular aspect of their existence whilst retaining their overarching personal identity and sense of self, 'substitution' essentially denotes the reverse; a minimal retention of the original self, whilst the individual embodies an otherwise entirely different identity. Here the critical emphasis is upon the desire to be someone or something other than oneself, as opposed to the escapist desire to discard or modify specific aspects of the self.

The difference between escapist and substitutive desire can be neatly demonstrated in digital games design, the former denoting a projection of the original self into the virtual world (the avatar and player share a single identity) whilst the latter replaces the self with a virtual alternative (the player embodies the personal identity of the avatar). In broad terms, it is the difference between building/selecting a character that is strikingly similar to yourself (but with a few minor tweaks for good measure) or one that is a complete and dramatic departure from you, both in terms of physical appearance and personality. Alternative digital games also pursue different VR desires by way of their approach to the relationship between the user and their avatar. For example, first person shooters such as *Doom* (Romero et al. 1993) typically prioritise escapism in their design by presenting an avatar with no characterisation for the purpose of evoking a sense in the player that they *are* the marine. By comparison, third person point and click adventure games, a good example of which would be *The Secret of Monkey Island* (Gilbert et al. 1990), typify substitution with an avatar who is fully rendered upon the screen and possesses an established physical (appearance, abilities, etc.) and psychological (personality, moral alignment, etc.) profile and backstory. Of course, each individual player is different in how they approach a game and the designer cannot be certain that their intention will always be realised, but these examples remain illustrative of key design differences as they relate to the designer's understanding of player's desire.

Substitution in this context is admittedly something of a neologism, but has connections to other, more established concepts. For instance, Francis (2015) differentiates between several definitions of nihilism, the philosophical perspective that argues against reputed meaning and value. His description of the term includes numerous nihilistic forms, including moral (concepts of right and wrong are artificial and nonsensical) and epistemological (a rejection of truth and knowledge). Francis' account of nihilism as it pertains to existentialism resonates most closely with the notion of substitution. For Francis, existential nihilism refers to: 'the belief that life has no intrinsic meaning or value' (p. 5). In terms of substitution, this description requires slight adaptation, to 'the belief that *you* have no intrinsic value'. This 'individual nihilism' refers to circumstances where the VR user (albeit temporarily one would hope) rejects their overarching self. They wish to abandon both their physical form and some aspects of their psyche, substituting their original self with that of someone or something completely different.

Sanders (2007) observes that identity can be divided into three discrete criteria: the body (biological/physiological form), the brain (a physical organ that, in fiction, can be transplanted into an alternate body to facilitate an identity-swap) and the mind (encapsulating personality, memory and continuity of experience). Therefore, total substitution would require the individual to abandon all three. Such total abnegation of personal identity is a notably absent occurrence in fiction, with certain instances in which multiple aspects of a person's identity are substituted, but not all. William Hartnell's Doctor introduced many viewers to the concept of regeneration in *The Tenth Planet* (Martinus 1966), in which the personality and physical appearance of the character changed almost entirely, but with the notable exceptions being retention of memories and continuity of experience. With regards to substitution, the other issue with regeneration is that it is not founded in desire, as the Doctor has no wish to die and it is often with great reluctance that he regenerates.

The desire to be someone else is explored outside of science fiction in a great number of 'body-swap' films. This recurring trope has become so popular in fact, that it has become a subgenre in its own right (see Beumers 2016). Cinematic depictions of substitution, released around the time of 1990s VR, include *Strange Days* (Bigelow 1995) and *Brainstorm* (Trumbull 1983). In both instances, the VR technology facilitates the multisensory re-living of an event originally experienced by another person. Users are typically aware of the identity that they are adopting, and other characters within the 'VR replay' interact with the user's assumed character, not the user themselves. Whilst both depictions are passive experiences in which the user has no agency with which to control events. One example that does enable users to interact within the virtual world and arguably presents one 1990s instance that embodies the substitutional desire of VR particularly closely brings us back to fictional representations in television. *Back to Reality*, a fifth series episode of the science fiction sitcom *Red Dwarf* (Bye 1988), first aired on television in 1992. The episode sees the four main protagonists seemingly die before awakening from a 'total immersion digital game' which, they are later informed, meant that their presumed identities were fictional constructs that they had chosen to 'play' as, preferring the idea of these characters to their own personal identities.

The substitution desire, unlike assimilation and escapism, has become a slightly more obscure aspect of VR fiction in recent years, with one of the only notable examples being a satirical sideswipe at the technology courtesy of *Rick and Morty* (Roiland and Harmon 2015). The second season episode *Mortynight Run*, features the VR experience 'Roy: A Life Well Lived", with which protagonist Morty lives out 55 years as Roy in a few short minutes of gameplay. During this time Morty forgets his given identity entirely, essentially becoming Roy, until the headset is promptly ripped from his head. One notable cinematic instance of substitution can be found in *The Congress* (Folman 2013), which depicts the cartoon aesthetic VR world of Abrahama City. This 'animated zone' enables individuals to become avatars which are initially cartoon likeness of themselves, but can be infinitely customised or based upon existing characters that are completely different to the individuals' prior physical and psychological constitution. Abrahama City residents therefore are able to swap out their entire identities for alternates, substituting their physical appearance, memories and entire personalities if so desired.

One possible reason for fiction not featuring many narratives in which the protagonist actively seeks to relinquish their entire self is due to such things being deeply connected to dissociative disorders and not typically acknowledged as healthy conscious desires. The consistent underlying 'message' of the body-swap subgenre is that the desire to be someone else is misguided and any realisation of that desire is ultimately unsatisfying.[11] Some may also argue that substitution is merely an alternate route to escapism and that the underlying desire of the individual is less the acquisition of an alternative identity and more the relinquishing and augmentation of the original self. It is fair to state that there is indeed some overlap between substitution, assimilation and escapism, but this is arguably because they are alternate sides to the composite desire to align the actual (perceived) self with the ideal self. As Sirgy (2012) posits, the conceptual distance between how we see ourselves and how we wish to see ourselves is inversely correlated to our self-esteem and life satisfaction. Where an imperceptibly small gap can evoke happiness and confidence, one that is more substantial has associations with shame and disappointment (Langsam 2007). What these visions of VR offer their users is a new alternate, the 'virtual self'. This could be an opportunity to escape from negative properties of the original self, retaining it overall but making adjustments to move it closer to the ideal. It could also be a means of substituting it for another, prepackaged identity that more closely resembles the ideal.

The difference between substitution and escapism is reinforced on our continuum in relation to their associations with virtuality and augmented virtuality respectively. Whilst escapism corresponds more to an augmented virtuality in which the original self is drawn into the virtual, substitution encourages the original self to be abandoned in favour of a fully realised virtual self. In terms of contemporary VR, substitution-related techniques include those that encourage dissociation from the physical world, such as noise-cancelling headphones, naturalistic interfaces that encourage avatar-embodiment and HMDs that fully occlude natural light. Such things bring the experience closer to a purist virtuality which, from a user's perspective, effectively swallows them whole and immerses them in a world and a role that is so cognitively, physically and emotionally engaging that their original self begins to slip away.

Assimilation

Whilst the VR desires relevant to escapism and substitution broadly describe a kind of identity subtraction, assimilation describes the opposite; an additive process in which the individual wishes to integrate themselves, building upon and thereby extending the self. For Johns (2016), assimilation is to 'embody one or a series of relations in order to see where those relationships go' (p. 6). This definition resonates with the meaning of the term within a VR context. The assimilation desire is to, by way of technology, establish new relations between ourselves and the wider world; expanding our reach, our breadth of experience, our abilities and our existence.

As with most of the concepts discussed within this section, assimilation possesses various definitions, some of which can contradict our interpretation of the term from within a VR context. For instance, Ghaffar-Kucher (2014) discusses various sociological forms of 'classic assimilation', of which one of the most prominent describes the term as subtractive. Assimilation in this context concerns cultural integration, specifically that of immigrants entering a different, but dominant, mainstream culture that causes them to lose their ethnic characteristics. Indeed, Tuan's (2000) notion of the inherent human desire to discard our 'intolerable uniqueness' (discussed earlier in this section) draws clear associations between escapism and assimilation, suggesting that the latter can also be a subtractive desire. However, despite some overlap between escapism, substitution and assimilation, the latter is distinguished in two ways. Firstly, assimilation describes a desire for the virtual self that is focused upon attaining the ideal self rather than escaping from elements of the actual self, thus making it different to escapism. Secondly, it differs from substitution by being a process of merging identity rather than replacing it.

Assimilation encapsulates various specific forms that are well established in fictional representations of VR. Such forms include transhumanism, human augmentation, telepresence and the collective network. Transhumanism is explored, albeit from a decidedly negative perspective, by Livingstone (2015) who describes the term as a 'movement', gaining increasing popularity of late, with ambitions of assimilating technology with the human form to 'augment human potential, and ultimately, to achieve immortality' (p. 5). Livingstone continues however, to note that the desires of transhumanism extend beyond immortality, to omnipotence (attributes of God), inclusive of invulnerability and omnipresence (the ability to be present anywhere/everywhere). In a related note, Livingstone also describes the transhumanist notion of 'the singularity', a predicted future event at which point it will be possible to decode the human brain and upload the mind to the Internet, granting immortality, before hiving the 'digital minds' together to form a collective consciousness; what Livingstone calls the 'global brain' (p. 6). A vision of the singularity appears in the contemporary film *Transcendence* (Pfister 2014), that depicts a fictive 'mind uploading' technology, in which the body and brain are discarded whilst the mind remains largely unchanged, but now operating from a global network platform.

A more restricted form of the transhumanist desire for omnipresence, namely telepresence, is also a common feature in fictive VR and has been a science fiction staple technology for decades. Where contemporary fiction differs from earlier works is largely in terms of visual fidelity and multisensory experience. For example, *Creative Control* (Dickinson 2015) presents *Augmenta*, a system that enables a person to project a holographic image of themselves to the AR glasses of anyone, anywhere. In addition to being able to move around in actual space, these projections are largely indistinguishable from physical people and Augmenta also enables a degree of haptic and tactile feedback, with individuals able to touch the holograms. Such a function is effectively teleportation, fulfilling the assimilation desire for omnipresence.

Returning to our continuum, fictive examples of assimilation in VR largely relate to human sensory/ability enhancement by way of some form of AR technology. As Tachi (2015) notes, human augmentation encapsulates AR functionality, by offering users enhancements to their'sensory and intellectual abilities' (p. 24) but within a physical-world context. Human augmentation does of course extend beyond AR, most notably into physical and mechanical forms of human augmentation. Indeed, reflections upon assimilation in both contemporary cinema and digital games tend to favour physical/mechanical forms, in examples such as *Elysium* (Blomkamp 2013) and *Deus Ex: Human Revolution* (Dugas et al. 2011). However, in line with our continuum, augmenting or merging one's self prescribes that AR, equally, extends beyond what is characteristically understood as human augmentation (i.e. augmenting the physical body and associated abilities) to encapsulate wider aspects of our lives. Essentially, assimilation by way of AR is akin to augmentation of the whole self, meaning we are not just extending our physical bodies, or even our minds, but everything that makes us, us. Interestingly, it is this form of assimilation that most frequently appears in fictive representations of VR technology.

One of the most well-known, contemporary realisations of human augmentation in mainstream cinema today is found within the Marvel Cinematic Universe, specifically in the relationship between central protagonist Tony Stark/Iron Man and augmentation technology. Across numerous canonically interlinked films that began with *Iron Man* (Favreau 2008), augmentation appears throughout in various forms, from the cybernetic augmentation and heads up display of the Iron Man suit, to the holographic AR displays used by Stark to interface with his JARVIS computer system. Similarly, the short film *Strange Beasts* (Barbe 2017) features an AR platform in which digital imagery is projected onto the retinas from a contact lens. In the film, this technology is specifically implemented to enable users to create bespoke virtual pets, from the domestic to the fantastic. The protagonist, Victor, demonstrates how integrating these virtual creatures into the actual world results in the building of virtual relationships that extend users' actual selves. Throughout the film, Victor is seen playing in the park with his young daughter, who has a strange beast of her own. In the closing scenes, Victor and his daughter are talking on a bench. Briefly, the AR overlay falls away and Victor is seen talking to no one. Using technology to mitigate a great loss is also a recurring theme in contemporary fiction, appearing in the Black Mirror episode *Be Right Back*, in which the data from social media profiles is used to create simulations of deceased individuals, enabling the grief-stricken to partially resurrect their loved ones, to continue their relationship and (in a macabre and uncanny sense) restore their self.

Strange beasts and Be Right Back both demonstrate the assimilation concept of the 'collective network'. In the former, multiple users can interact with each other's pets, whilst the pets themselves can also interact with one another. In the latter, our networked interactions have caused a significant proportion of our identities to leak into the digital collective, so much so that our entire persona can be digitally replicated to generate an independent virtual self; one that is almost indistinguishable from our physical self.

Another short film, *Hyper-Reality* (Matsuda 2016), presents another near-future vision of collective network assimilation, also in the guise of AR, in which our relationship with digital information has become something of an assault upon the senses. Games, news feeds, search engines, GPS, video chat and an abundance of garish advertising is thrust front and centre within the user's field of view. Underneath the myriad of digital overlays, Hyper-Reality presents an interconnected 'gamification of life', in which any conceivable activity is attributed a point-scoring system. This aspect of assimilation is itself a frequenter of contemporary VR representations, also featuring for example in the Black Mirror episode *Nosedive*, in which networked users continuously 'rate' (one to five stars) everyone they come into contact with. These ratings are so deeply embedded within the society that they become a kind of cultural currency, enabling those with average ratings above 4.0 to receive various perks (from access to exclusive housing to discounts on purchases).

Assimilation may have undertones of an underlying unhappiness with some aspect of the actual self, tying it to some extent with substitution and escapism, but the defining factor of the term is a priority in the mind of the user to extend and build upon their identity and their existence. However, it may have caught your attention that several the above representations appear to hold a decidedly negative opinion of these desires and the VR technologies that claim to fulfil them. This brings us onto the final part of this section, as we look at a distinctly opposing desire. A desire to avoid the virtual altogether.

Conservatism

True in both life and art, the fundamental quality of VR technology as a means of innovation, and ultimately change, can unsurprisingly trigger a counter-desire to preserve and enjoy that which is already present and available. Once again, we find rather limited existing terminology with which to cleanly express this desire within a VR context. This pushes us towards establishing another neologism but such a move would be counterproductive, obscuring the landscape rather than elucidating it. Consequently, a term is required that most closely resembles the meaning that this desire denotes. 'VR Conservatism' draws associations with Neo-Luddism (a scepticism towards modern technology—see Jones 2013). As Sale (1997) notes, the Neo-Luddite movement encapsulates a much broader remit, including antiglobalisation and primitivism (abandonment of all forms of mass production), however the most significant issues with this term is that it also suffers from inherently negative connotations. Jones (2013) observes that a prominent perception of the Neo-Luddite is a 'deluded technophobe', pointlessly pushing against the unstoppable march of technological progress. This focusses upon the emergence of doubts and fears in those that do not perceive value in the virtual promises of assimilation, substitution and escapism. There is no positive desire depicted here, merely a rejection of potential gains and a preoccupation with the possible risks and sacrifices; the proverbial dark side of the virtual moon.

Conservatism brings us closer to the specific phenomenon we are seeking to describe by being an arguably more positive term. For some, conservatism is primarily a political ideology purporting reactionary and hierarchical strategies of governance, with tradition and stability that which is to be conserved (see O'Hara 2011). Whilst this may be the most well-known meaning of conservatism, politics does not have dominion over the term. Technological conservatism is already an established expression, describing our partiality towards a system with we are experienced and comfortable over a new alternative, even if the

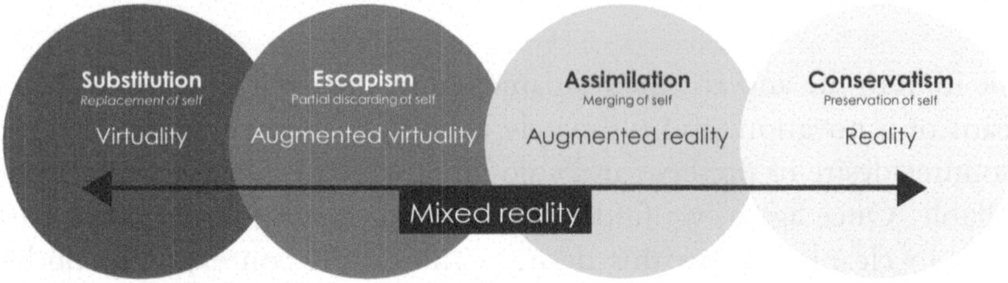

Fig. 5.1 Classifications of VR desire mapped along Milgram's continuum of mixed reality

latter offers us greater efficiency or functionality (Siracusa 2013). With regards to VR conservatism, we can observe how, for each of the previously discussed desires of VR, there is an opposing conservative attitude. For instance, conservative opposition to the assimilation desire is frequently addressed in terms of ethical considerations for human enhancement. Savulescu and Bostrom (2009) observe that a central ethical opposition to assimilation with technology is that it has the potential to undermine our human nature and it is this form of fundamental conservatism that resonates significantly with a VR context. As noted in the continuum at the end of this chapter, VR conservatism relates to the fundamental human desire to protect the original ('natural') self, or at least the components of the self that are perceived as ideal.

VR conservatism is a regular feature in fictional representations of virtual technology that appears in both diegetic and extra-diegetic forms. The diegetic form relates to the 'Evil Luddite' trope,[12] a character or group that works against the protagonist, often sacrificing their morality in pursuit of what they perceive to be a naturalistic greater good. The film Transcendence features an example of the Evil Luddite trope in RIFT, an anti-technology group that views artificial superintelligence as a great threat to humanity, branded as a terrorist organisation due to their use of kidnapping and torture. Likewise, films including The Congress (Folman 2013) and Surrogates (Mostow 2009) include renegade factions that oppose the central technology of the film and oppose the protagonist. Instances from contemporary digital games include the Purity First organisation in *Deus Ex: Human Revolution* and the Church of Yevon in *Final Fantasy X* (Kitase et al. 2002). These examples rally primarily against various forms of assimilation and their antagonistic portrayal supports the argument that, correspondingly, the desire for assimilation is generally perceived in mainstream fiction as 'good'. As an opposition to escapism and substitution however, VR conservatism within fiction appears more commonly as a trait of the protagonist. One of the most well-known examples of this appears in *The Matrix* (Wachowski and Wachowski 1999), as the titular simulated reality is perceived by the protagonists as a prison, chiefly because it is not 'real'. This is despite the virtual world of the Matrix arguably being a far more pleasant environment when compared to the desolation of the physical world.

In its extra-diegetic form, VR conservatism describes instances where the underlying message of the fiction is one of trepidation. Both Strange Beasts (Barbe 2017) and Hyper-Reality (Matsuda 2016) exemplify this. The latter in particular offers the viewer a few brief moments of un-augmented actuality amidst the cacophony of AR content as their system is reset. In this scenario, a bland supermarket aisle accompanied by the acousmatic sound of a baby crying is such a respite for the audience that it becomes a genuinely enjoyable moment and, as the audience is thrust back into the noise of the AR, they are led to experience the conservative desire for the world without digital augmentation. Dramatic predictions regarding our relationship with future technology underpin the tone and narrative of the Black Mirror series (Brooker et al. 2011), with individual episodes exploring the darker implications of VR systems that aim to fulfil our desires for escapism, substitution and assimilation. For instance, *White Christmas* depicts a networked AR built into irremovable contact lenses. The system enables users to significantly affect the sensory inputs of others (one user can 'block' another, causing the person who has been blocked to see them as a distorted greyish blur and hear their voice as no more than a muffled imperceptible noise). Whilst no character in the episode explicitly opposes the technology, the suffering it causes is plain to see.

VR conservatism is not a thoughtless and stubborn rejection of progress but should be thought of more as a provider of caution. It represents the ethical considerations of VR and encourages us to consider wider implications and potential dangers, helping us to avoid arrogantly charging ahead without due forethought. As Dr. Ian Malcolm once observed: '[y]our scientists were so preoccupied with whether or not they could that they didn't stop to think if they should' (Spielberg 1993). VR of course may not be an island of dinosaurs, but without care, there ahead be monsters.

A Continuum of Virtual Desires

One particular observation emerges when considering the above human desires in relation to the range of virtual forms along Milgram's continuum of mixed reality (see Chap. 2). Here, each position along the continuum seemingly reflects a particular desire. This is illustrated in the Fig. (5.1). As a primarily virtual construct but with key aspects drawn from the physical world, augmented virtuality has an arguable resonance with the desire for escapism. By comparison, virtuality is suggestive of the user's desire to completely abandon their identity for a complete existential substitution. Towards the other end of the continuum, the aspiration for some form of digital assimilation is notably reflected in AR, which fundamentally describes the merging of everyday experience with technology. At the far-right position upon the continuum, 'reality', we find one further human desire that is both highly relevant to VR and a recurring feature within fictional representations.

Pursuit of reality (as it pertains to Milgram's continuum) is indicative of an idealist, conservative desire based on an appreciation for what is perceived to be the supreme 'real' world. Inversely, a fear of change drawn from the assumption that technology will have an adverse, possibly even destructive impact. As shown above, this perspective is commonplace and almost consistent in its usage throughout fictional representations, but how it is used and its intended meaning for the audience varies widely.

Chapter Summary and References

Fictional worlds share a key function with virtual worlds, namely being means of escapism. Both facilitate travel into another place that one can become immersed in, potentially even favouring that world for our own, if only for a short while. Literature has proven itself an important component of what we understand to be VR. The fundamental aspect of VR as some form of world that is not quite our own has been shown to be something the written word has been exploring for thousands of years. Very virtual concepts have appeared in the literature of classical antiquity, predating the coining of the term 'VR' by millennia. Taking a broader overview of VR, inclusive of both technology and literary fiction, there is most certainly a bidirectional relationship in terms of influence. Twentieth-century science fiction has proven itself to be prophetic of several technological milestones, from HMDs to haptic gloves. However, depictions of VR in cinema, television and literature have consistently portrayed the technology in an idealistic frame. Fictive VR is lightweight, aesthetically and ergonomically designed. Its virtual worlds are convincing, immersive and responsive. It draws us into the extent that we cannot distinguish the physical from the virtual, or that we begin to perceive the virtual as a preferable form of existence.

In addition to the above, fiction's influence of VR technology extends beyond hardware design. The foundational elements of fictive worlds collectively forming a conceptual framework that has massively influenced the design and implementation of virtual worlds; from how they integrate fantastical elements to their handling of multiple realms and realities. Whilst sound may not enjoy pride of place in fictional worlds or virtual worlds, significant relationship effects between fiction and VR sound can be observed. Sound has also revealed value as a means of enhancing the perceived depth, dynamism and complexity of a world, be it fictive or virtual.

Lastly, the desires of VR reveal distinct overlap with various aspects of user experience and attributes of the hardware and software of VR. Assimilation, escapism and substitution all have connections to matters of identity and diegesis. Conservatism is a constant reminder to VR developers not to push users away from their 'naturalistic environment' too quickly, with numerous fictional representations on hand to provide a powerful 'how-*not*-to' guide. Here we have, yet again, only begun to uncover the massive network of components and connections that together form the aggregate from which VR emerges. That said, in doing so we are creating new opportunity for continued discovery. There really is so much more to be done.

Notes

1. Merriam-Webster Dictionary definition of Omniverse. https://www.merriam-webster.com/dictionary/omniverse.
2. Though the limitations of the procedural engine have been asserted to limit the actual number of distinct galaxies to 256.
3. For listings of the inspirations behind various films, games and literature, visit *Who Inspired?* http://www.whoinspired.com/wiki/Main_Page.
4. Though this is more often than not, the result of plot leaks or careless trailer creators giving away the twist.
5. Manus VR. Eindhoven, Netherlands. https://manus-vr.com.
6. 1992 box office figures. Box Office Mojo: http://www.boxofficemojo.com/yearly/chart/?yr=1992.
7. Empire Online, Strange Days Review: http://www.empireonline.com/movies/strange-days/review.
8. Specifications for Virtuality VR systems. Retro VR: http://www.retro-vr.co.uk/test/vr1000sd.html.
9. Ultrahaptics. https://www.ultrahaptics.com/.
10. A look at advertising for the Virtual Boy. Digital Spy: http://www.digitalspy.com/gaming/retro-gaming/feature/a562419/virtual-boy-retrospective-nintendos-disastrous-foray-into-vr.
11. As the internet meme insists: Always be yourself, unless you can be Batman, in which case, be Batman.

12. For a full explanation of Evil Luddites and list of examples: http://tvtropes.org/pmwiki/pmwiki.php/Main/EvilLuddite.

References

Abbott, E. A. (1884). *Flatland: A romance of many dimensions*. London, UK: Seeley and Co.

Adams, S. J. (1989). The noisiest novel ever written: The soundscape of Henry Roth's Call it Sleep. *Twentieth Century Literature, 35*(1), 43–64.

Ahrens, M. M., Veniero, D., Harvey, M., & Thut, G. (2017). P176 An EEG-TMS study investigating the oscillatory signatures underlying the time course of reflexive visuospatial attention. *Clinical Neurophysiology, 128*(3), e103.

Asimov, I. (1951). *Foundation*. New York, USA: Gnome Press.

Barbe, M. (2017). *Strange Beasts*. Red Knuckles. UK.

Barone, A. (Ed.). (1992). *Principles and applications of superconducting quantum interference devices*. Singapore: World Scientific.

Barrie, J. M. (1911). *Wendy and peter*. London, UK: Hodder and Stoughton.

Barrow, J. D. (2007). *Living in a simulated universe*. Online article: http://hinessight.blogs.com/files/living-in-a-simulated-universe.pdf.

Bobonich, C. (2011). *Plato's Laws: A critical guide*. Cambridge: Cambridge University Press.

Beumers, B. (2016). *A companion to Russian cinema*. NJ, USA: Wiley.

Bigelo, K. (1995). *Strange Days*. USA: 20th Century Fox.

Blomkamp, N. (2013). *Elysium*. USA: TriStar Pictures.

Bolshakov, A. et al. (2007). *STALKER: Shadow of Chernobyl*. Kiev, Ukraine: GSC Game World.

Bonnard, M., Chen, S., Carrere, M., Woodman, M., & Jirsa, V. (2017). Combined TMS-EEG to explore resting state brain dynamics and its transients. *Brain Stimulation: Basic, Translational, and Clinical Research in Neuromodulation, 10*(2), 459.

Bostrom, N. (2003). Are we living in a computer simulation? *The Philosophical Quarterly, 53*(211), 243–255.

Bradbury, R. (1950). *The Veldt*. In Bradbury, R. (1951). *The Illustrated Man*. New york, USA: Doubleday and Company.

Brooker, C., Jones, A., Reisz, B. et al. (2011). *Black Mirror*. House of Tomorrow. UK.
Bye, E. (1988). *Red Dwarf*. UK: Grant Naylor.
Calanchi, A. (2015). Searching for sounds in US literature: A multisensorial, multidisciplinary project. *European Scientific Journal, ESJ, 11*(6), 10–21.
Calleja, G. (2010). Digital games and escapism. *Games and Culture, 5*(4), 335–353.
Campanella, T. (2012). *La città del sole*. Newton Compton Editori.
Carroll, L. (1865). *Alice's adventures in wonderland*. London, UK: Macmillan.
Carter, S., Galt, T., Jenkins, P. et al. (2012). *The Darkness II*. Canada: Digital Extremes.
Conan Doyle, A. (1912). *The lost world*. London, UK: Hodder and Stoughton.
Churchill, W. (1931). If Lee had not won the Battle of New Gettysburg. In J. C. Squire (Ed.), *If It had happened otherwise*. London, UK: Longmans.
Cline, E. (2011). *Ready player one*. New York, USA: Random House.
Curry, J., & Pinchbeck, D. (2015). *Everybody's gone to the rapture*. Brighton, UK: The Chinese Room.
Dahl, R. (1982). *The BFG*. London, UK: Jonathan Cape.
Descartes, R. (1641). *Meditations on first philosophy*. In E. S. Haldane & G. R. T. Ross. (1978) *The Philosophical Works of Descartes*. Cambridge: Cambridge University Press.
Dhruv, N. T., Niemi, J. B., Harry, J. D., Lipsitz, L. A., & Collins, J. J. (2002). Enhancing tactile sensation in older adults with electrical noise stimulation. *NeuroReport, 13*(5), 597–600.
Dickinson, B. (2015). *Creative Control*. USA: Magnolia Pictures.
Désilets, P. et al. (2007). *Assassin's Creed*. Canada: Ubisoft Montreal.
Dick, P. K. (1962). *The man in the high castle*. New York, USA: Putnam.
Dick, P. K. (1966). *We can remember it for you wholesale*. The Magazine of Fantasy & Science Fiction. April 1996. Fantasy House.
Dick, P. K. (1968). *Do Androids Dream of Electric Sheep?* New York, USA: Doubleday.
Durango, J. (2015). *The leveller*. New york, UK: Harper Collins.
Dugas, J., Anfossi, F., Hamelin, S., et al. (2011). *Deus Ex: Human Revolution*. Canada: Eidos Montreal.

Dyess-Nugent, P., Handlen, Z., O'Neal, S., Robinson, T. & Semley, J. (2012). *The future won't look like this: 11 unintentionally ridiculous depictions of virtual reality*. A.V.Club: http://www.avclub.com/article/the-future-wont-look-like-this-11-unintentionally–84348.

Egan, G. (1994). *Permutation city*. London, UK: Millennium Orion Publishing Group.

Favreau, J. (2008). *Iron Man*. USA: Marvel Studios.

Fenlon, W. (2012). *How Futurist William Gibson is Still a Man of His Time*. Tested. http://www.tested.com/tech/3745-william-gibson-sci-fi/.

Fischer, D. (2011). *Science fiction film directors, 1895-1998*. McFarland.

Folman, A. (2013). *The Congress*. France: ARP Sélection.

Francis, A. M. (2015). *Nihilism: Philosophy of Nothingness*. Lulu.com.

Galouye, D. (1964). *Simulacron-3*. New york, USA: Bantam Books.

George, M. S., Wassermann, E. M., Kimbrell, T. A., Little, J. T., Williams, W. E. & Danielson, A. L. et al. (1997). Mood improvement following daily left prefrontal repetitive transcranial magnetic stimulation in patients with depression: A placebo-controlled crossover trial. *American Journal of Psychiatry*, *154*(12), 1752–1756.

Ghaffar-Kucher, A. (2014). Assimilation. In Phillips, D. C. (Ed.). *Encyclopedia of educational theory and philosophy*. Thousand Oaks, California: Sage.

Gibson, W. (1984). *Neuromancer*. New York, USA: Ace.

Gibson, W. (2007). *Spook country*. New York, USA: Viking Press.

Gilbert, R., Hammond, G., Schafer, T. et al. (1990). *The Secret of Monkey Island*. USA: Lucasfilm Games.

Gondry, M. (2004). *Eternal Sunshine of the Spotless Mind*. USA: Focus Features.

Graham, P. (2017). *Jaunt reviving cult classic The Lawnmover Man as a VR series*. VR Focus: https://www.vrfocus.com/2017/01/jaunt-reviving-cult-classic-the-lawnmower-man-as-a-vr-series.

Hanson, R. (2001). How to live in a simulation. *Journal of Evolution and Technology*, *7*(1).

Hardy, T. (1874). *Far from the Madding Crowd*. Cornhill Magazine. London, UK.

Harris, J. (1999). *Chocolat*. New York, UK: Doubleday.
Hart, S. M., & Ouyang, W. C. (2005). *A companion to magical realism* (Vol. 220). Suffolk: Boydell & Brewer.
Heinlein, R. (1941). *They*. In R. Heinlein. (1959). *The unpleasant profession of Jonathan Hoag*. USA: Gnome Press.
Henning, B., & Vorderer, P. (2001). Psychological escapism: Predicting the amount of television viewing by need for cognition. *Journal of Communication, 51*(1), 100–120.
Herbert, F. (1965). *Dune*. Philadelphia, USA: Chilton Books.
Howard, T. et al. (2008). *Fallout 3*. Maryland, USA: Bethesda Game Studios.
Hudson, C. (2007). *Mass Effect*. Edmonton, Canada: Bioware.
Huxley, A. (1932). *Brave new world*. London, UK: Chatto and Windus.
Johns, C. W. (2016). *Neurosis and assimilation: Contemporary revisions on the life of the concept*. UK: Springer.
Jones, S. E. (2013). *Against technology: From the Luddites to neo-Luddism*. London: Routledge.
Jousmäki, V., & Hari, R. (1998). Parchment-skin illusion: Sound-biased touch. *Current Biology, 8*(6), R190–R191.
Kitase, Y., Sugimoto, K., Katano, T. et al. (2002). *Final Fantasy X*. Japan: Square Enix.
Langsam, K. L. (2007). *Level of construal and the ideal self: Implications for attraction and support provision*. State University of New York at Stony Brook.
Lefaucheur, J. P., Drouot, X., Keravel, Y., & Nguyen, J. P. (2001). Pain relief induced by repetitive transcranial magnetic stimulation of precentral cortex. *NeuroReport, 12*(13), 2963–2965.
Lethem, J. (2009). *Chronic city*. New York, USA: Doubleday.
Leonard, B. (1992). *The Lawnmower Man*. USA: New Line Cinema.
Levine, K. (2013). *BioShock Infinite*. Novato, USA: 2K Games.
Longo, R. (1995). *Johnny Mnemonic*. USA: Tristar Pictures.
Lisberger, S. (1982). *Tron*. USA: Walt Disney Productions.
Livingstone, D. (2015). *Transhumanism: The history of a dangerous idea*. David Livingstone.
Maloney, D. (1976). *The Deadly Assassin*. Doctor Who. BBC. UK.

Marcus, L., & Nicholls, P. (2004). *The Cambridge history of twentieth-century English literature*. Cambridge, UK: Cambridge University Press.
Martinus, D. (1966). *The Tenth Planet*. Doctor Who. BBC. UK.
Matsuda, K. (2016). *Hyper-Reality*. Fractal. Columbia.
Mitchell, D. (2004). *Cloud Atlas*. Sceptre. UK.
Montgomery, L. M. (1908). *Anne of Green Gables*. USA: L. C. Page and Co.
More, T. (1975). 1516. Utopia. *Robert M. Adams (trans.)*. New York: Norton.
Mostow, J. (2009). *Surrogates*. USA: Touchstone Pictures.
Murray, S., Duncan, G., Doyle, R., & Ream, D. (2016). *No Man's Sky*. Hello Games. Guildford, UK.
Neveldine, M., & Taylor, B. (2009). *Gamer*. USA: Lionsgate.
O'Hara, K. (2011). *Conservatism*. London: Reaktion Books.
O'Shea, J., & Walsh, V. (2007). Transcranial magnetic stimulation. *Current Biology, 17*(6), R196–R199.
Pfister, W. (2014). *Transcendence*. Warner Bros. Pictures. USA.
Palmer, W. (2009). *The films of the '90s: The decade of spin*. Berlin: Springer.
Pascual-Leone, A., & Walsh, V. (2001). Fast backprojections from the motion to the primary visual area necessary for visual awareness. *Science, 292*(5516), 510–512.
Poe, E. A. (1839). *The fall of the house of usher*. USA: Burton's Gentleman's Magazine.
Pratchett, T. (1983). *The Colour of Magic*. Colin Smythe. Buckinghamshire, UK.
Pugh, W. (2015). *Dr. Langeskov, the tiger and the terribly cursed emerald: A whirlwind heist*. Crows Crows Crows. UK.
Pullman, P. (1995). *Northern Lights*. Scholastic Point. UK.
Raynall, F. (1992). *Alone in the Dark*. Infogrames. Lyon, France.
Reed, B. (1974). *Star Trek: The Animated Series*. Filmation. USA.
Rogasch, N. C., & Fitzgerald, P. B. (2013). Assessing cortical network properties using TMS–EEG. *Human Brain Mapping, 34*(7), 1652–1669.
Roiland, J. and Harmon, D. (2015). *Rick and Morty*. USA: Warner Bros. Television.
Romero, J., Petersen, S., Carmack, J. et al. (1993). *Doom*. USA: Id Software.
Roth, H. (1934). *Call it sleep*. New York, USA: Robert O. Ballou.
Rowling, J. K. (1997-2007). *Harry Potter*. London, UK: Bloomsbury.
Sadava, S. W., Thistle, R., & Forsyth, R. (1978). Stress, escapism and patterns of alcohol and drug use. *Journal of Studies on Alcohol, 39*(5), 725–736.

Sadler, R., & Dooly, M. (2014). Language learning in virtual worlds: Research and practice. In: Thomas, M., Reinders, H. & Warschauer, M. (Eds.) *Contemporary computer-assisted language learning.* 159–182.
Sale, K. (1997). *America's new Luddites.* Le Monde Diplomatique. http://mondediplo.com/1997/02/20luddites.
Sanders, S. (2007). *The philosophy of science fiction film.* Lexington: University Press of Kentucky.
Savulescu, J., & Bostrom, N. (2009). *Human enhancement.* Oxford: Oxford University Press on Demand.
Scott, R. (1982). *Blade Runner.* Burbank, USA: Warner Bros.
Sears, D. & Clayton, W. (2016). *We Happy Few.* Montreal, Canada: Compulsion Games.
Simeone, A. L., Velloso, E., & Gellersen, H. (2015). Substitutional reality: Using the physical environment to design virtual reality experiences. In *Proceedings of the 33rd Annual ACM Conference on Human Factors in Computing Systems* (pp. 3307–3316). ACM.
Singh, T. (2015). *Self/Less.* California, USA: Focus Features.
Singles, K. (2011). 'What If?' and beyond: Counterfactual history in literature. *The Cambridge Quarterly, 40*(2), 180–188.
Siracusa, J. (2013). *Technological conservatism.* Hypercritical. http://hypercritical.co/2013/04/07/technological-conservatism.
Sirgy, J. (2012). *The psychology of quality of life: Hedonic well-being, life satisfaction, and eudaimonia* (Vol. 50). Berlin: Springer Science & Business Media.
Spector, W. (2000). *Deus Ex.* London, UK: Eidos Interactive.
Spielberg, S. (1993). *Jurassic Park.* Universal City, USA: Universal Pictures.
Spinrad, N. (1980). *Songs from the stars.* London, UK: Orion Publishing.
St. Clair, J. (2013). *Sound and Aural Media in Postmodern Literature: Novel Listening.* Abingdon, UK: Routledge.
Stover, M. (1998). *Heroes Die.* Del Rey. USA.
Stephenson, N. (1992). *Snow Crash.* New York, USA: Bantam Books.
Strugatsky, A., & Strugatsky, B. (1972). *Roadside picnic.* Former Soviet Union: Macmillan.
Swift, J. (1726). *Gulliver's Travels.* London, UK: Benjamin Motte.
Tachi, S. (2015). Telexistence. In *Virtual Realities* (pp. 229–259). Springer International Publishing.
Tolkien, J. R. R. (1937). *The hobbit.* London, UK: George Allen and Unwin.
Tolkien, J. R. R. (1997). *The Monsters and the Critics and other Essays.* C. Tolkien (Ed.). New York, London: HarperCollins.

Toyama, K. (1999). *Silent Hill.* Tokyo, Japan: Konami Computer Entertainment.

Trollope, A. (1855). *The warden.* London, UK: Longman, Brown, Green and Longmans.

Trumbull, D. (1983). *Brainstorm.* Beverly Hills, USA: Metro Goldwyn Mayer.

Tuan, Y. F. (2000). *Escapism.* Baltimore: JHU Press.

Verhoeven, P. (1990). *Total Recall.* Culver City, USA: TriStar Pictures.

Von Schiller, P. (1932). Die rauhigkeit als intermodale erscheinung. *Z Psychol Bd, 127,* 265–289.

Wachowski, L. & Wachowski, L. (1999). *The Matrix.* California, USA: Warner Bros.

Warwick, D. D. (2015). *Hive Propolis.* Mythos Media [online-only publisher].

Weinbaum, S. G. (1935). Pygmalion's Spectacles. *Wonder Stories.* New York, USA: Thrilling Publications.

Williams, T. (1996). *City of Golden Shadow.* New York, USA: Legend Books.

Winton, C. (1960). Conversion on the road to Houyhnhnmland. *The Sewanee Review, 68*(1), 20–33.

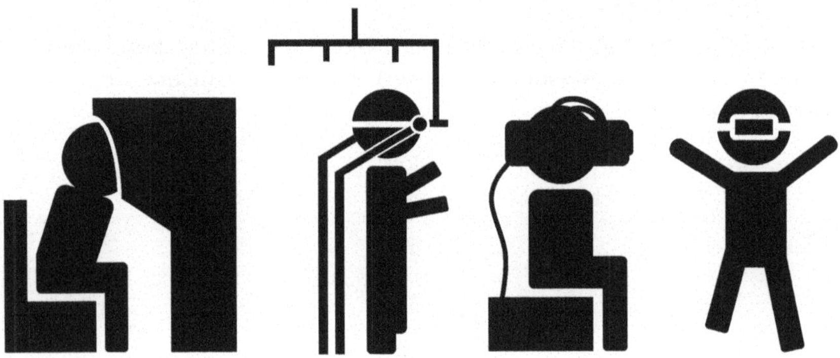

6

Technological History

So far in our exploration of VR and VR sound, we have considered various underlying concepts, points of user-experience and fictional representations in the literature, cinema and television. This chapter continues to uncover relevant matters by treading the historical pathway of VR from a technological perspective. Commencing with some of the concepts and artistic precursors of VR, dating back to the Medieval and Renaissance eras, this chapter focusses predominantly upon the VR developments between the 1960s and the closing of the twentieth century. As the forerunners of VR were introduced to the digital age, as the term 'virtual reality' was officially coined, and as the seemingly limitless potential of the computer began one of the technology's most crucial developmental milestones, we examine these periods with close attention paid of the role and value of sound. This chapter closes with a look at the history of 'VR sound' and posits that not only has sound been ever-present throughout the story of VR and its precursors, its technological and conceptual progress (particularly during the twentieth century) positions it as history's most successful aspect of VR technology.

T.A. Garner, *Echoes of Other Worlds: Sound In Virtual Reality*,
Palgrave Studies in Sound, DOI 10.1007/978-3-319-65708-0_6

Artistry and Mechanical Approaches to Early Virtual Worlds

The focus of this section is upon artistic and mechanical precursors to VR that reflect the most significant aspects of the technology as it exists today. Contemporary research has argued that stereoscopy, field of view (panorama) and synchronised multimodality are three of the most vital contributors to immersion and presence (Freeman et al. 1997; Hendrix and Barfield 1996) which, as we established in Chap. 4, are two of the most significant elements of VR experience. It is therefore these points that we focus upon here. Whilst there is certainly a great deal of further discussion that could be had regarding art, illusion and virtuality, there are already excellent texts available that cover this area comprehensively (see Grau 2003). Here, the topics are restricted to points in which art connects with concepts and mechanisms of VR and which also elucidate the nature of sound in VR's technological precursors.

Panorama

First officiated in 1787 when Robert Barker registered for a patent, panoramic art describes a sweeping image (including paintings, etchings, motion pictures and photographs) that displays a wide scene. Many panoramas extend across the curves of a cylindrical interior space, a rotunda typically referred to as a 'cyclorama'. The cyclorama positions the viewer in a central space so that they are surrounded by the image (Jacobs 2004). Interestingly, panoramic art of the eighteenth and early nineteenth century proved to be a controversial medium, specifically due to the illusory nature of its aesthetic. Critics would assert that the illusion yielded an inherent danger (conservatism making an appearance early on in the timeline of VR—see Grau 2003). This concern has proven to be a mainstay of VR with many people still unsure of the technology and, as the critics would argue, the uncertainty it presents.

Panoramic artistry of comparable form does predate the cyclorama, however, reaching back substantially further in time to examples such as Gu Hongzhong's *Night Revels of Han Xizai*, a painted scroll produced at an undetermined time between 937 and 975 CE. The principles of viewer-interaction in Hongzhong's piece reflect, to an extent, subsequent panorama and VR, specifically, in terms of the requirement for the viewer to actively move their heads to take in the entire scene. The turn of the sixteenth century brought with it Michelangelo's paining of the Sistine Chapel in Vatican City, possibly one of the most significant artistic accomplishments in all of history and one that is also an early instance of panorama. The work encompasses both the walls and ceiling of the chapel's interior, creating a panoramic piece that extends the field of view both horizontally and vertically.

Returning to the late eighteenth century, Robert Barker's rotunda was constructed in 1792, in the city of London's Leicester Square. Inside it housed a panoramic etching of the London skyline. Many years later, Franz Roubaud's *Battle of Borodino* (1911), depicting conflict during the Napoleonic invasion of Russia, is now regularly cited as an influential panoramic piece, further popularising design concepts that would eventually inspire VR. For Grau (2003), panorama represents the first significant instance of artistic illusion being generated by way of 'a pre-calculated outcome of the application of technological, physiological, and psychological knowledge' (p. 6). Panorama is recognised as one of the first instances of design integrating with the viewer and is inclusive of both physiological and psychological issues. Essentially, it is the eighteenth century's answer to user-centred design.

The fundamental quality of panoramic art is its encircling and surrounding of the viewer to varying degrees. The influence of panorama can be observed at several points during the years that followed, leading up to the present day. Cinéorama for example, refers to Raoul Grimoin-Sanson's experimental amusement ride, unveiled during the 1900 Paris Exposition. The piece utilised ten synchronised film projectors, each casting a moving image upon a nine-square metre screen. The arrangement of these screens formed a 360° cylindrical interior; now panorama was not only in-surround, but had also become in-motion. Later twentieth century offspring of panorama includes 'Expanded Cinema', a phrase coined by film-maker Stan Vanderbeek in the 1960s (Clark 2014). The term was later popularised by Youngblood and Buckminster

(1970), who explained the phrase as integration of additional artistic artefacts that spread beyond the limits of the flat screen.

Pushing forward to the present day and panorama certainly survives as an aspect of modern VR, with a fundamental quality of both HMDs and CAVE projection systems being to present the user with the image of a virtual world that surrounds them; it cannot be taken in fully without physical movement, requiring the user to engage with the world through interactivity. Modern VR also takes after its artistic precursors in terms of its experimentation with the type and degree of field of view, specifically panoramic photography. The roots of this art form can be roughly traced back to 1843 and another patent, this time for a bespoke wide-angle camera design by Joseph Puchberger (Vanvolsem 2011). Using these types of lenses, early panoramic photographs such as *San Francisco from Rincon Hill* (Behrmanx 1851) were able to capture much greater fields of view, specifically across the horizontal plane. Stitching together a series of such curved planar images could create up to 180° or 360° cylindrical images, much like the rotundas and cyclorama. However, by way of a technique known as inner-sphere panorama, modern photography is now able to extend both the vertical and horizontal field of view, like Michelangelo's work before it.

The assertion being presented here is that contemporary VR owes a debt to these panoramic forerunners of virtual worlds. Whilst computer-generated VR environments and digital games emulate the inner-sphere cyclorama techniques that enable the user to experience a fully enveloping image, VR photography applications such as *Cardboard Camera*[1] utilise a 360° horizontal field of view. At the time of writing, VR video is equally common in both 180° and 360° formats, often depending upon context and resolution requirements of the video (180° video requiring lower resolution relative to 360° for an equivalent image fidelity). For example, a video from an aeroplane cockpit would typically be in 180° format as all vital visual information exists within this field of view. That these panoramic options existed in alternative forms long before digital technology, reveal how they have been a significant influence on the varieties of contemporary VR today.

Across the various instances of panorama detailed above, any auditory component is arguably the subject of substantially less discussion and yet, none of them existed with sound in absentia. Michelangelo's frescoes for example are experienced in parallel to the soundscape of the Sistine Chapel, the materials of its architecture creating a highly reflective acoustic space able to bounce even very low amplitude sound waves repeatedly, creating a perceptual soundscape that is simultaneously dense and near-silent. Lewers and Anderson (1984) point out, in their analysis of the acoustical properties of St Paul's Cathedral, London, that spaces comparable to the Sistine Chapel significantly reduce speech intelligibility due to the reflecting sound waves obscuring one another. Such auditory properties connect sound inexorably to the visuals, arguably contributing to the overall multisensory experience and to our perception of the environment.

As new technologies began to emerge and panorama became more mechanical and less architectural, the soundscape demonstrates significant change. Grimoin-Sanson's *Cinéorama* for instance, that centres the viewer within a circle of film projectors, does not include any intentional audio content. However, being positioned next to ten active 70 mm projectors, each whirring away as the various mechanical components reposition and rotate, undoubtedly presented the Cinéorama's audience with a distinctive sonic environment that contributed to the overall experience.

Stereoscopy

As Baños and colleagues (2008) point out, stereoscopy is a technique that presents a different image to each eye to enhance our perception of depth in an image. Human visual perception has proven to be highly capable of discerning depth from a flat (or monoscopic) image, be it a printed photograph, a screen-rendered graphic on a computer, or a physical object (Koenderink et al. 1994). Monoscopic depth perception can be attributed to the phenomenon 'parallax'. Parallax as a term encompasses a range of meanings depending upon context and can be confused with wave-particle duality in physics and distinguishing the mind from the brain in neurobiology (see Žižek 2009, p. 7). [2] Sticking with visual perception, parallax describes the positional difference between our eyes that causes each eye to receive a slightly different image. Between our left and right eye, any object we direct our gaze upon will appear at a different position and angle. The brain then combines these two images in a way that grants us a sense of depth. Stereoscopy takes the principle of parallax to manufacture an 'enhanced' perception of depth by artificially extending the distance and viewing angle. The mechanism is simply the presentation of two subtly different images, one presented to each eye, depicting a single object or scene.

Underlying concepts that would drive the development of stereoscopy reach back substantially into ancient history. Sir David Brewster, a seminal figure in the mechanical development of the stereoscope, traces the foundational theory back to fourth century BCE and the mathematical theorems of Euclid (Brewster 1856). Leaping forward somewhat, illusions of depth in artistry begin to appear in the sixteenth century. The *Sala delle Prospettive*, by Italian architect and painter Baldassare Peruzzi, is often described as one of the first visual instances of parallax. Painted between 1516 and 1518 within the Villa Farnesina in Rome, Peruzzi's piece largely consists of numerous human characters engaging in various activities but also features imposing marble pillars standing in front of a panoramic landscape. The work extends across the four interior walls of the Villa's drawing room, creating for the visitor, a 360° viewing experience. Moreover, Peruzzi utilises mathematically determined perspective to scale objects within his painting in a way that evokes parallax to create an illusion of depth (Grau 2003).

The historical roots of stereoscopy also reach as far back as the early sixteenth century and have been observed in yet another masterpiece of Italian Renaissance art, Leonardo da Vinci's *Mona Lisa* (circa 1503–1503). Koenderink and colleagues (1994) note that artists including da Vinci would recommend viewers of their work to stand in front of the painting and view it with one eye closed, presumably to facilitate something akin to purity in the viewing experience. Whilst this implies that da Vinci possessed a working conception of parallax, it has been more recently asserted that he also may have knowingly developed a stereoscopic illusion by way of the Mona Lisa. Carbon and Hesslinger (2015) note that, during his creation of the original painting, da Vinci (or possibly one of his students) also produced a copy, entitled *La Gioconda* (presently located in the Prado Museum, Spain). Centuries later, the Louvre (in which the original is famously housed) approached conservators at the Prado Museum, asking to lend their copy for a special exhibition in 2012. During close inspection of the pieces prior to the event, analysts realised that the two paintings presented the same subject but from slightly different angles—and positioning the two pieces closely together created a stereoscopic illusion.

The stereoscope itself emerged in the mid-nineteenth century and is largely attributed to the work of Charles Wheatstone who revealed his invention in 1838, just prior to the unveiling of photography (Timby 2005). Wheatstone's device incorporated two adjacent mirrors at 45° angles that reflected two pictures, each positioned to the left and right sides of the person viewing the image. The viewer would place their head directly in front of the mirrors. There were no lenses, and the device was not encased. It was in the design of Wheatstone's competitor, the aforementioned Sir David Brewster, that what could be asserted to be the world's first HMD first appeared. Brewster's stereoscope grafted a small wooden box, the shape of a trapezoidal prism, onto eyepieces taken from opera glasses (Stafford et al. 2001). This design incorporated frosted glass at the front, enabling the stereoscopic images to be backlit, but otherwise created an enclosed viewing experience in which the viewer would only see the prescribed image, with all background visual noise blocked from their periphery. The stereoscope gained much exposure following demonstrations at the Great Exhibition of 1851 (Trotter 2004) and soon thereafter became a very popular and commonplace piece of technology.

Stereoscopic technology up to this point was a distinctively visual experience, with any auditory component depending upon the surrounding physical environment in which the stereoscope was used. It was not until 1881, when Clément Ader revealed two-channel stereo to the world that a sonic equivalent to the stereoscope emerged. However, this sonic breakthrough cannot be directly attributed to the stereoscope. As we shall see, it was discovered quite accidentally (Maconie 2002).

In the years that followed, stereoscopic technology would evolve from static to moving pictures with the Kaiserpanorama. Patented by August Fuhrmann in 1890, the Kaiserpanorama utilised a rotating set of slides to evoke a sense of motion and brought together panorama with stereoscopy (Luhmann 2004). However, it was not until 1961, when Morton Heilig patented the *Sensorama* that the path of technological development truly took a leap towards our contemporary vision of VR by championing multimodality.

Multimodality

The Sensorama is regularly cited as the first VR device (though still predating the term by many years) to present stimuli across multiple sensory modalities as a means of increasing presence (Boas 2013; Dinh et al. 1999). Helig's patent (1962) identifies some of the fundamental values of VR technology that remain equally prominent today. He notes that: 'There are increasing demands today for ways and means to teach and train individuals without actually subjecting the individuals to possible hazards of particular situations', citing the armed forces and industries dealing in hazardous materials as key application areas. Helig claims that VR (referred to in the patent as 'simulator apparatus') provides a solution to this demand, and he also presents rapid prototyping as a significant practical use of the technology. These statements were most certainly prophetic of contemporary VR and all of the above applications feature in a related discussion in Chap. 9.

The Sensorama itself is a device roughly equivalent in size to a video arcade cabinet. The user sits upon its integrated chair and leans into the device. A hood extending from the front to surround the head obscure the user's peripheral vision and the displayed moving images are both stereoscopic and panoramic. Several 'experiences' were created for the Sensorama and included driving a motorcycle through New York[3] or riding a helicopter over Los Angeles. Throughout the patent, Helig repeatedly advocates the importance of multimodal feedback, what he describes as 'developing sensations in a plurality of the senses'. Specifically, Helig's Sensorama design incorporates visual image projection, olfactory sensation (scent and the feeling of a breeze both delivered by way of an integrated ventilation unit), tactile content (via vibrations in the seat) and binaural audio.

In terms of auditory content, Helig's patent stresses the importance of directional sound as a means of evoking 'one's perception of reality' (1962). The Sensorama presents binaural recordings by way of two speakers built into the sides of the device's hood. Helig asserts that the binaural technique (in its most basic form, using two microphones to capture and two speakers to project—see Chap. 8) provides directional accuracy comparable to what would be achieved using the twenty-speaker set-up that would have been typical in a film theatre. In addition to the binaural playback, the auditory experience of the Sensorama featured sounds that by this point had become synonymous with simulation technology, though are rarely accounted for in the literature. Positioning the film projector in very close proximity to the viewer, with little attenuation afforded by the Sensorama's casing, a substantial amount of 'whirs' and 'clunks' emanated from within, often to the extent that they would compete with the binaural recordings for dominance in the overall soundscape.

The above section has outlined the development of artistic and technological precursors to VR, focussing upon aspects that have remained vitally important to VR theory and design across the twentieth century and into the twenty-first century. Moreover, sound has revealed itself to be a constant component of these precursors, irrespective of the designer's intentions. The subsequent section of this chapter continues along the VR timeline, picking up where the Sensorama left off, to explore the prominent VR developments that occurred across the latter half of the twentieth century.

Developments in VR Between 1960 and 1990

The representations and designs of VR prior to the twentieth century have characterised the concept as something primarily visual. The previous section of this chapter revealed how sound was an inexorable component of many a virtual experience, directly (if not intentionally) originating from the mechanics of the devices or indirectly, from the surrounding soundscape. As we shall discuss now, the subtle presence of sound did not change substantially in the years that followed.

That said, during the latter half of the twentieth century, two important observations are apparent. Firstly, whilst few early VR devices incorporated sounds, the conceptual work that accompanied these devices reveals how the individuals behind them did have an appreciation for multimodal content and particularly sound. Secondly, whilst the many developments throughout the century that are typically associated with the history of VR were occurring, a parallel line of research and development regarding sound was also taking place. This research, though not commonly described as 'VR sound', would nevertheless have a profound effect on contemporary VR when it eventually began to connect with VR development towards the end of the century. The history of VR between the 1960s and 1990s has received a good deal of attention (see Burdea and Coiffet 2003). Consequently, this section seeks to provide an alternative perspective by concentrating on the role and value of sound, before the chapter closes with a look at how research into sound technology progressed during the same period.

The 1960s: From the Sword of Damocles to the Ultimate Display

Whilst the Sensorama may have failed to achieve commercial success (despite being a revolutionary advance in VR innovation), it was not the singular endeavour of inventor Morton Helig. Two years prior to the patent of the Sensorama, Helig also patented the 'Stereoscopic-Television Apparatus for Individual Use' (Helig 1960). The device became known as the *Telesphere Mask*, though this name did not appear within the original patent. Whilst predated by Brewster's stereoscope, the Telesphere Mask has earned more official recognition for being the world's first HMD (Flores-Arredondo and Assad-Kottner 2015). Its design is discernibly similar to contemporary headsets, from its lightweight frame that covers the visual periphery, to lenses of adjustable positioning and focus. As with the Sensorama, Helig again reveals an appreciation for multimodality, with sound receiving notable attention within the patent. The Telesphere Mask includes a pair of speakers, built into the main positioning strap of the device, supporting stereo and binaural playback. Helig is specific, describing the earphones as 'so designed that the outer ear is completely free and untouched, thus allowing [them] to operate fully as sound focusing organs' (Helig 1960, p. 3). Helig also notes that the earpieces must be adjustable in the same way as the lenses, both for comfort and listening clarity.

Continuing onwards to 1961, the *Philco Headsight* brought us directional tracking of the head by way of magnetometers. By connecting the HMD to a close circuit television system, it is also one of the earliest instances of telepresence (Kiyokawa 2006). Computer-generated environments and the origins of (what we would recognise as) contemporary augmented reality followed with the work of Ivan Sutherland and the *Sword of Damocles*. Sutherland's device is an example of BOOM (Binocular Omni Orientation Monitor) technology, a HMD that tracks head direction and gaze by way of a heavy mechanical arm that is either counterbalanced against a weighted base unit or suspended from the ceiling. The Sword of Damocles is an example of the latter, a design aesthetic that earned its famous moniker (Mihelj et al. 2014). Neither the Philco Headsight nor the Sword of Damocles featured any auditory aspect, but Sutherland also made a theoretical contribution to VR with his concept for what he called the 'ultimate display'. In this vision, Sutherland emphasised the ideals of multimodal experience: 'If the task of the display is to serve as a looking-glass into the mathematical wonderland constructed in computer memory, it should serve as many senses as possible' (Sutherland 1965, p. 1). Sutherland's main assertion regarding sound was that it should be meaningful. He elucidates this point by describing what he calls a 'kinaesthetic display', a haptic-lead interaction between user and computer system that presents audio and visual material corresponding to the haptics, for more complete, multisensory feedback system.

The 1970s and 1980s

It wasn't until the 1990s that a significant number of commercial VR devices for entertainment appeared. Before this, however, the 1970s was a decade of behind-the-scenes progress made at academic institutions.

Whilst less well-known, 1970s VR had a substantial influence on that which would follow decades later. In one instance, Martin Krueger brought us the *Videoplace*, a process of digitising a user's silhouette then analysing gestural movements and manipulating the graphical display in response (see Reas and Fry 2007); arguably a significant precursor to the motion-tracking technology of the Microsoft *Kinect*. As an interesting aside, Krueger is also credited with coining the term 'Artificial Reality' (Muhanna 2015).

Towards the end of the 1970s, Eric Howlett presented the LEEP (Large Expanse Extra Perspective), an optical system facilitating a 110° horizontal field of view (55° vertical—see Kiyokawa 2006). LEEP optics found commercial application in the first HMD of the computing age, the *Eyephone* (Teitel 1990), and their design has persisted to the present day, across the entire range of contemporary HMDs. At around the same time, the forerunner of both *Google Street View* and *Google Earth* was being developed by Andrew Lippman and a research group at Massachusetts Institute of Technology. Their system, the *Aspen Movie Map*, took forward-facing photographs of Aspen's entire road network and integrated them into a computer program that would allow the user to 'travel' around the resort town. The Movie Map also incorporated a 'synthesised replica' (Lippman 1980, p. 32) of Aspen that could explor in much the same way. Additional data relating to the town and its locales was 'stored' in individual buildings, with users able to retrieve that information as they passed.

Throughout the 1970s and 1980s, technological developments in VR appear to largely overlook sound. However, as we shall see later in this chapter, important developments in VR sound technology and design were occurring throughout these decades, but as a separate strand of progress that would integrate with VR in the years that followed.

Wider VR Developments During the 1990s

It is a misnomer that the 1990s was the decade in which VR collapsed. Instead, it would be much more accurate to state that the final years of the millennium heralded a consumer market downturn, specifically with regard to the recreational application of the technology. Much like the 1970s, research and development for wider applications of VR continued, largely unaffected by the problems in the entertainment industry. Dissimilar to the 1970s, however, is an observable increase in the consideration of sound.

We begin in 1991, with the work of digital media artist Nicole Stenger and *Angels* (*Les Recontres Angeliques*). Acknowledged to be the first instance of both 'VR movies' and of a formal artist working in VR (Morie 2012), Angels is a largely linear sequence of 3D graphics with some user-interaction. Users could explore the space by way of head orientation-tracking within a HMD. They could also interact using gesture, by way of hand and finger-tracking with a Data Glove. Angels also present a musical soundscape consisting of a synthesised harp and a music box. The music is linear in composition, but primarily rhythmic and amelodic, so that it does not clash against the graphical transitions that are determined by the user.

A year later, Cruz-Neira and colleagues (1992) reveal the Cave Automatic Virtual Environment (CAVE), a cubic space onto which the four walls, ceiling and floors are projected images. It is likened to the *Omnimax* theatres and early flight simulators. CAVE systems have progressed through various developments but their original design did account for sound. Research using CAVE VR reveals that this system encouraged developments positional sound techniques. For example, Cruz-Neira and her colleagues (1993) note the difficulties in achieving convincing sound localisation with multiple speakers positioned in the four corners of the CAVE, due to the screens being highly reflective of the sound waves.

In 1995, contemporary artist Char Davies revealed *Osmose*,[4] a VR installation in which the user experiences several contrasting virtual renderings of physical environments (including a forest and an abyss) but also two more abstract spaces filled with floating text. The first space is described as the 'substratum' level where the user is surrounded by code used in the software itself. The second is the 'super-stratum' level, containing quotes from the artist and other texts related to the project. A retrospective by Davies and Harrison (1996) elucidates further, noting that their ambition was to generate full body immersion and that this aim guided many aspects of the design. To this end, head tracking is complimented with posture and respiration data (e.g., taking a deep breath would move the avatar upwards whilst exhaling would move it downwards). Davies and Harrison also note that the system's design paid particular attention to sound, specifically its affective and evocative qualities. Osmose utilises heavily edited samples of human voices that attempt to fall somewhere between literal and abstract interpretation. Throughout Osmose, synthetic audio sources are combined with acoustic recordings, and whilst some sounds are directly tied to changes in the visuals, or input from the user, others are not and the nature of the overall soundtrack is such that differentiating between the two is very difficult. Davies and Harrison also state that it was important for the soundscape to be flexible and support smooth transitions between the virtual environments. For this purpose, the sound was designed using techniques comparable to contemporary adaptive audio. Several soundscapes and sets of musical content were used, that were distinct from one another and reflected the individual environments, but also possessed shared qualities that would enable transition as determined by the user.

Davies elaborates on her notions regarding VR sound in her account of *Ephémère*,[5] a step forward from Osmose, but utilising broadly the same hardware and interfacing methods. In her retrospective on the project, Davies (2004) describes how the soundscapes in Ephémère utilise not just 3D positioning of sound objects, but also movement within the virtual space, acknowledging how the dynamic, fluctuating nature of sound can be exploited to evoke presence. For Davies, the fluctuating nature of sound offered VR to experience a great temporal dimension that could make users feel as though they were truly present in the virtual world.

1990s Consumer VR

In addition to these research innovations, 1990s VR systems developed for the entertainment industry are still deserving of a mention here, particularly as they provide early examples of audio design and technology becoming more integrated within VR. Released at the turn of the decade, the *Virtuality 1000* series was developed by Jonathan Waldern and W-Industries. The system utilised a modified Amiga 3000 computer system (marketed as the *Expality VR Computer*), running a bespoke operating system (*Animette VR*) and outputting the content of the virtual environments by way of a HMD (the *Visette* visor). The Virtuality system was a pioneer of multiplayer VR, with several systems able to communicate, enabling users to share a single environment. A quadraphonic audio system is employed, with four speakers integrated within the Visette HMD. In 1992, Waldern himself published a conference paper, largely publicising his Virtuality technology, but also discussing pertinent VR issues that included a section on sound. Here, Waldern discusses positional audio and notes that the properties of VR technology are (as of 1992) inherently supporting of 3D auditory localisation techniques. Waldern also makes predictions concerning voice detection and speech recognition, asserting that these shall become key features within future VR systems (Waldern 1992).

The year 1993 saw the official reveal of the *Sega VR*[6] system at the annual Consumer Electronics Show. The device itself did not progress from the prototype stages of development and was never officially released. The design of the prototype was as a peripheral extension of Sega's *Genesis* (*Mega Drive* in the UK), a HMD that incorporated two display screens and a stereo audio configuration that positioned two speakers over the ears. Although failing to reach consumers, the technology of the Sega VR did feed into the arcades in Sega's *VR-1* simulator system. We shall return to this in Chap. 7, where we look in greater detail at the relationships between VR and digital games (both home computing and the arcades).

Moving forward to 1995 and we reach the pariah of VR, the *Virtual Boy* by Nintendo. Developed as a standalone console rather than a peripheral for an existing system, the Virtual boy HMD housed both the display hardware and all the computing components (processor, RAM, etc.). Consequently, the rather heavy HMD was stand mounted, with the user in a fixed, typically seated position. The Virtual Boy's graphical display was a collaboration between Nintendo and Reflection Technology Inc. (see Steinicke 2016) that resulted in a red/black colour scheme. The device delivered 16-bit stereo audio by way of two speakers positioned at the outer left and right edges of the headset. Although it is admittedly rather difficult to source more detailed technical information, details of the Virtual Boy's audio chipset do exist in a few choice locations online.[7, 8] These sources reveal that Nintendo developed a bespoke audio processing unit for the Virtual Boy, entitled the VSU. The chip supports four stereo wavetable synthesis (WTS) channels, with the left and right volumes independently adjustable and graded signal enveloping that together enable a certain degree of audio source localisation. A fifth WTS channel is included that additionally features frequency-over-time sweeping (pitch bending) and modulation (e.g. tremolo) effects. The final sixth channel is a basic noise generator. Interestingly, comparing this unit against Nintendo's home console of the time, the *Super Nintendo Entertainment System* (SNES, released in 1991) reveals that the Virtual Boy audio chipset was largely inferior to that of its older relative. By contrast, the SNES boasted 8 channels of audio and utilised pulse code modulation technology, which enabled sampled recordings of physical sound waves to be presented alongside synthesised sounds.[9]

The history of the Virtual Boy system now stands as a powerful illustration of Gartner's Hype Cycle for emerging technologies (Fenn and LeHong 2011). As with the Virtuality 1000 and Sega VR systems before it, the Virtual Boy was riding high on the peak of inflated expectations, the point on the cycle where the consumer base is fully hyped about the product, but their anticipations are significantly removed from the reality of its functionality and quality. A retrospective by Zachara and Zagal (2009) splits the blame for failure across multiple points, including a disengaging visual display and a socially isolating game experience. Whilst it would be a stretch to assert that the failure of the Virtual Boy is directly connected to its inferior audio, there is a broader related point of which the sound is a factor. Specifically, that what derailed the system (and commercial VR in general during the 1990s) was an overreliance on the unique novelty aspects of VR, to the extent that maintaining quality in other areas (including audio, but also graphics and gameplay) was deprioritised. The hope essentially being that the consumer's desire to experience a realisation of the VR fantasy was so intense that they would overlook these qualities. As we shall explore further in the next chapter, the reality of virtual reality proved to be something of a stern lesson.

A Brief History of (and a Case for) VR Sound

This chapter closes with a look at the historical road less travelled in the context of VR that of sound design audio technology. Much like the visual-centred developments documented above, advances in sound also stretch back centuries, with similar flurries of development across the nineteenth and twentieth centuries and sustained progress as we reach the present day. We begin with a look at early developments in audio recording and synthesis before moving on to a brief history of multichannel audio and closing with something of an argument, about headphones.

Electronic Sound and Recording Technology

The dawn of the electronic era of sound technology is commonly traced back to 1874, and Elisha Gray's accidental discovery that tuned materials (such as reeds or metal bars), when stimulated with electromagnetic current, would oscillate at the frequency of that material and that oscillation could be transmitted over a telephone wire. This discovery leads to Gray developing the first electronic music instrument, a single note oscillator entitled the *Musical Telegraph* (Salazar-Palma et al. 2009). Writing for the North American Review in 1888, Thomas Edison himself recounts his own work developing the cylinder phonograph in 1877, a few short years following Gray's Musical Telegraph. Initially, using paper tape and then tin foil, Edison's cylinder phonograph signified the beginning of mechanical sound recording (Edison 1888). The mechanism of recording had been realised decades prior with Édouard-Léon Scott's 1857 invention, the phonautograph, a means of translating acoustic energy into visual representations of sound waves

(Brock-Nannestad and Fontaine 2008). However, it was Edison's phonograph that facilitated both recording and playback of sound.

Citing Edward Johnson's article *A Wonderful Invention* (1877), Roy (2016) describes how the phonograph marked a turning point in the very meaning of scientific progress: 'Science had eventually left the cold realm of reason and the intellect to enter the domain of emotions' (p. 2). What particularly sparked Johnson's attitude towards the technology was undoubtedly something foundational to the meaning of the virtual. As Roy elucidates, it was the notion that the phonograph could replay the voices of those who had died that moved Johnson that individuals who were not present in the actual environment could be experienced as if they were. In the years that followed, sound technology continued to develop, experimenting with various materials, mechanisms and devices, from the gramophone to Guglielmo Marconi's discovery of radio transmission (Schoenherr 2005). Towards the end of the First World War, a sound recording that became known as the 'Gas Shell Bombardment' was produced. The recording itself was of an artillery attack that took place near Lille in October 1918. The recording was reproduced and commercialised, with many people replaying it in their own homes as a virtual means of experiencing the nature of the war (Roy 2016).

All the above developments come after those made in panorama and stereoscopic technology, with the visual precursors of VR able to both capture the physical world through photography and create worlds beyond our own through art. Auditory precursors of this time may have been preoccupied with capture, but in terms of virtual experience, offered something the stereoscope could not: motion and time. At this point in history, most visual experiences of a virtual nature were static, whilst sound could present a fully dynamic affair. This is not to say that moving images did not exist at this point, the phenakistoscope (a cardboard disc attached to a handle that is spun to animate a series of images) and zoetrope (a panorama of still images within a cylinder that is spun and viewed through slits) being two notable examples. However, with the release of technology, such as the gramophone, that was able to suitably amplify playback, sound waves were able to extend into physical space, filling the room and surrounding the listener. In this instance, audio technology had revealed a particularly important virtual quality of sound that we shall discuss below.

Surround(ing) Sound

As a concept, the precursors of multichannel audio[10] can be traced back as far as the sixteenth century, with the spatial separation technique of antiphonal music in the medieval period (Holman 2014). Within the electronic era of sound, the development of multichannel audio as a considered technique dates back to the late nineteenth century. One of the first public appearances of multichannel audio is noted to be Clément Ader's experimental use of stereo telephony. As was briefly mentioned earlier within this chapter, 1881 saw Ader successfully demonstrate the audio transmission of operatic and theatre performances from the Palais Garnier opera house to the Paris Electrical Exhibition by way of two telephone lines. Théberge et al. (2015) make an interesting observation concerning the commercial offspring of Ader's design, the *Théâtrophone*. They note that whilst comparisons were drawn between the Théâtrophone and the stereoscope, Ader himself distanced his invention from aspects of innovation and novelty, by focussing the public's attention towards its core function of enabling listeners to receive a more 'intelligible impression of the performance' (Ader 1882, p. 1). This exemplifies a common thread that runs throughout the remainder of this chapter. Whereas visually based virtual technology willingly associated itself with aggrandisement and public notions of innovation, audio-based technology largely avoided the hype and prioritised function.

Existing research (see Holman 2014; Théberge et al. 2015) already provides us with detailed accounts of multichannel audio and its origins. This section will therefore not provide a comprehensive timeline of such technological and design developments, but rather pick out a few select items that most resonate with notions of VR. Of particular noteworthiness is Holman's account of the very first multichannel audio project. In 1933, Bell Labs engineers presented a three-channel (front left, centre and right speaker configuration) system that, whilst not actually surround in terms of speaker positioning, was able to generate an effective enveloping experience by way of sound waves reflecting off the rear wall of the listening space. The concept of utilising multichannel audio as a means of enhancing motion pictures began in the early 1940s, with Walt Disney's *Fantasia* (Armstrong et al. 1940). Disney himself wished for sections of the film to include sound objects that would perceptually dart around the auditoriums. To this end, a bespoke 'Fantasound' system was produced, a five-channel (three front facing, two rear facing) soundtrack projected by way of 54 individual speakers (Miller 2004).

After the Second World War ended, a great deal of military technology was cannibalised by the motion picture industry, including permanent magnets that improved the amplification of theatre speakers, and magnetic film for higher quality recording of audiovisual material (Holman 2014). In the 1950s, Cinerama and Cinemascope technologies were two consecutive attempts by the industry to counter losing ticket sales due to the impact of the television. These systems were based heavily upon surrounding the audience to better immerse them in the film. As Riva and colleagues note, experiencing Cinerama 'was often likened to being transported to other lands'. The listener could 'escape from reality, do anything one may desire to do, and go anywhere one wishes' (2003, p. 2). Both systems relied heavily upon multichannel audio, using various alternative speaker configurations to achieve this desired effect.

Throughout the latter half of the twentieth century, and continuing to the present day, multichannel audio systems and techniques have consistently been utilised as a means of enhancing the experience of cinema's visuals, specifically by way of surrounding the audience in sound, positioning them in the centre of the film's diegetic space and heavily encouraging them to feel present within it. As a final example, it would be amiss not to mention *Dolby Stereo* and its popularisation in *Star Wars* (Lucas 1977). The film's opening scene, in which an Imperial Star Destroyer descends upon the screen as if passing audience from above, utilised Dolby Stereo to evoke the starship's movement, as the great rumbling sound of the engines passes from the rear of the theatre to the front. Later, retrospectives on this opening scene would assert that the effect it had upon the audience was so intense, that it was largely responsible for the Dolby system subsequently being installed in theatres across the world (Kerins 2010).

With a specific focus upon digital games, Chap. 7 continues our timeline of multichannel audio's theoretical and technological progress. But even before we consider digital game sound, the consistent success of stereo and other forms of multichannel audio across multiple forms of media raises a key point. Latour (1999) notes that the ubiquitous uptake and consistent functionality and quality of stereo sound have led to it being 'made invisible by its own success' (p. 304). This effect is noted again in the closing section of this chapter, as we observe how sound hardware has been facilitating VR effectively and efficiently for years, but very few people appear to have noticed.

Headphones: The VR Headset of Consistent Design and Popularity

With its nineteenth-century origins beginning with the earpiece receiver of a telephone (Bell 1876), headphones are in many ways a head-mounted (auditory) display, a means of experiencing VR by holding a device to your ears rather than to your eyes. Following on from the telephone came dual-earpiece audio courtesy of Ader's Théâtrophone in 1881, then the *Electrophone* service in 1894 launched the concept of a listening subscription service. Charging £10 a year, Electrophone Ltd. gave very well-to-do customers[11] a set of four dual-earpiece receivers through which listeners could enjoy live performances from eighteen London Theatres and several churches across the city (Pain 2011). The Electrophone hardware consisted of two earpieces, connected by a downward curving bar, attached to a pole that the listener would hold. This design was essentially an inversion of what became traditional headphones, which were realised in the early twentieth century, when Nathaniel Baldwin (1915) patented a 'head-band for telephone-receivers'. Baldwin's design became a template from which the headphone has yet to significantly stray, with subsequent developments focussing more upon subtle and gradual improvements as opposed to heavy experimentation or paradigm shifts. This set of circumstances could go some lengths to explaining why, once the technology overcame certain social hurdles, became so ubiquitous and unassumingly low profile.

Headphones were the primary means of listening to the radio until the late 1920s, when integrated loudspeakers began to override headphone's popularity. However, this was not due to issues of audio fidelity or realism, but rather that the headphones were less able to integrate into everyday life (Ehardt 2014). In terms of cultural reception, headphones predate a conservative concern levied at visual HMDs during the 1990s, more than seventy years later, by first raising the issue of virtual technology socially isolating individuals within a household (Théberge et al. 2015). Loudspeakers, on the other hand, facilitated a shared experience and the ability to receive audio content whilst also engaging in most normal household and working activities. Here quality came second to functionality.

Despite this setback for headphones, development in the technology continued, and in 1954, Regency released the *TR-1*, the world's first transistor radio and the origin of portable audio (Kimble and Wang 2012). Enabling listeners to select between the integrated speakers and an earpiece, the TR-1 did not exactly mark a turning point in the popularity of headphones but was the progenitor of Sony's 1979 portable audio cassette player, the *Walkman*. During the 1980s and 1990s, the Walkman may have dominated the portable audio market, but the market itself was substantial, worth $1 billion globally by the mid-1990s (Sanderson and Uzumeri 1992). The Walkman design, much like many competing devices, exclusively utilised headphones for audio playback. This signified a great culture shift towards headphones as the function of portability overcame the stigma of antisocial use. As of 2015, market researchers have stated the value of the global earphone/headphone market as $8.7 billion.[12] To put that in context, the global consumer VR market for the same year of 2015 was valued at $1.21 billion[13] and short-term projections still position VR far behind headphones, with an estimated future value of $5.2 billion in 2018.[14]

In addition to being a commercial success and a ubiquitous global technology, looking into developments in headphones also provides us with indications as to likely future design directions for VR-HMDs. In

2003, Rick Alden founded Skullcandy Inc.,[15] launching a range of colourful and patterned headphones. The year 2006 saw the unveiling of *Beats by Dre*, headphones co-developed and endorsed by Andre 'Dr Dre' Young, who also co-founded the company Beats Electronics LLC.[16] Celebrity endorsement proved to be a substantially profitable endeavour. As of 2012, US sales of Beats by Dre headphones accounted for 64% of US market share for headphones priced over $100, with this market of higher priced units expanding year on year (Neate 2013). Certain products also look to provide additional functionality, such as wireless connectivity and waterproof/underwater devices, the latter proving commercially popular in sporting applications. Contemporary HMD manufacture is already indicating that these trends shall influence their designs, potentially even in the same order, with different colour options now available for purchasers of the Google *Daydream View* headset.[17] If the pattern continues, we should be expecting to see visually customisable headsets, designer/celebrity-endorsed designs and additional functionalities targeting niche markets in the not-so-distant future.

The humble headphones have, during their development, presented us with several points from which we can identify them as a 'virtual technology'. Fundamentally, headphones facilitate the experiencing of virtual worlds, from telepresence through playback of both live broadcast and pre-recorded real-world soundscapes (akin to 360° video) to fully electronically generated material by way of audio synthesis (comparable to a digital game). The positioning of headphones as VR hardware can also be further supported by at least three features of the technology, namely binaural audio, 3D audio/head-related transfer function emulation and interactive audio-based gaming.

The term 'binaural' can be used to describe our natural mechanism of listening but also methods of audio recording and/or signal processing. Literally meaning 'two ears', binaural listening describes the psychoacoustic effects of receiving acoustic sound waves by way of our head and our two ears (specifically, effects pertaining to source localisation—see Blauert 2013). The precedence effect (see Chap. 3) is a good example of binaural source localisation, describing the phenomenon of our identification of a source's position from which ear first received the relevant sound wave and the delay of reception between the ears (inter-aural time difference). Stereo processing can recreate this effect in headphone listening by simply delaying the signal of one channel, a technique and effect that is exceptionally similar to that of the stereoscope, which presents each eye with the same image but with one's position slightly adjusted, also to evoke a perception of three-dimensional space. Binaural audio, however, is able to produce a significantly more convincing sense of localisation. It can be created by way of specific recording techniques or simulated by way of digital signal processing (see HRTF below). As Møller (1992) explains, basic binaural recording can be quite simply two microphones in a relative position comparable to human ears, recreating inter-aural time difference, but can also be more complex. To more fully recreate the effect, binaural recording may employ a 'head simulator', an artificial reproduction of the human head with two microphones built into the ears. This method of binaural recording more accurately reproduces the acoustic effects in actual listening, as the recorded sound waves have interacted with the (artificial) head, face, pinnae and ear canals of the head simulator before reaching the microphones. This technique is particularly powerful as a means of creating positional audio, with the listener able to perceive source position and movement in three dimensions, giving them a sense of relative position and presence in the virtual sound space. Møller (1992) also notes that binaural recordings function best when played back by way of headphones rather than speakers, which add crosstalk as both ears can hear both the left and right binaural channels, thereby cancelling out the effect.

Although we shall return to this in Chap. 7, positional audio is worth a brief mention here as it describes an audio processing technique that substantially resonates with VR and works most effectively with headphones. In games and virtual environments, head-related transfer function (HRTF processing, often described in gaming as 'positional audio' or '3D audio') describes a means of audio processing that positions sound objects and events in virtual space around the avatar of the player (Stevens and Raybould 2013). With HRTF in games, sound objects can be placed across both the horizontal (around) and vertical (above and below) axes. In practical terms, audio processing by way of an HRTF algorithm means that the signal will be affected in different ways depending on its frequency components. For example, a sound wave composed of mid-high frequencies will typically sound heavily muffled if the source moves behind the listener (just as in natural listening, when the sound waves hit the back of the pinnae). HRTF is most commonly utilised in first-person perspective games, where the position and orientation of the avatar determines the position of the sound to create an effective sense of relative position in virtual space. Turn your (avatar's) head to the left and a sound that was previously straight ahead now buzzes in your right ear. Walk towards a sound object and that sound becomes louder; walk away and it fades. This technique of employing algorithms so that the listener's movements and orientation determine the perceived relative position of content within the virtual space, arguably epitomises the central function of a modern VR-HMD. You simply visuals for audio. The effect remains largely the same.

Finally, and yet another area that we shall explore in greater depth in Chap. 7, we can consider interactive audio games that are specifically facilitated by headphones. Audio-only digital games are gradually becoming more commonplace, particularly in the mobile games market. *Vanished* (see Farokhmanesh 2013), for instance, uses a smartphone's internal compass and accelerometer to control avatar orientation and movement as the player explores a horror-themed virtual world built entirely of sound. Another example is *Zombies, Run!* (Alderman et al. 2012), a mobile 'exergame' (fitness game) that utilises GPS tracking and an entirely auditory interface to direct the player to physical locations to pick up virtual supplies, rescue virtual survivors and evade virtual zombies, all as part of an actual exercise routine.

From the above, we can observe how strikingly similar headphones are to HMDs. This is in terms of their hardware design (they are effectively head-mounted *auditory* displays) and their facilitation of multichannel and positional audio systems that evoke a sense of spatial presence within virtual worlds. Furthermore, we can see that headphones have developed largely in parallel with visually centred VR technology but we can observe a significant difference between their histories. Whilst VR progressed consistently in research but in fits and starts commercially, its auditory cousin enjoyed steady improvement in both arenas, arguably due to the relatively minor changes made to the overarching design of headphones over the decades. The final point to note here is that, despite being arguably a very virtual technology, headphones and positional audio have yet to be appreciated as such. This is partially as a result of comparably less interactive audio applications being available, but also due to the general unassuming ubiquity and effectiveness of the technology. It seems we're more likely to notice technology that doesn't work than that which does.

Chapter Summary and References

Contemporary VR has deep roots stretching back many centuries and has directly descended from artistic, mechanical and electrical precursors. This chapter has traced developments in both VR and sound technology, two lines of progress that have largely been identified as separate, but that are brought together here by their shared virtual natures. Whilst early virtual machines such as the Cinéorama did not utilise sound intentionally, they nevertheless provided inherently sound-filled experiences. Subsequently, multimodal developments including the Sensorama and the Sutherland's concept for the Ultimate Display heavily advocated sound with their designs. As the timeline nears the end of the twentieth century, sound appears to become increasingly downgraded, particularly during the first generation of VR entertainment in the 1990s. However, whilst this generation of VR suffered a dramatic failure, its auditory equivalent continued on, steadily making progress as we approach the present day and, as we shall explore in subsequent chapter, the integration of sound into both digital games and contemporary VR as it makes its second bid for commercial success.

Notes

1. Google Cardboard Camera. https://googleblog.blogspot.co.uk/2015/12/step-inside-your-photos-with-cardboard.html.
2. Or indeed the DC Comic supervillain and living embodiment of fear. Marz, R. (1994). *Green Lantern*, 3:50, DC Comics.

3. Different articles appear to disagree over whether this experience was specifically Manhattan island or Brooklyn.
4. Osmose. http://www.immersence.com/osmose/.
5. Ephémère. http://www.immersence.com/publications/char/2004-CD-Space.html.
6. Sega VR. http://segaretro.org/Sega_VR.
7. Details on audio chipset for Virtual Boy. http://chipwiki.ru/wiki/Nintendo_VSU.
8. Details on audio chipset for Virtual Boy. http://www.planetvb.com/modules/dokuwiki/doku.php?id=audio_overview.
9. SNES chipset details. http://emureview.ztnet.com/developerscorner/SoundCPU/spc.html.
10. This term is used throughout the chapter although some sources prefer the term 'surround sound'. 'Multichannel audio' is used instead as, in Chap. 7, surround sound is contrasted with '3D sound', both of which were built from the historical progress discussed here.
11. As of 2017 currency, its equivalent to paying roughly £1150 annually.
12. Global headphone market stats. http://www.grandviewresearch.com/industry-analysis/earphone-and-headphone-market.
13. VR market stats. https://www.mordorintelligence.com/industry-reports/virtual-reality-market.
14. Future VR market expectations. http://www.kzero.co.uk/blog/consumer-virtual-reality-market-worth-13bn-2018/.
15. For details on Skull Candy headphones. http://www.skullcandy.com/.
16. Interestingly, between 2010 and 2012 Beats Electronics were majority owned by HTC, who also manufacture the *Vive* VR headset. The company is currently a subsidiary of Apple.
17. Colour options on Google Daydream HMD. https://vr.google.com/intl/en_uk/daydream/headset/.

References

Ader, C. (1882). Telephonic Transmission of Sound from Theatres. *U.S. Patent No. 257,453*. Washington, DC: U.S. Patent and Trademark Office.

Alderman, N., Hon, A., Chan, E. & Levene, R. (2012). *Zombies, Run!* London, UK: Six to Start.

Armstrong, S., et al. (1940). *Fantasia*. USA: RKO Radio Pictures.

Baldwin, N. (1915). Head-band for telephone-receivers. *U.S. Patent No. 1,127,161*. Washington, DC: U.S. Patent and Trademark Office.

Baños, R. M., Botella, C., Rubió, I., Quero, S., García-Palacios, A., & Alcañiz, M. (2008). Presence and emotions in virtual environments: The influence of stereoscopy. *Cyber-Psychology & Behavior, 11*(1), 1–8.
Behrmanx, M. (1851). *San Francisco from Rincon Hill.* USA: Library of Congress.
Bell, A. G. (1876). Telephone. *U.S. Patent No. 174,465*. Washington, DC: U.S. Patent and Trademark Office.
Blauert, J. (Ed.). (2013). *The technology of binaural listening*. Berlin, Germany: Springer.
Boas, Y. (2013). Overview of VR technologies. In *Interactive Multimedia Conference 2013*.
Brewster, D. (1856). *The Stereoscope; its history, theory and construction, with Its application to the fine and useful arts and to education*. etc: John Murray.
Brock-Nannestad, G., & Fontaine, J. M. (2008). Early use of the Scott-Koenig phonautograph for documenting performance. *Journal of the Acoustical Society of America, 123*(5), 3802.
Burdea, G. C., & Coiffet, P. (2003). *Virutal Reality Technology*. New Jersey, USA: John Wiley & Sons.
Carbon, C. C., & Hesslinger, V. M. (2015). On the nature of the background behind Mona Lisa. USA: Leonardo.
Cruz-Neira, C., Sandin, D. J., DeFanti, T. A., Kenyon, R. V., & Hart, J. C. (1992). The CAVE: Audio visual experience automatic virtual environment. *Communications of the ACM, 35*(6), 64–73.
Cruz-Neira, C., Sandin, D. J., & DeFanti, T. A. (1993). Surround-screen projection-based VR: The design and implementation of the CAVE. In *Proceedings of the 20th Annual Conference on Computer Graphics and Interactive Techniques* (pp. 135–142). USA: ACM.
Clarke, G. (2014). How 'expanded cinema' rethinks the film screening. *Tate.* http://www.tate.org.uk/context-comment/articles/how-expanded-cinema-rethinks-film-screening.
Davies, C. (2004). Virtual space. In: Penz, F., Radick, G. & Howell, R. (Eds.). *Space: In science, art and society.* New York, USA: Cambridge Unviersity Press. 69–104.
Davies, C., & Harrison, J. (1996). Osmose: Towards broadening the aesthetics of VR. *ACM SIGGRAPH Computer Graphics, 30*(4), 25–28.
Da Vinci, L. (c.1503–1506). *Mona Lisa*. Paris, France: Museé du Louvre.
Dinh, H. Q., Walker, N., Hodges, L. F., Song, C., & Kobayashi, A. (1999, March). Evaluating the importance of multi-sensory input on memory and the sense of presence in virtual environments. In *VR, 1999. Proceedings., IEEE* (pp. 222–228). IEEE.

Edison, T. A. (1888). The perfected phonograph. *The North American Review, 146*(379), 641–650.
Erhadt, C. (2014). Phones, horns, and "audio hoods" as media of attraction. In: D. Morat, (Ed.), (2014). *Sounds of modern history: Auditory cultures in 19th-and 20th-century Europe*. USA: Berghahn Books.
Farokhmanesh, M. (2013). Experimental horror gam Vanished uses only audio. Polygon. http://www.polygon.com/2013/9/29/4783934/experimental-horror-game-vanished-uses-only-audio.
Fen, J. & LeHong, H. (2011). *Hype cycle for emerging technologies*. Stamford: Gartner. https://www.gartner.com/doc/1754719/hype-cycle-emerging-technologies-.
Flores-Arredondo, J. H., & Assad-Kottner, C. (2015). VR: A look into the past to fuel the future. *The Bulletin of the Royal College of Surgeons of England, 97*(10), 424–426.
Freeman, J., Avons, S. E., Davidoff, J., & Pearson, D. E. (1997). Effects of stereo and motion manipulations on measured presence in stereoscopic displays. *Perception, 26*(1_suppl), 144–144.
Grau, O. (2003). *Virtual Art: From illusion to immersion*. Cambridge, England: MIT press.
Heilig, M. L. (1960). *U.S. Patent No. 2,955,156*. Washington, DC: U.S. Patent and Trademark Office.
Heilig, M. L. (1962). *U.S. Patent No. 3,050,870*. Washington, DC: U.S. Patent and Trademark Office.
Hendrix, C., & Barfield, W. (1996). Presence within virtual environments as a function of visual display parameters. *Presence: Teleoperators & Virtual Environments, 5*(3), 274–289.
Holman, T. (2014). *Surround sound: Up and running (second edition)*. Burlington, USA: Focal Press.
Jacobs, C. (2004). Introduction. In *Interactive Panoramas* (pp. 1–7). Berlin and Heidelberg: Springer.
Johnson, E. H. (1877). A Wonderful invention: Speech capable of indefinite repetition from automatic records. *Scientific American, 37*(20), 304.
Kerins, M. (2010). *Beyond Dolby (stereo): Cinema in the digital sound age*. Bloomington, Indiana, USA: Indiana University Press.
Kimble, C., & Wang, H. (2012). Transistors, electric vehicles and leapfrogging in China and Japan. *Journal of Business Strategy, 33*(3), 22–29.
Kiyokawa, K. (2006). An Introduction to Head Mounted Displays for Augmented Reality. In Haller, M. (Ed.), *Emerging technologies of augmented reality: interfaces and design* (pp. 43–63). IGI Global.

Koenderink, J. J., van Doorn, A. J., & Kappers, A. M. (1994). On so-called paradoxical monocular stereoscopy. *Perception, 23*(5), 583–594.
Latour, B. (1999). *Pandora's hope: Essays on the reality of science studies.* Cambridge, USA: Harvard University Press.
Lewers, T. H., & Anderson, J. S. (1984). Some acoustical properties of St Paul's Cathedral, London. *Journal of Sound and Vibration, 92*(2), 285–297.
Lippman, A. (1980). Movie-maps: An application of the optical videodisc to computer graphics. In *Acm Siggraph Computer Graphics* (Vol. 14, No. 3, pp. 32–42). NY, USA: ACM.
Lucas, G. (1977). *Star Wars: A New Hope.* USA: Twentieth Century Fox.
Luhmann, T. (2004). A historical review on panorama photogrammetry. *International Archives of the Photogrammetry, Remote Sensing and Spatial Information Sciences, 34*(5/W16), 8.
Maconie, R. (2002). *The second sense: Language, music, and hearing.* Plymouth, UK: Scarecrow Press.
Mihelj, M., Novak, D., & Beguš, S. (2014). *VR technology and applications.* Springer.
Miller, M. (2004). *The History of Surround Sound.* Que Publishing. http://www.quepublishing.com/articles/article.aspx?p=337317.
Møller, H. (1992). Fundamentals of binaural technology. *Applied Acoustics, 36*(3–4), 171–218.
Morie, J. F. (2012). Female artists and the VR crucible: Expanding the aesthetic vocabulary. In *IS&T/SPIE Electronic Imaging* (pp. 828908–828908). Bellingham: International Society for Optics and Photonics.
Muhanna, M. A. (2015). VR and the CAVE: Taxonomy, interaction challenges and research directions. *Journal of King Saud University-Computer and Information Sciences, 27*(3), 344–361.
Neate, R. (2013). Dr Dre Beats valued at more than $1bn following Carlyle deal. *Guardian Online.* https://www.theguardian.com/music/2013/sep/27/dr-dre-beats-1bn-carlyle-sale.
Pain, S. (2011). *Farmer Buckley's exploding trousers: & other events on the way to scientific discovery.* UK: New Scientist.
Peruzzi, B. (c.1516–1518). *Sala delle Prospettive.* Rome, Italy: Villa Farnesina.
Reas, C., & Fry, B. (2007). *Processing: A programming handbook for visual designers and artists* (No. 6812). Cambridge, USA: MIT Press.
Riva, G., Davide, F., & IJsselsteijn, W. A. (2003). Being there: The experience of presence in mediated environments. *Being there: Concepts, effects and measurement of user presence in synthetic environments, 5.*
Roubaud, F. (1911). *Battle of Borodino.* Moscow: Borodino Battle Museum.
Roy, E. A. (2016). Worn Grooves: Affective connectivity, mobility and recorded sound in the First World War. *Media History,* 1–20.

Salazar-Palma, M., Sarkar, T. K., & Sengupta, D. (2009). A brief chronology of the origin and developments of wireless communication and supporting electronics. In *Applied Electromagnetics Conference (AEMC), 2009* (pp. 1–4). IEEE.

Sanderson, S. W., & Uzumeri, V. (1992). Industrial design: The leading edge of product development for world markets. *Design Management Journal (Former Series), 3*(2), 28–34.

Schoenherr, S. (2005). *Recording Technology History*. Audio Engineering Society. http://www.aes.org/aeshc/docs/recording.technology.history/notes.html.

Steinicke, F. (2016). *Being Really Virtual: Immersive Natives and the Future of VR*. Switzerland: Springer.

Stevens, R., & Raybould, D. (2013). *The Game Audio Tutorial: A practical guide to creating and implementing sound and music for interactive games*. Burlington, USA: Focal Press.

Stafford, B. M., Terpak, F., & Poggi, I. (2001). *Devices of Wonder: From the World in a Box to Images on a Screen*. Los Angeles, USA: Getty Publications.

Sutherland, I. E. (1965). *The ultimate display*. Multimedia: From Wagner to VR.

Théberge, P., Devine, K., & Everrett, T. (Eds.). (2015). *Living Stereo: Histories and Cultures of Multichannel Sound*. USA: Bloomsbury Publishing.

Teitel, M. A. (1990). The Eyephone: A head-mounted stereo display. In *SC-DL tentative* (pp. 168–171). Bellingham: International Society for Optics and Photonics.

Timby, K. (2005). Colour photography and stereoscopy: Parallel histories. *History of Photography, 29*(2), 183–196.

Trotter, D. (2004). Stereoscopy: Modernism and the 'haptic'. *Critical Quarterly, 46*(4), 38–58.

Vanvolsem, M. (2011). *The art of strip photography: Making still images with a moving camera* (Vol. 11). Universitaire Pers Leuven.

Waldern, J. (1992). VR the Applications and Commercialisation. In *Proceeding - AUUG 92 Conference on Maintaining Control in an Open World*.

Youngblood, G., & Buckminster, R. (1970). *Expanded cinema* (p. 41). New York: Dutton.

Zachara, M., & Zagal, J. P. (2009). Challenges for success in stereo gaming: A Virtual Boy case study. In *Proceedings of the International Conference on Advances in Computer Enterntainment Technology* (pp. 99–106). New York, USA: ACM.

Žižek, S. (2009). *The parallax view*. Cambridge, USA: MIT Press.

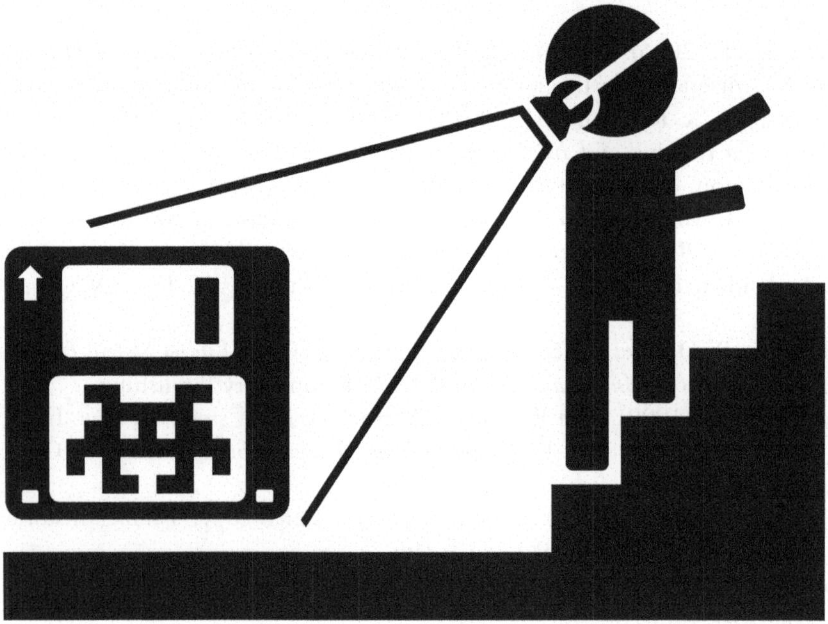

7

Reality Check

So far, our journey across the timeline of VR has investigated a progression of relevant concepts and technologies. At this point, we reach a fork in the road that separates VR in terms of consumer and commercial applications. The technological developments made during the latter part of the twentieth century created opportunities for virtual systems to have notable value across a wide range of industries. Consequently, the history of VR diverges between the commercial and consumer domains, creating two different (but still connected) aspects of contemporary VR to explore. In Chap. 9, these wider commercial applications are discussed (including education/training, simulation, visualisation, etc.). Before that, we examine consumer VR, which broadly incorporates gaming and recreational applications. This chapter revisits the significant moments that brought VR into our living rooms by way of its integration with personal computers and home console technology. We also return to the expectations that were largely set by fictional representations (see Chap. 5) to explore the gap between expectation and reality that lead to what was effectively the fall of VR.

The central assertion that this chapter seeks to raise is that, whilst the 1990s crash undoubtedly plunged consumer technology into several years of obscurity, VR includes a much broader range of technologies

T.A. Garner, *Echoes of Other Worlds: Sound In Virtual Reality*,
Palgrave Studies in Sound, DOI 10.1007/978-3-319-65708-0_7

and underlying concepts. These not only survived the crash, but also continued to advance throughout the decade and into the twenty-first century, largely by becoming of substantial value to digital games. Here we examine several of these VR concepts and technologies as they evolved within digital games, before being reintegrated into contemporary VR as it exists today. Throughout this chapter the treatment of sound during this period is a focus of discussion, reinforcing the notion of sound as a fundamental component of VR, from sound design techniques and auditory interfaces to the integral sonic aspects of various VR hardware devices.

The Failure to Launch of Consumer VR in the 1990s

In the years leading up to the advent of contemporary VR, feeling towards the technology raised much scepticism indicative of a continuing disillusionment (Fox et al. 2009). As we briefly discussed in Chap. 6, endeavours into 'consumer VR' (i.e. systems developed for the general market with a primarily recreational function) failed considerably, temporarily banishing the technology into darkness. This section examines the problems encountered by VR in the 1990s, specifically looking at the great division that arose between consumer expectation and the actual functionality and quality of VR. This discussion retains a focus upon sound as it is revealed that, although the visually-orientated systems typically understood as encapsulating VR (i.e. HMDs) were a commercial failure, various other VR technologies existed during this period that are commonly not thought of as VR. These technologies not only embodied principles of VR but also managed to evade the 1990s crash, maintaining relatively stable commercial success throughout the decade and into the new millennium.

The Reality of Virtual Reality

In more recent years, revived interest in contemporary VR has prompted a wave of late 1990s retrospectives asking 'where did it all go wrong' and attempting to extract learning opportunities for present day and future developments. These articles focus primarily upon HMDs, which epitomised VR throughout the decade, but do extend to a limited number of other VR technologies. Citing the pariah of VR, Nintendo's *Virtual Boy*, Betters (2013) highlights the fundamental mismatching of player's expectations and their actual experience. Betters notes that the Virtual Boy did not support head tracking (the display therefore did not respond at all to head movement) or multi-player, two features that consumers widely anticipated but did not receive. Betters also notes that the Virtual Boy launched with only three games (its entire catalogue consisting of 22 titles, split across different regional markets) which resulted in players being offered a distinctly small number of gameplay hours before all available content was thoroughly exhausted. Charara (2015) raises the additional issue of VR-side effects, referring to the Virtual Boy's use of a varifocal mirror display (a single image is projected to the eye via a rapidly oscillating mirror to create a 3D effect; see Hughes et al. 2014) that caused eyestrain and headaches following prolonged use.

Difficulty in integrating VR into broader games culture proved another sticking point. McFerran (2014) observes how the technology proved problematic at the video game arcades due to being an 'unknown quantity' when compared to more established games and systems. Arcade VR expected players to don an offensively sweaty and pungent headset that would require an attendant to hoist each player into the VR cabinet then remain with them for the duration to help manage various usability issues. Cinema's representations of VR during this time are also frequently charged with raising consumer expectations far beyond the capabilities of the technology (Betters 2013; Charara 2015). Additional reasons identified as contributing to consumer VR's failure include the lack of design quality standardisation (with devices rushed to market ahead of testing to any significant breadth or depth) and that it added very little to the experience of its primary functionality, playing games.

In addition to HMDs, 1990s consumer VR included accessories that brought together human-computer interaction devices with VR technology, one of which performed equally badly upon consumer release. Retailing at the beginning of the decade, Nintendo's *Power Glove* was derived from the *Data Glove* (Zimmerman and Lanier 1991) and combined a traditional gamepad interface (taken from the Nintendo Entertainment System [NES]—the console for which the controller was built) with sensors embedded within the glove to track hand orientation across a single axis of movement (roll). During its heavily limited commercial release period, a grand total of two games were developed for the device, though it was intended to be backwards-compatible with existing NES games. Though not a HMD, retrospectives on the Power Glove cite striking similar issues to those mentioned above. Khoury (2015) states that difficult set-up, ineffective calibration and a poor range of games were contributing factors to its commercial failure. Khoury also asserts that the broad but central problem was, again, the gap between expectation and reality for consumers. Hype resulting from films, television programmes and magazine advertising resulted in an imagined user experience that the technology could not replicate in either functionality or quality. Control was deemed less intuitive than a traditional controller, and players had anticipated a full six degrees of freedom (meaning three axes of movement and three of orientation) tracking that enabled fluid control and accurate feedback, much like its fictional representation. It is therefore unsurprising that they were largely disappointed by a temperamental and inaccurate control system that only tracked a single axis of rotation.

For 1990s HMDs and peripherals such as the Power Glove, poor reception came not only from comparisons drawn against ideal representations in fiction, but also from traditional digital games technology. With a direct history dating back to 1947 and the filling of a patent for a 'cathode-ray tube amusement device' (Goldsmith and Ray 1948), the digital games industry had been steadily increasing in refinement for many years and had become firmly established and trusted. In many retrospectives on 1990s VR, this would be where the story ends, until 2013 and the re-emergence of consumer VR in its contemporary form. However, this would neglect various other conceptual aspects of VR and other VR technologies that successfully evaded the hype-trappings suffered by HMDs and the Power Glove, to generate both a stable lineage of development over the years and also consistent, if modest, commercial success.

VR's Conceptual Survival Through Digital Games

With a focus upon several aspects of game design, this section examines how central concepts of VR survived and advanced within the consumer domain by way of digital games. Beginning with a look back at the lineage, evolution and diversification of first-person perspective games, the connection between VR and games of this nature is established. We then proceed to examine how key VR mechanics, aesthetics and user-experience intentions were retained and evolved before the second wave of contemporary VR arrived to reap the benefits. This section also charts the developments of VR sound, including a review of techniques in positional audio, avatar communication and auditory intelligence.

The First-Person Perspective

In literature, the differences between first-, second- and third-person narrative devices are wrapped up in a complex blend of diegetic meaning and identity (see van Peer and Chatman 2001). Whilst such issues are still relevant to these terms in the context of digital games, the meaning of first/second/third person is more straightforwardly associated with viewing the perspective of the virtual environment. The third-person perspective typically presents the game's visuals either from a top-down/isometric position such as *The Sims* (Wright et al. 2000), or from just behind the graphically rendered avatar (commonly referred to as 'over the shoulder' perspective), an example of which would be *Gears of War* (Fergusson et al. 2006). A much rarer approach by comparison, second-person perspective games utilise a view of the game environment very similar to the third person in which the avatar is rendered and visible to the player from a distance. The key difference is that in second-person games the player's perspective is portrayed as a separate diegetic character, such as in *Super Mario 64* (Miyamoto et al. 1996), in which the player's view is through the camera of a cloud-riding reptile called Lakitu. Finally, first-person games, as their name would imply, utilise a perspective that extends from the eyes of the player-controlled avatar. Essentially, the avatar and player share the same viewpoint with the intention being to forge a closer connection of identity between the two.

In addition to the viewing perspective, first-person games characteristically present three-dimensional virtual environments, with the player able to explore the lengths, breadths and depths of the levels. As Wolf (2012) observes, first-person games have origins in numerous strands of digital game (and other) developments. Digital games utilised the first-person perspective within several genres, from the shooting at carnival fairgrounds and light gun games to experimental wireframe first-person games such as *Maze War* (Colley et al. 1974) and Atari's tank shooter *Battlezone* (Rotberg et al. 1980). First-person racing games were introduced in *Night Driver* (Michon et al. 1976), an arcade racer later ported to the Atari 2600 and Commodore 64 systems. Role-playing games appeared several years later, in the mid-1990s with titles such as *The Elder Scrolls: Arena* (Lakshman et al. 1994). However, during that decade one particular genre came to prominence, becoming the prime example of games featuring a first-person perspective; the first-person shooter (FPS).

Prior to the 1990s, the forerunners of what was to become the FPS genre underwent various developments. What became the template for FPS games as we recognise them today was first realised in early 1990s titles such as *Wolfenstein 3D* (Hall et al. 1992) and *Doom* (Romero et al. 1993). Whilst Wolfenstein 3D stands as the progenitor of the modern FPS template, it was Doom that built distinct standards of design quality into that template to create a work that developers of many future games would cite as a significant inspiration. The game tasks the player-as-protagonist with battling through hordes of countless demons, aided by an array of weapons all rendered on screen and held in the outstretched arms of the avatar. In addition to reducing enemies to disembowelled puddles upon the floor, Doom's gameplay also involved navigating the level to find the exit point, avoiding environmental hazards (toxic slime pits, lava, etc.) and searching for key-cards to unlock doors. These elements all continue to be used regularly in modern FPS titles, including many of those that are developed specifically for contemporary VR platforms. Because of the head-tracking functionality of the HMD, the first-person perspective largely defines contemporary VR, with FPS games in particular taking up several positions on current 'best VR' lists online.[1]

Across the various game genres in which it is employed, the first-person perspective has particular relevance to user-experience elements, specifically immersion and presence. Horrowitz and Looney (2014) quote Alexander Brandon, sound designer for 'deathmatch/arena-style' FPS *Unreal Tournament 3* (Morris et al. 2007). Brandon asserts that FPS games are 'all about immersion. Otherwise, don't bother. The player will be encountering things through the same eyes as the [avatar], so you want them to feel that experience first and foremost' (p. 58). Numerous research studies utilise immersion as a key qualitative measure of user experience in digital games (McMahan 2003; Nacke and Lindley 2008) and 'an immersive experience' regularly features in online articles discussing most important qualities of FPS game design.[2, 3] As we discussed in Chap. 4, immersion and presence are a pivotal aspect of VR user experience, which is hardly surprising when we consider the fundamental similarities between VR and first-person perspective games.

Overall, the above draws general parallels between first-person/FPS games and VR, both in terms of design and intentions for players' user experience. However, whilst 1990s consumer VR was suffering at the hands of its own hype, the FPS was making both significant and ongoing commercial progress. As Pinchbeck (2013) notes, Doom itself had an extremely impressive launch, becoming a game that seemingly the majority of PC owners had on their systems. The game enjoyed high praise from consumers and critics alike, setting the stage for an abundance of imitators and innovators. Throughout the 1990s, the FPS genre has delivered numerous titles now revered as seminal games of the decade[4] with notable titles including *Quake* (Romero et al. 1996), *Half Life*

(Newel et al. 1998) and *Unreal Tournament* (Bleszinski et al. 1999); all of which were hugely successful games that expanded into ongoing series'.

Positional Audio

In addition to realising the FPS template in terms of gameplay and visual aesthetic, Doom is also celebrated for its contribution to sound design, with several features that still have great relevance to contemporary VR a quarter-century later. One of these features is positional audio. As noted in Chap. 6, multichannel sound designs and technologies have enjoyed consistent progress and stable market success, impacting upon our experience of cinema, television and radio to name a few. We shall address the hardware that facilitated multichannel sound later in this chapter, but here our attention is on the related sound design techniques that progressed with thanks owed to digital games. More commonly referred to as 'positional (or 3D) audio' in a games context (see Collins 2008), sound design techniques that contribute towards the sensation of a truly three-dimensional virtual environment have flourished over the years. Tied in with the first-person perspective and requirements for immersive environments that evoke presence, they are also highly relevant to contemporary VR.

In a retrospective of Doom by Pinchbeck (2013), sound designer Bobby Prince's work on the game is detailed. As Prince describes, the underlying principle of the sound design for Doom was integration of sound objects within the interactive, three-dimensional virtual space. Careful attention to the mix of the soundscape meant that certain sounds could reliably be discerned over others. The stereo output was exploited to approximate three-dimensional positioning of sound objects, with the movements and actions of the player reflected in auditory feedback. Sounds made by certain game objects would get louder as the player approached. They would become muffled if a door closed that separated the player from an object. If the player faced an object then rotated to one side, the sound of that object would become louder in the opposite ear. The sound design for Doom was not true positional audio but instead more akin to surround sound. As noted in Chap. 6, surround sound is, alongside positional audio, a form of multichannel audio. Though whilst positional audio describes three axes of spatialisation (i.e. sound objects can be placed both around and above/below the listener), surround sound is only across the two horizontal axes (i.e. left-right and front-back).[5] That said sound objects in early FPS games such as Doom did effectively simulate vertical axis positioning by way of the relative loudness of objects within the mix of the soundscape. FPS games such as Doom also achieved simulated positional audio by taking advantage of the first-person perspective, where visual objects could be clearly placed above and below the player, creating an impression of vertical sound by way of synchresis (the coupling of sound and image and its perceptual effects, see Chion and Murch 1994).

As Collins (2008) observes, the primary function of positional audio is localisation, relating to player's ability to establish the position of their avatar, other game objects and non-player characters. In terms of localising their avatar, positional audio provides the player with a multimodal illusion of depth that immerses them within the virtual space as the audio and visuals reinforce one another. With regard to localising other objects within the virtual space, positional audio is a means of integrating sound by making it a core aspect of gameplay. In games that use positional audio this way, players are required to listen carefully to the soundscape and ascertain the position and movement of enemies, thereby generating ludic immersion (see Chap. 4) as the player becomes more engaged in the game world through their actions. This technique is particularly prominent in first-person stealth games such as *Dishonored* (Colantonio et al. 2012), in which one of the central mechanics is to avoid detection by enemies, many of whom are often hidden from view behind walls or around corners. Third-person equivalents (*Splinter Cell* [Coulon 2002] for example) typically enable players to exploit their camera-view to see around obstacles, essentially giving them a superhuman field of vision. In contrast, the first-person perspective limits the accessibility of visual information and, consequently, the player must listen for positional audio cues (such as footsteps and speech) to establish the location of enemies and evade them successfully.

In a retrospective on positional audio, Chase (2016) asserts that VR inherently prioritises positional content by way of its stereoscopic, parallax and head-tracking functionalities. Of these three points, the first two are the primary visual techniques for evoking a perception of depth in a three-dimensional virtual space, whilst the latter enhances the principles of the first-person perspective by positioning the player right in the centre of the action. User experience in VR is deeply connected to localisation. With visual content first establishing this user-experience quality, auditory content is therefore required to reciprocate or else cause a multimodal mismatch that risks severely breaking immersion (Lalwani 2016). For many recent commentators (see Chase 2016; Lalwani 2016; Nguyen 2016), this represents a mandate for positional audio to reclaim its past glory and advance even further.

Auditory Intelligence

As one of the most well-known progenitors of various sound design innovations, Doom also integrated basic elements of auditory perception into their artificial intelligence. Characters in the game were reactive to the sound of the player's gunfire and, if close enough, enemies would exploit the sound to search for the player by moving to the location the player was in when the gunshot was fired. As well as being prominent examples of games that utilise sound cues to evoke ludic immersion, first-person stealth games also commonly imbue enemies with simulated listening abilities. One of the most well-known (and well-implemented) examples of this is *Thief: The Dark Project* (LoPiccolo et al. 1998), in which enemies are highly sensitive to noises. This requires the player to carefully consider sound throughout the game. Moving too quickly or treading upon hard surfaces raises the loudness of your footsteps, requiring you to tread gently and spread moss upon tiled flooring and grates to avoid detection.[6] The player can also recruit the enemy's responsiveness to sound to their advantage, by firing 'noisemaker arrows' to distract enemies, a mechanic that has been repeated frequently in the numerous stealth games that followed.

More recent games' titles reveal that consideration of game sound in terms of ludic immersion is not only of continued interest, but also a design technique that encourages ongoing development and innovation.

For example, *Alien Isolation* (Hope et al. 2014) combines the principles of simulated listening with machine learning, pitting the player against a Xenomorph antagonist that, on top of being acutely responsive to the sight and sound of the player's avatar, can analyse the player's behaviour patterns and undermine frequently used strategies. As with Thief: The Dark Project, Alien Isolation allows players to distract enemies with sound. It even requires the player to control their avatar's breathing and soften their footsteps to avoid detection. The game's musical score is either minimal or absent throughout much of the gameplay, accentuating the soundscape so that it can be exploited by the player to make progress. Drawing sound and gameplay even closer together, the player can also use a motion tracker to detect the Xenomorph's movement, but the device itself emits noise that will draw the creature to their location. As testimony to the increasing value and regard of sound in digital games; amongst the plethora of awards Alien Isolation received, four were for sound design, including its accolade at the 2015 British Academy Games Awards.

The Sound of Fear

The relationship between VR and the elicitation of fear is highly relevant in non-gaming contexts, such as the treatment for phobias and post-traumatic stress disorder by way of VR-based exposure therapies (see Chap. 9). There is also, however, a distinct lineage of horror-themed digital games that flows directly into contemporary VR. As we shall see, these games have contributed significantly to the nature of contemporary consumer VR, and a significant aspect of both their aesthetics and their mechanics is sound.

The survival horror genre of digital games describes yet another facet of Doom (Romero et al. 1993), one which the game successfully popularised, inspiring a significant number of contemporary VR games and experiences. As Fahs (2009) explains, prior to Doom, digital games already had an established relationship with the macabre. Horror aesthetics and narratives were employed in several text-based adventure games before Atari's *Haunted House* (Andreasen 1982) established the foundational gameplay mechanics of the genre (an underpowered avatar, labyrinthine levels and unstoppable enemies that the player could only run and hide from). In his tracing of the genre's origins, Fahs (2009) reveals a consistent popularity in survival horror gaming throughout the 1970s and 1980s, partially due to the various Hollywood film tie-in games released during this time. Then in the early 1990s, the seminal *Alone in the Dark* (Raynal et al. 1992) set the template for contemporary survival horror. For many years following Alone in the Dark, the lineage of survival horror mainly consisted of third-person perspective gameplay. However, many of the more recent survival horror games have since transitioned to first-person perspectives, primarily as a means of increasing their capacity to evoke fear. The thinking behind this transition relates to the use of the first-person perspective in stealth games such as Dishonored (Colantonio et al. 2012), specifically that it removes the enhanced field of view that third-person perspectives afford. In first-person, if you can see the enemy, then your enemy can potentially see you. This creates a continuous sense of vulnerability and was the reasoning behind the developers of Alien Isolation (Hope et al. 2014) changing the game's perspective from third- to first-person mid-development (Gera 2015). This steady progression towards first-person survival horror reveals how the genre has flowed into VR, with numerous games of this type being ported into VR or built for VR from the ground up.

Sound is of great significance to games that intend to evoke fear in their players, with many of the most established 'terror-inducing' sound design techniques linking closely with the prior subsections of this chapter. Positional audio, for instance, can be exploited in a survival horror context by inverting its typical usage to disorientate and undermine the player. Specifically, a 'delocalisation' effect (see Grimshaw 2009) can be achieved by either masking the position of a potential threat (e.g. by making the materials of the virtual environment acoustically reflective so that the reverberations negate the player's ability to localise the source of the sound) or contradicting established perceptual connections (e.g. the first four times the player hears a 'hissing' it is air escaping from a nearby vent, the fifth time it is something a little more hungry).

Similarly, Kromand (2008) presents the notion of 'diegetic masking' in a game's soundscape as a means of inducing uncertainty and fear. Broadly speaking, diegetic masking describes the process of building a soundscape comprising of diegetic sounds (that are tied to in-game enemies) and extra-diegetic sounds (that play randomly irrespective of other game events or player actions). Making these sounds similar, in terms of timbre and associated meaning, causes the player to confuse the diegetic status of the sounds, thereby limiting their ability to discern if a certain sound is indicative of a threat, or part of the underlying soundtrack and therefore safe to ignore. This technique has not only been implemented in several modern VR titles but also developed conceptually further in VR games like *Alone*,[7] a 'meta-VR' experience in which you sit in a virtual living room playing a survival horror game on a virtual home console. As the game continues, much of the sound is revealed to be trans-diegetic, changing in diegetic status during play. The perceived objects from which the sounds originate transitions over time from the 'game-within-a-game' to the virtual living room, causing the player to question where the spooky sounds are coming from, and what their meaning is.

Garner and Grimshaw (2011) assert that the fundamentals of how we typically experience fear in a more general sense can be exploited by sound designers for survival horror games. Their framework describes a series of 'fear states', each of rising intensity compared to the last, and connected by way of a perpetual 'priming' effect of which sound cues can be powerful instigators. To explain, *Silent Hill* (Toyama et al. 1999) is a noteworthy example of 'fear priming' by way of sound. As the player begins to explore the town of Silent Hill, the subtle drone of the musical score and the echoing of their avatar's footsteps emphasises a broader sense of unnerving silence. As they continue, a distant air raid siren rolls across the landscape, raising the unease by indicating that a threat (whatever it may be) is getting closer. The tension arc is completed as the player is ambushed by several demons, all of whom approach from behind and first reveal themselves by way of their tortured screams. Whilst a consistent and orderly increase in tension has arguably become a design trope, experimental implementation of this foundation epitomises sound design in contemporary VR horror games. *Lost in the*

Rift,[8] for example, begins with a discordant bed of strings alongside the sounds of heavy rainfall. However, rather than continue to increase the intensity of the soundscape to induce greater tension, the soundscape intensity is instead reduced almost to zero, becoming silent in all but for the trembling breaths of the avatar. In this instance, whilst the intensity of the soundscape has decreased, the player's fear-intensity will have undoubtedly increased.

A final, but significant point to mention here is regarding acousmatic sound (when a sound can be heard but its presumed source cannot be seen, see Kane 2014). Once again Doom (Romero et al. 1993) stands as an early example, utilising acousmatic sound as a central method of raising tension. As Pinchbeck (2013) observes: 'When the player steps out of the antechamber in E2M8 and first hears the pounding, slamming, giant footsteps of the Cyberdemon before they even see the thing, that job of scaring the living hell out of people is already done' (p. 55). As with trans-diegesis, priming and delocalisation, acousmatic sound design continues to be experimented with today in both traditional games and those built for VR. Collectively, these techniques reveal how survival horror sound embodies several distinct facets of VR. These facets have been advanced upon in the years between the two generations of VR and continue to contribute to the nature of VR as it exists today.

The Silent Protagonist Paradox

In the years and decades leading up to the early 1990s, nearly all game sound content was generated by synthesis. Audio samples were used very sparingly due to the limitations of audio processing units and storage memory at the time. Synthetic audio was unable to generate realistic (or often even comprehensible) human speech and, consequently, few games endowed their characters with voices (save for a few grunts perhaps). However, as Collins (2008) notes, the fifth generation of consoles came with optical discs, largely replacing the ROM cartridges that came before. This change enabled game developers to include, amongst other things, relatively large quantities of pre-recorded speech in their games. Characters could now talk at length and this could include the player's avatar. This development presented a conundrum for game developers with regard to giving their avatar a voice. In an examination of dialogue in digital games, Domsch (2017) observes that games striving to present a detailed fictional world inherently raised an expectation amongst players for dialogue that would draw them into the narrative. At the same time, however, the first-person perspective implies that the player and avatar possess a shared identity as a means of heightening immersion (you are your avatar). Consequently, hearing the voice of another person as the avatar could create a severe incongruity that breaks immersion. Paradoxically, making the avatar silent creates a scenario in which all other characters in the game can speak (often directly to the protagonist) whilst the protagonist cannot. As Domsch points out, the silent protagonist technique is therefore 'seen as a weak solution for the implementation of narrative into digital games' (p. 253).

The silent protagonist issue has relevance to VR as the head tracking combined with a first-person perspective in VR games inherently enhances the identity connection between user and avatar. In VR, a voiced avatar would likely sound particularly jarring, with someone else's voice seemingly emanating from the player's own mouth. Fortunately, as first-person games have developed, various titles illustrate the means of successfully addressing this issue. Half Life (Newel et al. 1998) for instance, presents carefully crafted dialogue that subtly characterises the protagonist as 'the quiet type' and avoids creating scenarios in which speech from the avatar would make better sense. The mechanics of FPS arena-shooter games such as Quake (Romero et al. 1996) and Unreal Tournament (Bleszinski et al. 1999) take this effect to its extreme by minimising narrative and focussing almost entirely upon gameplay. Alternative solutions include voicing the protagonist minimally and only during scripted sequences (see *Bioshock* [Levine 2007]) whilst first-person role-playing games such as *Fallout 3* (Howard et al. 2008) enable speech without sound, by way of multiple choice text options. These techniques further support the assertion that first-person perspective games took fundamental VR concepts and kept them relevant whilst also making steady progress managing various issues (both in general and in terms of sound design) that are central to contemporary VR.

VR's Technological Survival Through Digital Games

The previous section of this chapter outlined how design techniques that embody VR principles have been built upon over the years and have become integrated into contemporary VR. This section continues this argument, but with a focus upon hardware technologies including various home console and personal computer peripherals and accessories. The following is certainly not a comprehensive documentation of every VR-related game peripheral,[9] and examples of technology related to interactive music (guitar controllers, dance mats, etc.) are omitted as there are already excellent and detailed books in this area (see Collins 2008). Instead, we focus upon devices that have significant relevance to sound and those that, despite embodying VR principles, are not typically appreciated as VR.

Multichannel Audio Technology

One particular assertion raised in Chap. 6 was that multichannel systems (inclusive of headphones, speaker arrays and auditory processing units [APUs]) were essentially VR technologies: the auditory equivalents of the HMD or 'HMDs for your ears'. The sales trajectory of consumer HMDs featured a steep initial rise due to inflated expectations but then an even steeper fall resulting from the realities of the technology being very poorly received by consumers. By comparison, multichannel systems propagating VR sound enjoyed a slower but steadier progression that largely avoided inflated hype and its unfortunate pitfalls. Chapter 6 makes this point in a more general context, broadly crossing the whole of the twentieth century, but it remains equally true in the more specific terms of digital games technology as it developed throughout the 1990s, up to the present day.

As an understated 1990s home console peripheral, headphones themselves were hardly appreciated as a VR technology. Scouring the various books on video game history returns only very brief mentions of headphones. Despite this, we can still confidently infer that the use of headphones in 1990s gaming was commonplace, particularly when we consider portable games systems. With release dates between 1989 and 1991, roughly similar to those of 16-bit (fourth-generation) era home consoles, portable games platforms such as Nintendo's seminal *Game Boy*, Atari's *Lynx* and Sega's *Game Gear* all featured headphone audio sockets, with all but the Lynx outputting in stereo. The early 1990s also brought us the flagship home consoles of the 16-bit era, including Nintendo's *Super Nintendo Entertainment System* (SNES) and Sega's *Mega Drive*. Both used bespoke APUs that introduced stereo audio to home consoles. Two audio channels facilitated basic spatial sound design techniques such as stereo panning, the effect of which could be greatly enhanced with headphones. A small number of titles for the SNES, including *Jurassic Park* (Beard et al. 1993) went further, utilising an early iteration of *Dolby Pro Logic* processing to emulate 'surround sound' through the stereo output.[10] Even at this early stage, digital games could now successfully envelope players in three-dimensional virtual soundscapes, and without many of the drawbacks that plagued HMDs. The APU hardware was integrated within a product that had much more to offer as a traditional gaming system. Likewise, the surround sound encoding programs (such as Pro Logic) came integrated with the ROM cartridge of the games in which it featured, whilst headphones were an established peripheral, compatible with numerous devices, that the majority of consumers already owned. The overall effect of these points, from the perspective of the consumer, was that VR sound both served a purpose and brought enhanced quality to an established and trusted product. It was accessible, required nothing from them and its enhancements were subtle but effective.

Towards the end of the decade, 1990s home consoles continued to embrace surround sound even further with the release of the first sixth-generation system[11] and Sega's underappreciated swansong, the *Dreamcast*. Unlike previous consoles, the Dreamcast presented players with hardware that supported real-time multichannel audio[12] meaning that, depending upon their speaker configuration, sounds could now be more realistically auditioned around and behind the player. Later consoles of this generation (Sony's *PlayStation 2*, Nintendo's *GameCube* and Microsoft's *Xbox*) all featured surround sound but were released after the turn of the new millennium so the Dreamcast stands as the most advanced VR sound console of the 1990s. Surround sound in games was not limited to home consoles however. Predating the Dreamcast, soundcards for personal computers began to enable multichannel audio in games, with Ensoniq's *AudioPCI*. Released in 1997, the AudioPCI soundcard was one of the first to present listeners with surround sound gaming, supported by the Microsoft *DirectSound3D* system (discussed below) which enabled quadraphonic (4-speaker) audio playback.

The somewhat colourful history of multichannel audio in personal computers takes us back to 2006. As Chase (2016) explains, many multichannel computer audio systems in the early 2000s generated true positional audio and could generate convincing and detailed three-dimensional soundscapes by way of bespoke APUs such as Aureal's *A3D*,[13] which utilised head-related transfer function (HRTF, see Chaps. 6 and 8) algorithms to simulate the relative distance of sound objects and their positon across both the vertical and horizontal axes. A3D chips were also capable of simulating first-order acoustic reflections (i.e. the first point at which the sound wave reflects from a surface before reaching the ear, see Self 2009), thereby dramatically enhancing the perceived realism of a game's soundscape. Chase (2016) continues, observing how a questionable legal challenge in 1998 leads to Aureal's competitor, Creative Technology, gaining monopoly over positional audio. Creative themselves made the pivotal decision to build almost all their developments for the next eight years upon the same architecture first used by Ensoniq's AudioPCI card, DirectSound3D. However, Microsoft's audio interface was, unfortunately for Creative, a foundation that was later pulled out from underneath them in 2006, when Microsoft decided to discontinue DirectSound3D. As Chase describes, positional audio was subsequently dealt a substantial blow, reverting back to two-dimensional surround sound and simplified acoustic modelling processes for several years. The discontinuation of DirectSound3D had a profound effect on positional audio in the early 2000s but it did not stop the technology entirely. At the same time as DirectSound3D's departure, the home console market saw Sony release their *PlayStation 3*, which supported up to 7.1 surround sound by way of its HDMI output connectors.[14] Sony also released a proprietary 'sound bar' surround speaker (compatible with Dolby and DTS formats), specifically for use with their console. To this day, home consoles continue to support positional audio. Although personal computers now utilise relatively basic integrated audio chips, more powerful dedicated soundcards are widely available and are becoming increasingly popular due to greater consumer interest in signal fidelity and more accurate positional audio.[15]

For some, it may be a contentious point to designate headphones and multichannel audio (whether surround or positional) as 'virtual' or 'VR sound'. However, as we discussed in the previous chapter, these technologies embody some of the defining qualities of VR. Like their visual counterpart, headphones provide a direct connection to the virtual environment by occluding sensory information from the actual world. Furthermore, by way of positional sound techniques, headphones can present the listener with an immersive three-dimensional soundscape that enhances user experience by evoking greater depth of presence. In identifying such sound technologies as part of the VR family, we can also observe how VR sound managed to avoid the struggle that HMDs endured. The underlying reason for the relative success of VR sound within the general consumer market is, ironically, the result of consumers *not* perceiving it as VR. Without that label, VR sound was not explicitly represented in television or motion pictures, nor was it subject to the same juxtaposition between market hype and product reality. Expectations were by contrast, far more conservative and their realisation more achievable.

Haptics

With the wider public understanding of VR 'tethered' to HMDs throughout the 1990s, haptics describes a further set of devices that fall under the conceptual umbrella of VR yet were not commonly regarded as such. Like VR sound, haptic devices also enjoyed relative commercial success in the general consumer market and are particularly relevant to sound, as they highlight a certain aspect of interactive audio that is often overlooked.

Haptics is a central aspect of VR technology. They have been used to represent virtual environments for visually impaired individuals (Colwell et al. 1998), supported movement rehabilitation within VR therapy for post-stroke patients (Broeren et al. 2004) and are a popular mechanism for VR surgical training (see Van der Meijden and Schijven 2009) to name a few. Haptic devices have until recently been dissociated from VR with regard to the consumer market, despite being a quintessential VR technology that has enjoyed consistent commercial success. Haptics incorporates several related concepts that include tactile sensation and force feedback. Both concepts are of great relevance to consumer VR of the 1980s and 1990s, with tactile sensation describing the feel of the hardware interface (which could include a traditional gamepad/mouse/keyboard but, as a VR technology, also include naturalistic game interfaces such as light guns, flight simulator joysticks and racing wheels/pedals) and force feedback referring to vibration stimuli delivered by way of the hardware. As with VR sound technology, haptics defied the trajectory of 1990s HMDs, managing to retain a relatively stable popularity and, to this day, continues to feature in various contemporary forms. Here, we take a look at haptic hardware, specifically the light gun. Although alternative haptic interfaces such as racing wheels/pedals and flight simulator joysticks have gathered and retained comparable success, it is the light gun peripheral that has the most relevance to sound in VR.

Within a games context, the term 'light gun' describes a physical representation of a gun (of some kind—be it a pistol, rifle or rocket launcher) that utilises the emission and detection of light to register the 'hit' as players aim the device at a screen and depress the trigger. The development of light gun technology dates back to the 1920s and the work of William Gent, which led to the release of *Ray-O-Lite* by the Seeburg Corporation in 1936 (Wolf 2012). Ray-O-Light was a virtual duck shooting arcade game in which players 'shot' at a mechanically moving duck. At the centre of the duck was a photo-emissive cell that detected the light emanating from the rifle. Throughout its technological history, the principles and mechanics of the light gun have remained relatively unchanged. In 1966, the light gun became the basis of *Periscope*, the very first game manufactured by Sega (Pettus et al. 2013). Light gun hardware was first integrated with home consoles in 1973 with the advent of the Shooting Gallery peripheral for the *Magnavox Odyssey* system (see Wolf and Perron 2014) but it was the 1980s and the 8-bit (third-generation) era of home console entertainment that first saw light gun games achieve commercial prominence. They were compact, sleek and science fiction-inspired devices that included Nintendo's *Zapper* (compatible with the NES), Atari's *XG-1* (working on the Atari *2600*, *7800* and *XEGS* consoles) and Sega's *Light Phaser* (for the Sega *Master System*). Their operation reversed that of Ray-O-Lite, with the gun now receiving (rather than emitting) light from the television. Light guns of this generation were designed to function primarily on cathode ray tube (CRT) televisions, either by way of 'sequential targeting' or 'cathode ray timing'.[16] Within the consumer market, the light gun continued to be modestly successful in the subsequent 16-bit and 32-bit console eras, throughout the duration of the 1990s, but with some variation in success between products. The year 1994 saw the release of the Sony *PlayStation* and a year later, Namco presented the first iteration of the *Guncon* light gun, a PlayStation exclusive peripheral released alongside *Time Crisis* (Sawano et al. 1995) a popular game from the 'rail shooter' genre (first-person perspective games built around the light gun interface, in which movement is largely controlled by the computer whilst the player's main interaction is simply to shoot). Namco continued to manufacture Guncon controllers for subsequent iterations of the PlayStation and whilst it is difficult to obtain precise market statistics, light gun peripherals and rail shooter games have been consistently popular with home console consumers, up to the present day. As a VR technology, light gun peripherals ran the gauntlet of the 1990s, and survived.

In terms of sound, devices such as the Zapper and Light Phaser did not produce sound by design. Pulling the trigger and hitting a target in *Duck Hunt* (Yokoi et al. 1984), for example, would instigate a sound (typically a short burst generated via the hardware's noise channel) in a much the same way as would a game utilising a traditional controller. Use of the light gun however, presented an entirely new way of interacting with game sound. Rather than the noise burst being kinaesthetically linked to a button press, now it was fed back from a more embodying action as the user was required to physically raise their arms and aim the light gun towards the screen in a movement more closely connected to the activity it was representing. Furthermore, unlike gamepad button presses, which were largely silent, each pull of the Zapper's trigger caused a loud mechanical 'click' which became an integrated part of the overall experience and created a mixed reality soundscape by providing the user with simultaneous auditory feedback from both the physical world and the virtual world. In their seventh-generation console, the *Wii*, Nintendo more intentionally recreated this effect in their *Wii Zapper* peripheral, which is essentially a casing that houses the system's *Wii Remote* motion controller. Because the motion controller itself featured integrated speakers, games such as *The House of the Dead: Overkill* (Pritchard et al. 2009) could play gunfire samples, matching the user's actions and the specific type of in-game weapon that they were using.

In addition to being naturalistic interfaces, the above hardware also incorporated force feedback in at least some of its designs. The origins of force feedback can be traced back to the early twentieth-century research that sought to draw connections between human psychology and physiology. The result of this research became known as 'machine haptics', the design and development of 'mechanical devices that replace or augment human touch' (Orozco et al. 2012: p. 217). El Saddik and colleagues (2011) posit that force feedback technology in digital games enhances the overall experience by establishing a more accurate imitation of the actual artefacts and activities that the game is representing, thereby deepening immersion by 'creating a more realistic physical feeling when playing a game' (p. 36).

One of the first digital games to feature force feedback, *Moto-Cross* (later rebranded as *Fonz*), was an arcade game that featured physical handlebars mounted to a cabinet. These handlebars facilitated control of the virtual motorbike and would vibrate whenever the player collided with another vehicle. Home consoles did not employ force feedback into their systems until 1997, with the arrival of Nintendo's *Rumble Pak*, a removable peripheral for the *Nintendo 64* that connected to the underside of the console's controller. Depending upon the game, the Rumble Pak would vibrate in response to various in-game events (characteristically explosions and earthquakes) or user actions such as firing a gun or forcefully colliding with an object.

As with some of the haptic VR devices described above, force feedback had (and continues to have) a significant relationship with sound. This is not at all surprising due to the physical nature of the sound wave as 'a natural feature of the mechanical impacts and vibrations that we experience […] everyday' (Adelstein et al. 2003, p. 73). In research, so-called audio-haptics describe techniques in which sound wave and vibration cues are combined as a means of enhancing user experience. Audio-haptics have been integrated into mobile phone interfaces (Chang and O'Sullivan 2005) and used to improve user experience in digital audio editing software (Chu 2007), with various other studies asserting the value of sound as a means of enhancing haptic interfaces (Adelstein et al. 2003; DiFranco et al. 1997; Lederman et al. 2002).

In many video game examples, force feedback provides informational cues that align more closely with sound than with visuals. For example, in the now-classic Nintendo 64 shooter *Goldeneye 007* (Hollis et al. 1997), the rapid rate of fire for automatic rifles as displayed by the sound cue and the vibrations of the Rumble Pak are closely matched, whilst the accompanying graphical muzzle flash is not. Various games, that include Goldeneye 007, frequently match force feedback rumbles to auditory in-game events such as explosions or earthquakes, with such events often occurring outside the player's field of view. This kind of scenario highlights points of gameplay in which force feedback can be separated from graphics but unless game audio was muted entirely, sound and haptics are continually linked. Furthermore, the physical correlation of sound waves and vibration inevitably means that all force feedback devices themselves emit a certain degree of sound.

The commercial success of the Rumble Pak has led to the device being recognised for establishing the value of force feedback to user experience (specifically as a means of enhancing immersion) and paving the way for the many devices that followed (Watts 2013). In the same year, Microsoft revealed their first implementations of force feedback, which they integrated into both their flight simulator joystick and racing wheel controllers. As a direct competitor to Nintendo, Sony also released force feedback hardware, the *Dual Shock* controller for their *PlayStation/PSX* console. As with the Microsoft peripherals, force feedback was integrated into the Dual Shock controllers, a design template that remains largely unchanged in the current eighth generation of games consoles, with force feedback presently a feature of all major console controllers. The success of haptics in digital games, as a VR technology, reveals how such peripheral devices fit very well into the central argument of this chapter. Whilst the HMD may have been synonymous with 1990s VR, hardware and design principles pertaining to haptics were (and continue to be) fundamentally virtual. As with the other concepts and technologies discussed above, haptics also circumvented the commercial failure of 1990s HMDs as a direct result of being more closely integrated with digital games whilst avoiding expectation/reality gap.

Motion Tracking and Gestural Interfaces

Premaratne (2014) recalls the development of hand gesture recognition, dating back to 1977 and the *Sayre Glove* (Defanti and Sandin 1977) which used photocells and flexible tubing to measure finger flexion. The tubing ran along the thumb and each finger, with small light emitters at each apex. Each digit was also affixed with a photocell array for detecting light. As the user flexed their fingers, the amount of light hitting each photocell would change and the computer would use this information to calculate finger movement. Premaratne (2014) continues, noting several advances in hand gesture recognition across the 1980s and 1990s, including the replacement of light tubes with fibre optics and the implementation of accelerometers to enable hand orientation tracking in addition to finger movement. Also acknowledged by Premaratne is Zimmerman and Lanier's (1991) *Data glove*, which employed an ultrasonic system. In the design of the Data Glove, ultrasonic emitters and detectors were placed across the hand and the delay between emission and reception was used to calculate finger flexion. This device is noteworthy for being the precursor to the first gesture recognition device built specifically for the home console market, Nintendo's *Power Glove*. The Power Glove is discussed earlier in this chapter (and in Chap. 9 we shall return to such devices in terms of their various non-gaming applications). Despite not being a commercial success, the Nintendo Power Glove was the forerunner of the various games control devices that would follow, with a legacy in the hardware of contemporary VR.

In the years that followed, Nintendo's broader commercial success enabled them to embark on riskier ventures in the name of innovation (McFerran 2016). Outcomes included their *Super Scope* light gun and the first use of an analogue thumb stick within their *Nintendo 64* controllers. Whilst not always profitable, Nintendo's ventures proved the company to be a great contributor to VR-related devices, in terms of both their technical development and their maintaining of consumer interest. This became most apparent with the release of the *Wii* console in 2006. This home console featured the Wii Remote (or Wiimote), a wireless motion controller that brought the players a whole new approach to interacting with games. The Wii Remote was not the world's first motion sensing games controller. That honour that goes to Datasoft's *Le Stick* controller, which used a set of mercury tilt switches to track orientation. Nintendo did however bring the technology to the masses and was the first to exploit modern inertial measurement units (IMUs: electronic position and orientation tracking devices typically using a combination of accelerometers, magnetometers and gyroscopes).

Initially built using a three-axis accelerometer and infrared optical sensing technology, the Wii Remote facilitated the measuring of gross motor movements and more precise pointing actions, respectively. As with all IMU-based devices, the Wii Remote tracking was limited to three DoF (pitch, yaw and roll) with broader movements, as displayed on screen by the avatar, effectively simulated from the orientation data. Nintendo later increased the overall accuracy of the system by way of the *MotionPlus* sensor, a gyroscope module extension that could be attached to the controller (Jones 2016). Nintendo's own published data states that the Wii sold 101.63 million units by the final quarter of 2016,[17] making it the company's best-selling home console to date and evidencing the significant popularity of the motion control interface. A recent retrospective by Bailey and Oxford (2016) observes that the motion control offered by the Wii was by no means free from problems. The accelerometer could not discern fine movements, making certain interactions difficult. Also, the mapping of player to avatar-movement was not relative, meaning feedback from the game didn't accurately reflect the player's gestures. Despite this, Bailey and Oxford note that the uniqueness of the interface, nevertheless 'got a lot of people into gaming. Mothers, fathers, grandmothers, grandfathers—everyone was doin' it'.[18]

Whilst motion sensor-based controllers are not necessarily the sole inspiration of interaction within VR, the progression of relevant technology leading into contemporary VR interface hardware casts little doubt as to its great influence. HTC's *Vive*, Oculus' *Rift* and Google's *Daydream* all utilise motion controllers, though the tracking approaches do vary between systems. Both the Rift[19] and the Vive[20] use infrared tracking to facilitate both positional and orientation tracking (6DoF), similar to the Wii's pointing mechanism. This provides greater accuracy to facilitate finer movements, a lesson arguably learned from criticisms levied at the Wii Remote. To preserve portability and ease of use, the Daydream and other smartphone-based VR devices rely upon IMU-based tracking[21] and, much like with the Wii, evoke the sensation of making broad motor actions (swinging a virtual golf club, throwing a virtual bowling ball, etc.) by mapping the orientation input of the controller into a virtual action visually rendered in the movements of the avatar. As the central selling point of a hugely successful home console, the Wii Remote has earned its place in history as the device that first acquired mass appeal for motion sensing technology in digital games. That said, games devices that enable gesture-based interactions are not limited to IMUs and infrared emission. Furthermore, the first home console accessory to offer this type of functionality predates the Wii Remote by several years.

Released in 2003, the Sony *EyeToy* was a simple USB-connected camera not too dissimilar to a webcam (indeed with a little ingenuity, the device could function as such on most major operating systems[22]) that facilitated simple but relatively robust body-tracking (primarily arms and head movement) across a two-dimensional plane. The basic mechanism of the tracking is based upon detection of changes in the pixels of the moving image and the implementation of the device is notable for projecting the image of the player into the game itself, essentially presenting one of the first modern mixed reality gaming experiences. In its first iteration, no additional hardware (other than the camera itself) was required to facilitate tracking, though later versions did incorporate hand-held controllers that the camera could track more accurately, including the *PlayStation Move* system. Kim (2008) notes that the immediate and corresponding feedback of the player's moving image within the game made use of the EyeToy highly intuitive and accessible. Despite being the progenitor of home console motion sensing technology, the EyeToy appears to be condemned to being an afterthought, which would arguably be rather inappropriate for several reasons. Firstly, the sales of the device were relatively strong (particularly for a peripheral device sold separately to the main console), with over 10.5 million units sold within the first five years of its release (Kim 2008). Secondly, the legacy of the EyeToy extends to present day with its most recent iteration continuing to advance the specifications of the product.[23] Finally, the technology, implementation and commercial success of the EyeToy is arguably a direct influence to Microsoft's well-known 2010 entry in motion sensing market, the *Kinect*.

As Borenstein (2012) explains, the Kinect broadly takes the functionality of the EyeToy and extends it into three dimensions. Whilst the EyeToy is solely a light receptor, the Kinect utilises a tracking system highly comparable to that which is utilised in contemporary VR, both emitting and receiving infrared light then analysing the received light to interpret depth information.[24] As with the EyeToy, the Kinect had a solid but modest commercial impact, selling 24 million units in the first five years.[25] A new iteration of the device currently remains available as a peripheral for Microsoft's current system, the *Xbox One*.

Motion sensing devices such as those above reveal clear connections to contemporary VR in terms of their technology, implementation within gaming and their marketing. With regard to design and functionality, the Wii Remote's legacy can be observed in the controllers for various mobile VR systems. Equally, so too can the influences of the Kinect and EyeToy be seen in the underlying tracking technology of the HTC Vive and Oculus Rift. Additionally, the history of these devices reminds contemporary VR of several important lessons that arose during the 1990s era of VR, such as the need for accurate feedback, relative affordability, ease of use and integration within game mechanics. As we noted above, the Wii faced criticism for feedback that corresponded poorly to player's physical actions (Bailey and Oxford 2016).

With regard to the Kinect, critics noted that too few games were made specifically for the device and games that used it as an optional extra were deemed wholly unenhanced by the additional control interface (Weinberger 2015). These issues resonate almost eerily with those raised in the earlier discussion in this chapter, concerning the 'reality' of 1990s VR. What this suggests is that, for contemporary VR to enjoy ongoing commercial appeal, it must present consumers with genuine added value by offering them something that is innovative enough to excite, whilst familiar enough to be accessible. It must present an experience that largely meets consumer expectations and it must integrate itself into the wider games and entertainment industry as much as possible.

At present, motion sensor interfaces are arguably missing a significant opportunity for enhancement by way of better sound design. Presently, auditory feedback to player-gestures notably limited and typically absent throughout the back catalogue of games that support motion control. This issue is certainly worth mentioning as it highlights the opportunity for sound to participate in future gesture-based interfaces, something that we shall discuss in Chap. 8. However, this is not to say that the above technology did not often incorporate a significant sound element in their functionality, as we shall explore below.

Speech Recognition and Voice Commands

Our final foray into VR technologies looks at speech recognition and its implementation within digital games, specifically as an auditory interface that facilitates voice commands. Of the various motion sensing systems released throughout the first few years of the twenty-first century, the majority also incorporated some form of microphone hardware and speech recognition software. As with much of the VR technology discussed earlier in this chapter, a voice-based computer interface takes cues from fiction. A notable example of this can be observed within the television series *Star Trek: The Next Generation* (Roddenberry 1966). In numerous episodes, crew members of the Starship Enterprise frequently accessed information and made requests to the Enterprise's computer using voice commands. This fictive example became so well known that it even provided the vocal template for issuing actual voice commands, with nearly all voice command systems requiring an identifier prefix before the instruction.[26] Jean-Luc Picard's renowned prefacing of his commands to the Enterprise's computer with 'computer, …', was so popular in fact that in early 2017, Amazon updated their *Echo* device with an option for users to change the default preface, 'Alexa', to 'computer' (Pullen 2017).

Voice command technology falls under the umbrella of speech processing research, which is itself a branch of artificial intelligence. It encompasses speech coding and signal enhancement, which are methods of compressing incoming sound waves to extract only the 'speech signals' necessary for the subsequent processes of speech recognition and speaker identification (see Ogunfunmi and Narasimha 2010; Sigmund 2003). The final aspect under this umbrella is speech synthesis, which is essentially a reconstitution of the processes behind speech recognition as a means of shaping synthetic sound. Prior to CD-ROM games formats, synthetic speech did feature in various digital games but was used sparingly due to being computationally expensive. Since being replaced by sample-based speech, synthesis is not a common feature of digital games, but recent breakthroughs in the quality of the technology have seen synthesis regularly paired with speech recognition across the range of personal assistant software presently available, making it a likely technology to be integrated into future games systems.

Sigmund (2003) provides a concise history of speech processing, dating back to the work of physicist and engineer Christian Gottlieb Kratzenstein, who in 1779 successfully reproduced five vowels mechanically. Early twentieth century technology progressed to electrical speech synthesis with Homer Dudley's *Vodor* (Dudley 1939), and the 1950s saw the first attempts to implement speech analysis theory to construct an automatic speech recognition machine. Speech synthesis and speech analysis formally separated into discrete research fields in 1960. The latter years of the twentieth century saw speech synthesis integrate with home computing, most notably as a form of assistive technology that enabled visually impaired individuals to access digital content by translating text into synthesised speech. As Pinola (2011) notes, automatic speech recognition software for personal computers arrived in 1990 with the release of *Dragon Dictate*, a brand which continues to be associated with speech recognition.[27]

Speech recognition as a voice command interface first appeared during the sixth generation of games consoles. Historically, the *Dreamcast* has proven itself to be a truly innovative console, from being the first console to implement an operating system to first that featured a second screen into the controller.[28] In line with this pattern, Sega's system also stands as the forerunner of voice command technology within home console gaming. Released just before the turn of the century, *Seaman* (Saito et al. 1999) came bundled with Sega's proprietary microphone and put players in the rather surreal situation of speaking to a deeply pythonesque human-fish hybrid. Several years later, voice command gameplay resurfaced in the *PlayStation 2* survival horror title *Lifeline* (Nishizawa et al. 2003). The game utilised a third-person perspective, and the player was required to direct the avatar through the game but could only do so by way of voice commands. Whilst some packages of Lifeline included a microphone headset, the game would function with various third party USB-compatible sets. Whilst Sony's EyeToy (discussed earlier) featured an integrated microphone, its function was to enable Internet video chat. Voice commands did not feature in any of the games developed specifically for the EyeToy, with the gameplay priority being on exploiting its motion sensing functionality. Rival to Sony's console, the Nintendo *GameCube* briefly explored voice command interfacing with their *GameCube Microphone* peripheral. As with Sony, Nintendo was clearly experimenting cautiously with the technology and only a handful of games were compatible with the device, the most notable of which being *Mario Party 6* (Nishiya et al. 2004) which was often bundled with the microphone peripheral. Consisting of various mini-games, Mario Party 6 players used the microphone to carry out various actions, with the system able to recognise a limited lexicon ('bombs to fire bombs, 'lasers' to fire lasers, etc.). Whilst games were very rarely developed around voice command as a central gameplay mechanic, the technology made more frequent appearances during this console generation as a non-required enhancement. Early examples of this include *Ghost Recon 2* (Allen et al. 2004), a tactical shooter title in which players use a traditional game controller for all actions but can supplement this with voice command 'shortcuts' to strategically direct non-player characters.

As with the legacy of motion sensing, the early use of voice command in consoles has yielded modest but stable commercial success and has continued to be implemented in all major consoles released up to the present day. Later iterations of the technology have also been developed further, with both functionality and quality moving ever closer to the fiction-derived ideals. As Luo and Yang (2014) explain, the Kinect hardware includes an array of four microphone capsules that enable sound waves to be detected from most points around the room in which the device is situated. The Kinect's processing software examines incoming audio, separating it into discrete audio streams and tracking the loudest stream by default (Giori 2013). This enables the device to distinguish voice input from background noise and, when combined with language processing, the Kinect enables players to control various *Xbox 360/One* functions, from 'Xbox, broadcast' (starts screen-capturing gameplay and streams it live to the Internet) to 'Xbox, call' (sets up a Skype call).[29] It is worth noting that voice functionality for the Kinect is not primarily intended for use in actual games, but rather to support wider applications of the console's operating system. Again, we can see here how the priority is largely on subtle enhancements and functionality; making an existing and trusted system more flexible and intuitive but as an optional extra that doesn't take the player too far from their comfort zone.

Critical reception of auditory interface technology has, in line with its commercial success, been modest but stable. Reviews for the latest iteration of the Kinect for instance largely praise the functionality of the voice command interface, often citing it as the most impressive aspect of the device.[30] That said, detractors note that the voice recognition can be somewhat unreliable, to the extent that repeated failures to discern instruction often causes users to stop using voice commands and revert to using the traditional gamepad interface. This highlights a now rather familiar contrast between expectation and reality, the former being raised by the perfectly functioning fictional representations such as the Enterprise computer in Star Trek. However, whilst this does signify a similarity between voice command technology and the various

VR devices that suffered the expectation/reality gap, the former has arguably not suffered same the fate of languishing in the 'trough of disillusionment' (Gartner's Hype Cycle, see Chap. 6). Whilst the precise reason for this remains unclear, it is highly likely that one significant factor is something we have discussed previously, regarding the progression of haptics and headphones. Unlike HMDs, which were central to their related experience, auditory interfaces were in most cases supplementary and optional. As with headphones, they did not dictate how players should interact with the game but rather presented a method of interaction that could potentially enhance gameplay, but in a game that would function perfectly well without it. With enhancement as its central value proposition, the quality of the technology was not a pivotal factor in the player's overall enjoyment of the game. Any issues were unlikely to evoke intensely negative experiences and therefore equally unlikely to provoke aggressive criticism. In a final, related point, virtual technologies such as auditory interfaces evade the damnation levied at 1990s VR by way of their cost. Whilst presenting consumers with something that under-delivers on their expectations is one thing, requiring them to invest substantially, based on that false expectation, is something quite different. VR sound interfaces, much like many of the devices detailed throughout this chapter, largely avoided this issue by being either low-cost peripherals (often bundled with games to increase value further) or integrated technology that shipped with the core games hardware.

As technology is very much akin to VR, it is unsurprising that auditory interfaces enabling voice commands have made their way into contemporary VR systems. At the time of writing, the most recent version of the Oculus store now utilises *Oculus Voice*, enabling voice commands for navigating around the store.[31] VR tools such as *Modbox* (a building tool that enables users to create VR environments from within a virtual environment) also facilitate voice commands to make their overall interface more intuitive.[32] Various VR games now provide the option for voice command, including very popular titles such as *Elite Dangerous* (Braben 2014). With a focus upon increasing accessibility and operational efficiency, the user-experience intentions of contemporary VR tools and games that implement voice command interfaces closely reflect those of the digital games systems that preceded them.

Chapter Summary and References

As an idealised representation of the functionality and quality of VR, fictional representations from cinema, television and literature certainly bear some of the responsibility for the deeply inflated expectations that contributed to the downfall of 1990s consumer VR, from fully multimodal, high fidelity virtual worlds that are perceptually indistinguishable from the actual world to extraordinary functions such as the ability to re-experience actual events from recorded memories. With such fantasy put upon screen and written in books, all reinforced by actual VR advertising, the inflated expectations of the consumer had little chance of being met by a fledgling technology. That said, it would arguably be unfair to suggest that fictional representations should take full responsibility. Firstly, without such ideals, both sources of inspiration for developers and the interest of consumers would likely have been significantly lower and fictional forms of VR continue to inspire future VR development. Secondly, whilst genuinely innovative and conceptually fascinating, the reality of 1990s VR was clunky and uncomfortable, expensive and difficult to operate, lacking in games (particularly third party), poorly standardised between systems, brought with it a host of physical side effects and delivered feedback that was inconsistent with player actions.

Consumer VR had suffered a notable setback but, contrary to common understanding, did not disappear. Instead, consumer VR continued in various, more subtle forms, largely driven by developments in digital games. The first-person perspective embodied various design concepts that both embodied and advanced VR principles. It prioritised user-experience qualities such as a perception of presence within virtual space. It also introduced various new game genres like the FPS and first-person survival horror that now make up a large proportion of current VR titles. Developments in digital game sound also have made a significant contribution by way of advances in positional audio techniques, intelligent games systems that respond more realistically to sound in support of ludic immersion, experimentation with auditory perception to undermine player expectation to evoke fear in survival horror and addressing issues of player identity and narrative immersion by way of the silent protagonist paradox.

In addition to conserving and building upon VR-related designs, digital games provided the same service for VR hardware. From light guns and IMU motion control, to multichannel audio hardware for positional audio and auditory interfaces for voice command interaction, personal computer and home console peripheral devices maintained consumer interest in VR. Whilst marketed with much less ceremony than 1990s HMDs, these devices were unmistakably virtual, commercially successful and contributed significantly to contemporary VR. The next chapter concludes our timeline and brings us up to now, with a review of VR sound as it pertains to contemporary systems.

Notes

1. Example of current (at time of writing) top VR games. https://www.wareable.com/vr/best-steam-vr-games.
2. Value of immersion in FPS. http://www.gamersdecide.com/pc-game-news/10-things-make-fps-awesome.
3. Qualities of FPS games. Gamasutra.com. http://www.gamasutra.com/view/news/170721/.
4. Popularity of FPS games. Wow247. http://www.wow247.co.uk/2015/05/19/greatest-90s-video-games.
5. Difference between surround sound and 3D audio. http://www.electronicproducts.com/News/Surround_sound_vs_3D_sound.aspx.
6. How sound works in Thief. Old PC Gaming: http://www.oldpcgaming.net/thief-the-dark-project-review/.
7. Alone article. https://www.destructoid.com/alone-is-a-horror-game-within-a-game-for-oculus-rift-262394.phtml.
8. Lost in the Rift article. VRWiki. https://vrwiki.wikispaces.com/Lost+in+the+Rift.
9. Of notable absence is the truly glorious *Dreamcast Fishing Rod* peripheral which is certainly worth a look (see: https://www.giantbomb.com/dreamcast-fishing-controller-support/3015-4057/).
10. How the SNES used surround sound. Gamnesia. https://www.gamnesia.com/articles/throwback-thursday-top-5-uses-of-surround-sound-on-snes.
11. Using bits to distinguish generations became superfluous at this point as console processors became too complex.
12. Sega Dreamcast technical specifications. Digital Extremes Online. http://dextremes.com/dc/specs.html.

13. Chase (2016) provides a link to an Aureal A3D demo video. https://www.youtube.com/watch?v=-oSlbyLAksM.
14. PS3 user's guide. http://manuals.playstation.net/document/en/ps3/current/settings/audiooutput.html.
15. Value in dedicated soundcards. Techspot. http://www.techspot.com/article/751-should-you-buy-a-sound-card/.
16. In sequential targeting (without getting too technical), when the trigger is depressed, the computer flashes a single frame image on the screen, displaying only the target but as a block of white light. If the gun detects light at this point, a hit is registered. Cathode ray timing by comparison exploits how CRT screens refresh by an electron beam that scans from left to right and top to bottom, pixel by pixel and row by row for each frame. Here, when the trigger is depressed, the beam brightens the screen for a single frame. When the gun detects the lighter pixel, it calculates the time from the trigger to the detection to determine a hit.
17. Nintendo hardware and software sales units. https://www.nintendo.co.jp/ir/en/finance/hard_soft/index.html.
18. In line with the deeply competitive nature of the home console market, arch-rival Sony very quickly implemented motion-sensing technology by way of accelerometers into their official *PlayStation 3* console controller, the *Sixaxis*. Whilst previously trailing behind Nintendo with motion-sensing technology, Sony managed to make the first move in terms of combining the accelerometers with a gyroscope for increased tracking accuracy with the release of their *DualShock 3* controller.
19. Oculus Touch controller details. https://techcrunch.com/2016/12/05/review-oculus-touch-motion-controllers/.
20. HTC Vive controller details. https://www.vive.com/eu/support/category_howto/720333.html.
21. Google Daydream controller details. http://www.twistedreality.tv/daydream-controller-comprehensive-guide/.
22. Using the EyeToy as a webcam. http://www.instructables.com/id/Turn-an-Eyetoy-into-a-USB-Webcam/.
23. Subsequent iterations of the EyeToy include the PlayStation Eye (https://en.wikipedia.org/wiki/PlayStation_Eye) and the PlayStation Camera (https://en.wikipedia.org/wiki/PlayStation_Camera).
24. As Borenstein (2012) notes, the specific technique used by the Kinect is a 'displacement calculation'. The system is programmed with an 'understanding' of how each beam of light will appear if reflected at a certain

distance. The system then measures the difference in the received image from the calibrated 'standard' image to determine distance.

25. Kinect 5-year sales. https://www.gamespot.com/articles/kinect-sales-reach-24-million/1100-6403766/.
26. Though it is acknowledged that whilst this may have been inspired by Star Trek, it was implemented so that the system could confirm that the speaker was addressing them.
27. Current Dragon speech recognition software. http://www.nuance.co.uk/dragon/index.htm.
28. Dreamcast innovations. http://www.gameinformer.com/b/features/archive/2015/01/21/10-gaming-innovations-the-world-wasnt-ready-for-but-is-now.aspx.
29. List of popular voice commands for the Kinect/Xbox. http://news.xbox.com/2014/05/02/xbox-one-kinect-voice-tips/.
30. Xbox One Kinect review. Engadget. https://www.engadget.com/products/microsoft/xbox/one/kinect/.
31. Voice commands in Oculus store. https://www.vrheads.com/how-use-oculus-voice-your-rift.
32. Voice commands in Modbox. https://uploadvr.com/modbox-bringing-voice-commands-vr/.

References

Adelstein, B. D., Begault, D. R., Anderson, M. R., & Wenzel, E. M. (2003, November). Sensitivity to haptic-audio asynchrony. In *Proceedings of the 5th International Conference on Multimodal Interfaces* (pp. 73–76). ACM.

Allen, C., Brown, B., Salta, T. et al. (2004). *Ghost Recon 2*. Cary, USA: Red Storm Entertainment.

Andreasen, J. (1982). *Haunted House*. New York, USA: Atari Inc.

Bailey, K., & Oxford, K. (2016). Ten years later: US Gamer debates whether the Nintendo Wii was a success or a failure. *US Gamer*. http://www.usgamer.net/articles/ten-years-later-usgamer-debates–nintendo-wii-was-a-success-or-a-failure.

Beard, J. H., Kerry, C., et al. (1993). *Jurassic Park*. Manchester, UK: Ocean Software.

Betters, E. (2013). Virtual reality: Lessons from the past for Oculus Rift. *BBC News Online*, http://www.bbc.co.uk/news/technology-23877695.

Bleszinski, C., Schmalz, J., Sweeney, T. et al. (1999). *Unreal Tournament*. Cary, USA: Epic Games.

Borenstein, G. (2012). *Making things see: 3D vision with kinect, processing, Arduino, and MakerBot*. California, USA: O'Reilly Media, Inc.

Braben, D., Brookes, M., Sammarco, S. et al. (2014). *Elite Dangerous*. Cambridge, UK: Frontier Developments.

Broeren, J., Rydmark, M., & Sunnerhagen, K. S. (2004). Virtual reality and haptics as a training device for movement rehabilitation after stroke: a single-case study. *Archives of Physical Medicine and Rehabilitation, 85*(8), 1247–1250.

Chang, A., & O'Sullivan, C. (2005, April). Audio-haptic feedback in mobile phones. In *CHI'05 extended abstracts on Human factors in computing systems* (pp. 1264–1267). ACM.

Charara, S. (2015). Virtual reality: Then and now—why it won't fail this time. Wareable. https://www.wareable.com/vr/virtual-reality-then-now-why-it-wont-fail-this-time.

Chase, M. (2016). How VR is resurrecting 3D audio. *PC Gamer Magazine*. http://www.pcgamer.com/how-vr-is-resurrecting-3d-audio.

Chion, M., & Murch, W. (1994). *Audio-vision: Sound on screen*. New York, USA: Columbia University Press.

Chu, L. L. (2007). *U.S. Patent No. 7,208,671*. Washington, DC: U.S. Patent and Trademark Office.

Colanonio, R., Smith, H., Tardiff, H. et al. (2012). *Dishonored*. Lyon, France: Arkane Studios.

Colley, S.,& Thompson, G. et al. (1974). *Maze War*. Colley.

Collins, K. (2008). *Game sound: An introduction to the history, theory, and practice of video game music and sound design*. MA, USA: MIT Press.

Colwell, C., Petrie, H., Kornbrot, D., Hardwick, A., & Furner, S. (1998, January). Haptic virtual reality for blind computer users. In *Proceedings of the third international ACM conference on Assistive technologies* (pp. 92–99). ACM.

Coulon, F., Ferland, M., Schneider, R. et al. (2002). *Splinter Cell*. Montreal, Canada: Ubisoft Montreal.

Defanti, T., & Sandin, D. (1977). Final report to the National Endowment of the arts. *US NEA R60-34-163, University of Illinois at Chicago Circle, Chicago, Illinois*.

DiFranco, D. E., Beauregard, G. L., & Srinivasan, M. A. (1997, November). The effect of auditory cues on the haptic perception of stiffness in virtual environments. In *Proceedings of the ASME Dynamic Systems and Control Division* (Vol. 61, pp. 17–22). New York City, USA: American Society of Mechanical Engineers.

Domsch, S. (2017). Dialogue in video games. In J. Mildorf & B. Thomas (Eds.), *Dialogue across Media*. Amsterdam: John Benjamins Publishing Company.

Dudley, H. (1939). Remaking speech. *The Journal of the Acoustical Society of America, 11*(2), 169–177.

El Saddik, A., Orozco, M., Eid, M., & Cha, J. (2011). *Haptics technologies: Bringing touch to multimedia*. Berlin, Germany: Springer Science & Business Media.

Fahs, T. (2009). IGN Presents: The history of survival horror. *IGN Online*. http://uk.ign.com/articles/2009/10/30/ign-presents-the-history-of-survival-horror.

Fergusson, R., Bleszinski, C., Brown, J., Perry, L. et al. (2006). *Gears of War*. NC, USA: Epic Games.

Fox, J., Arena, D., & Bailenson, J. N. (2009). Virtual reality: A survival guide for the social scientist. *Journal of Media Psychology, 21*(3), 95–113.

Garner, T., & Grimshaw, M. (2011). A climate of fear: Considerations for designing a virtual acoustic ecology of fear. In *Proceedings of the 6th Audio Mostly Conference: A Conference on Interaction with Sound* (pp. 31–38). ACM.

Gera, E. (2015). Alien Isolation looks a lot different in third-person. *Polygon*. http://www.polygon.com/2015/3/5/8152555/alien-isolation-looks-a-lot-different-in-third-person.

Giori, C. (2013). *Kinect in Motion–Audio and Visual Tracking by Example*. Birmingham, UK: Packt Publishing Ltd.

Goldsmith, J. T. T., & Ray, M. E. (1948). *U.S. Patent No. 2,455,992*. Washington, DC: U.S. Patent and Trademark Office.

Grimshaw, M. (2009). The audio Uncanny Valley: Sound, fear and the horror game. In: *Audio Mostly 4th Conference on Interaction with sound*, Glasgow, 2nd–3rd September.

Hall, T., Romero, J., Carmack, J. et al. (1992). *Wolfenstein 3D*. Richardson, USA: Id software.

Hollis, M., Hilton, K., Smith, A., Jones, B., Doak, D. et al. (1997). *Goldeneye 007*. Twycross, UK: Rare.

Hope, A., Court, J., Smith, O. et al. (2014). *Alien Isolation*. Horsham, UK: Creative Assembly.

Horowitz, S., & Looney, S. R. (2014). *The essential guide to game audio: The theory and practice of sound for games*. Boca Raton, USA: CRC Press.

Howard, T., Carter, G., Pagliarulo, E. et al. (2008). *Fallout 3*. Rockville, USA: Bethesda Softworks.

Hughes, J. F., Van Dam, A., Foley, J. D., & Feiner, S. K. (2014). *Computer graphics: Principles and practice*. London, UK: Pearson Education.

Jones, S. E. (2016). Controller. In H. Lowood, & R. Guins (Eds.), *Debugging game history: A critical Lexicon* (pp. 80–88). Cambridge, USA: MIT Press.

Kane, B. (2014). *Sound Unseen: Acousmatic sound in theory and practice*. New york, USA: Oxford University Press.

Khoury, A. (2015). You'll love the Power Glove documentary. It's so bad. Digital Trends. https://www.digitaltrends.com/gaming/nintendo-power-glove-documentary/.

Kim, T. (2008). In-depth: Eye to eye—The history of the EyeToy. *Gamasutra*. http://www.gamasutra.com/view/news/111925/InDepth_Eye_To_Eye__The_History_Of_EyeToy.php.

Kromand, D. (2008). Sound and the diegesis in survival-horror games. *Audio Mostly*. Pitea, Sweden. October 22nd–23rd: 16–19.

Lakshman, V., Lefay, J., Heberling, E. et al. (1994). *The Elder Scrolls: Arena*. Rockville, USA: Bethesda Softworks.

Lalwani, M. (2016). For VR to be truly immersive, it needs convincing sound to match. *Engadget*. https://www.engadget.com/2016/01/22/vr-needs-3d-audio/.

Lederman, S. J., Klatzky, R. L., Morgan, T., & Hamilton, C. (2002). Integrating multimodal information about surface texture via a probe: Relative contributions of haptic and touch-produced sound sources. In *Haptic Interfaces for Virtual Environment and Teleoperator Systems, 2002. HAPTICS 2002. Proceedings. 10th Symposium on* (pp. 97–104). IEEE.

Levine, K., Hellquist, P., Kline, C. et al. (2007). *Bioshock*. 2K Games, USA/Australia.

LoPiccolo, G., Gilby, J., Randall, J. et al. (1998). *Thief: The Dark Project*. Cambridge, USA: Looking Glass Studios.

Luo, Q., & Yang, G. (2014). Research and simulation on virtual movement based on Kinect. In R. Shumaker & S. Lackey (Eds.), *Virtual, Augmented and Mixed Reality: Designing and Developing Augmented and Virtual Environments: 6th International Conference, VAMR 2014, Held as Part of HCI International 2014, Heraklion, Crete, Greece, June 22–27, 2014, Proceedings* (Vol. 8525). Berlin, Germany: Springer.

McFerran, D. (2014). Reality crumbles: Whatever happened to VR? *Eurogamer Online*. http://www.eurogamer.net/articles/2012-04-12-reality-crumbles-whatever-happened-to-vr.

McFerran, D. (2016). How we got to the switch: A brief history of Nintendo controllers. *Tech Radar*. http://www.techradar.com/news/how-we-got-to-the-switch-a-brief-history-of-nintendo-controllers.

McMahan, A. (2003). Immersion, engagement and presence. *The Video Game theory Reader, 67*, 86.

Michon, T., Shepperd D., Milner, R. et al. (1976). *Night Driver*. Hudson, USA: Micronetics Inc.
Miyamoto, S., Koizumi, Y., Tezuka, T. et al. (1996). *Super Mario 64*. Kyoto, Japan: Nintendo.
Morris, J. F., Capps, M. V., Polge, S. et al. (2007). *Unreal Tournament 3*. Cary, USA: Epic Games.
Nacke, L., & Lindley, C. A. (2008). Flow and immersion in first-person shooters: Measuring the player's gameplay experience. In *Proceedings of the 2008 Conference on Future Play: Research, Play, Share* (pp. 81–88). ACM.
Newel, G. et al. (1998). *Half Life*. Bellevue, USA: Valve.
Nguyen, T. (2016). 3D audio is back, and VR needs it. *PC Gamer*. Bath, UK: Future Plc. http://www.pcgamer.com/3d-audio-is-back-and-vr-needs-it/.
Nishiya, S., Ikeda, A., Outouge, S. et al. (2004). *Mario Party 6*. Tokyo, Japan: Hudson Soft.
Nishizawa, M., Kobayashi, Y., Fujisawa, T. et al. (2003). *Life Line*. Japan: Sony Computer Entertainment Inc.
Ogunfunmi, T., & Narasimha, M. (2010). *Principles of speech coding*. Boca Raton: CRC Press.
Orozco, M., Silva, J., El Saddik, A., & Petriu, E. (2012). The role of haptics in games. In *Haptics Rendering and Applications*. Rijeka: InTech.
Pettus, S., Munoz, D., Barroso, I. & Williams, K. (2013). *Service Games: The Rise and Fall of Sega*.
Pinchbeck, D. (2013). *Doom: Scarydarkfast*. Ann Arbor, USA: University of Michigan Press.
Pinola, M. (2011). Speech recognition through the decades: How we ended up with Siri. *PC World Magazine*. http://www.pcworld.com/article/243060.
Premaratne, P. (2014). Historical development of hand gesture recognition. In *human computer interaction using hand gestures* (pp. 5–29). Singapore: Springer.
Pritchard, S., Dyke-Wells, T., Crooks, B., McEwan, N, et al. (2009). *The House of the Dead: Overkill*. London, UK: Headstrong Games.
Pullen, J. P. (2017). Star Trek fans will love Amazon's newest Echo feature. *Time Magazine*. http://time.com/4645187/amazon-echo-star-trek-computer-voice/.
Raynal, F., Bonnell, B., Girolami, F. et al. (1992). *Alone in the Dark*. New York, USA: Infogrames.
Roddenberry, G. (1966). *Star Trek: The Next Generation*. Santa Monica, USA: CBS Television Distribution.

Romero, J., Petersen, S., McGee, A., Green, S., Hall, T. et al. (1993). *Doom*. Richardson, USA: Id Software.
Romero, J., Petersen, S., McGee, A., Carmack, J. et al. (1996). *Quake*. Richardson, USA: Id Software.
Rotberg, E., Rubin, O., & Hector, R. (1980). *Battlezone*. New York, USA: Atari.
Saito, Y., Hirose, Y., Ito, S. et al. (1999). *Seaman*. Japan: Vivarium.
Sawano, K., Takashi, S., Kami, H., Satsukawa, T. et al. (1995). *Time Crisis*. Japan: Namco.
Self, D. (2009). *Audio engineering explained*. Burlington, USA: Focal Press.
Sigmund, M. (2003). *Voice recognition by computer*. Tectum Verlag DE.
Toyama, K., Kitao, G., Yamaoka, A. et al. (1999). *Silent Hill*. Japan: Konami.
Van der Meijden, O. A. J., & Schijven, M. P. (2009). The value of haptic feedback in conventional and robot-assisted minimal invasive surgery and virtual reality training: a current review. *Surgical Endoscopy, 23*(6), 1180–1190.
van Peer, W., & Chatman, S. B. (Eds.). (2001). *New perspectives on narrative perspective*. Albany, USA: SUNY Press.
Watts, M. (2013). Taking a look back at the Nintendo 64 Rumble Pak. *Nintendo Life*. http://www.nintendolife.com/news/2013/07/feature_taking_a_look_back_at_the_nintendo_64_rumble_pak.
Weinberger, M. (2015). The downfall of Kinect: Why Microsoft gave up on its most promising product. *Business Insider UK*. http://uk.businessinsider.com/why-microsoft-xbox-kinect-didnt-take-off-2015-9.
Wolf, M. J. (2012). Battlezone and the Origins of First-Person Shooting Games. In G. A. Voorhees, J. Call, & K. Whitlock (Eds.), *Guns, grenades, and grunts: First-person shooter games* (pp.25–40). New York, USA: Bloomsbury Publishing.
Wolf, M. J., & Perron, B. (Eds.). (2014). *The Routledge Companion to Video Game Studies*. New York, USA: Routledge.
Wright, W., Martin, J., Russo, M., Casey, K. R. et al. (2000). *The Sims*. USA: Maxis.
Yokoi, G., Kiyotake, H., Tanaka, H., et al. (1984). *Duck Hunt*. Kyoto, Japan: Nintendo.
Zimmerman, T. G., & Lanier, J. Z. (1991). *U.S. Patent No. 4,988,981*. Washington, DC: U.S. Patent and Trademark Office.

8

Current Status

After a long road that stretches far into the past, we now reach the present state of contemporary VR. We start, however, with a look to the future. There is certainly no shortage of speculation and commentary regarding the anticipated future of VR. In a recent white paper published by the Internet Advertising Bureau UK,[1] it is asserted that the near future of VR will primarily feature its widespread commercial uptake across multiple industries. The technology will facilitate efficient prototyping, virtual product testing, VR-based marketing and skills training to name a few. Increasing normalisation of the technology (see Chap. 2) and greater brand focus on VR content for marketing are equally likely near-futures. Ilif (2016) posits that fashion and retail will likely have a significant influence on VR, with a priority on personalised 'designer' hardware. This point was raised in Chap. 6, in which aesthetic individualisation of headphones was shown to be very likely to reappear in future HMDs.

Dating back to August 2013 (and still accessible at time of writing), the Guardian newspaper has amassed a substantial archive of VR-related articles, with notable examples including a VR penalty shootout simulator to help England transcend 'quarter-finals purgatory' (Riach 2013)

T.A. Garner, *Echoes of Other Worlds: Sound In Virtual Reality*,
Palgrave Studies in Sound, DOI 10.1007/978-3-319-65708-0_8

and a recent VR-based cinema exhibit in which users are immersed in a refugee crisis as they embody immigrants attempting to cross the Mexico-US border (see Bradshaw 2017). A review of these articles highlights several consistent trends in VR over the last few years that point to likely continuing trajectories for VR in future. These include the use of VR to experience physical world locations and scenarios with the intention of deepening understanding, efforts made by VR companies to move away from hype and deliver immediate and practical applications with concrete value, a push towards negating persistent VR side-effects so that the hardware can be used throughout the day, and decentralising interaction in VR applications to encourage more holistic improvement of the experience.

Some of these themes appear throughout this chapter as we explore some of the cutting-edge positions in VR and VR sound across both industry and research, with the occasional bit of speculating on likely future trajectories. Various strands of VR technology and design appear to be gradually converging. Therefore, this chapter begins with a more general proposition on the integrative trajectory of VR, with a discussion regarding the role of sound in this probable future direction. Subsequently, we review contemporary audio hardware as it pertains to VR, including both input (microphones, arrays) and output (headphones, loudspeakers) technologies, before examining the wide range of VR sound design and implementation tools that are currently available, from source development kits and plug-ins to game engines and middleware. This chapter closes with a look at the aspects of contemporary sound design that are most relevant to VR, before speculating on how such techniques will advance in future.

The Integrated Future of VR

Whilst future advances in VR will undoubtedly be broad and varied, one principle distinguishes itself as an overarching theme of development that extends beyond VR to describe how all human technology is coming together. This section explores this principle of integration as it pertains to our increasing connectivity, with objects through the

Internet of Things and with people by way of social media and collaborative virtual environments.

Integrating Things

The meaning of 'technological singularity' arguably extends beyond the transhumanist vision, of uploading a digitised brain and the dawn of superintelligence, to something altogether different and appearing much sooner over the horizon. The Internet of Things (IoT) describes a network technology that enables various everyday electronic devices to communicate. IoT is built upon a paradigm that 'promises to make "things", including consumer electronic devices or home appliances [...] part of the internet environment [and] enables the realisation of smart cities, infrastructures and services for enhancing the quality of life' (Buyya and Dastjerdi 2016, p. xxi). IoT features are certainly varied, with seemingly limitless functionality that can emerge from combining any number of 'smart' devices. Very soon we could all be 'telling our barista coffee machine at home to prepare us a chocolate fudge Café Cubano from five miles away, using a bespoke interface in our car as we're driving home' (Garner 2017a).

The principles of IoT have two significant (and related) ramifications for VR, one practical and the other more conceptual. In terms of the practical, VR technology stands poised to become the central informational hub and direct human interface with IoT systems. Several years prior to consumer VR's re-emergence, mixed reality systems were already being touted as logical means of interfacing with IoT (Macias et al. 2011). More recent research often points to augmented reality (AR) as a powerful partner to IoT, displaying the information being output by smart devices as an easily digestible visual overlay (Lappänen et al. 2014). Furthermore, research has now begun to think big, with experimental systems integrating mixed reality interfaces and IoT with geographical data as means of creating entire 'smart-cities', enhanced with digital content. This kind of research aims to bring the virtual and the physical together in a kind of 'trans-reality', a mixed reality ecosystem in which large numbers of virtual objects, physical objects and human users all interact with one another (Martin and Laviola 2016).

The second, more conceptual implication of IoT for VR relates to integration as an underlying principle of technology. As noted later in this chapter, making VR tools that are interoperable (from game engines and source development kits to third-party plug-ins and content design software) is rapidly becoming a significant trend. These tools are diversifying in terms of functionality, creating a modular production pipeline where VR developers and designers need to use a variety of tools to create content, with each individual tool providing specialist functions. Such diversification may present advantages for larger development teams, who can designate specialists to each 'module' within the pipeline, but it also raises crucial issues if compatibility and communication between modules are poor. Solving such issues with integration can help developers to quickly and easily transition content from one platform to another, or utilise multiple toolsets within a single interface for efficient production. The value of such integration is clearly being recognised already by manufacturers of VR tools, but with the principle becoming ever more pronounced in the fundamentals of technology, this trend is very likely to gather momentum throughout the coming years.

As Greengard (2015) notes, a comprehensive integration of technology by way of IoT is an exceptionally complex task. It requires robust standardisation of data transfer and communications, backwards compatibility with legacy hardware/software and, critically, effective privacy and security to protect users. For VR, this implies that various wider technological issues need to be addressed before such integration, currently seen in experimental research, filters through to the consumer market. In the immediate future, therefore, we are unlikely to be able to sit in our living rooms, with a digital representation of our fridge in front of us (with its entire contents on display), so that we can order our groceries online without setting foot in our kitchen for a stock check. That said, although such a scenario is unlikely to be mainstream and widely available in the next couple of years, it is most certainly on the horizon. With regards to sound in this trajectory of VR's evolution, three key matters outline how sound design for IoT applications should progress. These are enhancement (the potential of sound to improve the user experience of integrated IoT/mixed reality applications), problem mitigation (countering adverse auditory effects) and accessibility (how sound content can provide individuals who cannot rely on visual feedback the opportunity to engage with this developing technology).

Matters of enhancement and accessibility relate this discussion to part of this chapter and auditory perception research into sonification systems, the underlying principles of which largely concern ways in which sound can present a listener with information that is multilayered and temporally relevant. Whereas visuals can provide explicit and focussed information, sound favours that which is broader and more implicit (the wood as opposed to the trees). A novel example of research in this area is presented by Lockton and colleagues (2014) and their development of *Powerchord*, a sonification system that displays electricity consumption data as birdsong. As more household objects are turned on, the birdsong becomes louder and more intense. This gives the listener clear auditory feedback regarding their total power usage, but additionally encourages an affective response that promotes conservation of energy. In another example, Chernyshov and colleagues (2016) highlight how sound can enhance IoT by presenting continuous information to complement discrete visual information. Their system allows users to select their own sounds to represent each networked smart device, then combines the sounds into an ambient soundscape, the characteristics of which sonify the devices' various ongoing processes and variables (wireless signal strength, battery power, CPU utilisation, etc.). Chernyshow and colleagues also note that people can extract useful information from complex and subtle sound even when engaged in cognitively demanding tasks. Therefore, sonification offers enhancement in IoT by facilitating multitasking and more efficient interaction with the technology. In a review of IoT technology for disabled individuals, Domingo (2012) observes how AR interfaces that utilise sonification systems can enable a greater number of people to access IoT technology, such as the blind or partially sighted who cannot rely upon visuals.

Mitigation of adverse auditory effects connects IoT to sound studies by way of acoustic ecology, the fundamentals of which we also discussed in Chap. 3. As Williamson (2017) observes, a lacking in foresight with regards to acoustic ecology in IoT could lead to our homes being filled with a cacophony of smart objects desperately vying for our attention, a next-generation loudness war. This has particular relevance to VR as a hub for IoT. In this future scenario, a single VR/AR interface would give the user access to all IoT-related content, and therefore, audio would not be emitted directly from the smart objects themselves but from a single pair of headphones or loudspeaker. Here, methods supporting positional audio would be in a much stronger position to present listener with multiple streams of incoming sound, with the listener able to differentiate and attune to individual streams because of the spatialisation of sound objects. A single loudspeaker (as is currently the arrangement with personal assistant IoT hubs), however, would severely limit this ability, and IoT sound information would in this instance need to be significantly more linear. If not, this presents an even greater risk of sound being intrusive, difficult to interpret and damaging to the overall experience. The effect of this for the user would almost certainly be undesirable, an auditory equivalent of the AR assault on the senses portrayed in *Hyper-Reality* (Matsuda 2016).

The above points reinforce the need to better understand acoustic ecology in the context of VR-mediated IoT, but also acknowledge its complexity. At present, very little research exists regarding auditory perception within IoT, with the primary interest in IoT sound concerning the use of sound waves to transmit encoded information between smart objects (commonly referred to as 'the Internet of Sound', see Bergel 2015). However, it is anticipated that the next few years will likely see greater investment in IoT sound research and development, specifically concerning how sound can be utilised effectively in ways that address some of the above issues. If not, IoT interfaces are likely to be condemned to silence, with users persistently reaching for the mute button (Williamson 2017).

Integrating People

The pivotal social aspect of VR is profoundly highlighted when we consider why a social media platform like Facebook would invest $2 billion in acquiring Oculus. At the time of acquisition, Dredge (2014) observed how the prominent explanation for this move by Facebook was that it could be attributed to CEO Mark Zuckerberg's youthful susceptibility to novelty. This notion was quickly dismissed, however, with Facebook revealing that their strategy was to obtain control over a technology that had significant potential to be the next major platform for social media (see Parkin 2014). Furthermore, in the IAB white paper (see Note 1) mentioned briefly in the opener of this chapter, a key assertion raised is that events in which large numbers of people gather together 'could be the key to [unlocking] mass market VR'. The argument here is that, with physical entry to popular events (music festivals, live theatrical performances, sporting events, etc.) becoming increasingly difficult and expensive due to capacity constraints, VR has the potential to provide virtual entrance to an almost limitless number of attendees. If the above assertions are correct, VR now seems poised to be the central hub of our digitally augmented social lives.

Such visions for VR are not unexpected, with analogous ideas appearing frequently in fictional representations such as *Snow Crash* and *Neuromancer* (see Chap. 5), in which VR meets with the global computer network, enabling users across the world to interact irrespective of physical distance within a shared virtual space. Such projections draw clear associations with the fundamental VR desire for assimilation, the result of which has stimulated resistance from the counter-desire of conservatism. Social media in particular has faced significant conservative opposition from detractors who assert that such virtual communications reduce face-to-face interactions, thereby devaluing social interaction in general (see Drago 2015).

With regards to relevant research, collaborative virtual environments (CVEs) possess both a relevance to VR-mediated social integration and a history that far predates contemporary VR systems. Broadly akin to multiplayer functionality in digital games, Churchill and colleagues (2012) describe a CVE as 'a computer-based, distributed, virtual space [in which] people can meet and interact with others, with agents or with virtual objects' (p. 4). They elaborate, noting that CVEs are inclusive of mobile technology, wearable devices (e.g. biometrics) and various forms of graphical spaces (3D, 2D, 2.5D). On the surface, there is overlap between this explanation of CVEs and the principles of integration in IoT. The key difference, however, is that whilst IoT is primarily concerned with our integration with objects, CVEs prioritise our integration with other people. Massively multiplayer online video games such as *Second Life* (Rosedale 2003—see Chap. 9), where the most significant game mechanics centre upon person to person interactions, represent some of the best illustrations of the CVE concept from that particular form of media. As Churchill and colleagues (2012) assert, the virtual space in question does not have to be a fully realised 3D world, making Facebook and all forms of social media, established instances of CVEs. Collaborative function is undoubtedly a key aspect of cutting-edge VR games, with one of its key benefits being that it can recreate the experience of local multiplayer gameplay (several players in the same physical space) with networked players potentially thousands of miles apart from one another (Porter 2016). The current value and future potential of multiple-user VR is significant and extends beyond games applications to enhance and facilitate numerous commercial applications. Consequently, it is also highly probable that developments in this area of VR shall be an important and continuing trend within future VR.

Whilst discussing binaural processing, Kohlrausch and colleagues (2013) present a quotation from Helen Keller that traces back to Kantian philosophy: 'Blindness separates us from things but deafness from people'. This statement illustrates the role sound plays in social communication and relationships. With regards to VR, it can also be interpreted to suggest that any social function of VR would be best-served by way of sound. Sonification of big data sets, such as those that are generated by social media, presents interesting ways in which sound can facilitate technology-mediated social interaction. For instance, Hermann and colleagues (2012) present *tweetscapes*, a system that sonifies message streams from Twitter to give listeners sound-mediated insight into how groups of people are behaving, both in general and in response to a given event. Sound is also a significant component of CVEs, particularly in terms of real-time interpersonal conversation within a virtual space (see Benford et al. 1995; Waters et al. 1997). However, the application of sound does extend beyond the capture and transmission of actual speech to include insights regarding how a single virtual soundscape can present a unique listening experience to every user within the CVE. This returns us to concepts presented in Chaps. 3 and 4, specifically those concerning virtual acoustic ecology.

As a noisy virtual object moves away from one individual and towards another within the same virtual space, it stands to reason that the sound of the object should be attenuated for the first person but amplified for the second. This effect is referred to as 'tele-diegesis'. Processing sound in this way supports suspension of disbelief, encouraging users to feel as though they are present within the virtual space and that their virtual companions are only a few feet in front of them and not half way around the world.

Technology is, in a way, bringing us closer together but through an increasingly virtual lens. One interesting observation regarding the above technologies concerns how our interaction with objects is becoming increasingly similar to our interaction with people. This resonates heavily with the opposing desires of assimilation and conservatism discussed in Chap. 5. Greater integration is obscuring the traditional lines that once separated technology from nature, virtual worlds from physical worlds and people from objects. The ethics of this issue fall outside the scope of this chapter and are not something that will be addressed here. Instead, the following section explores the cutting edge and likely future directions of technologies and design approaches that pertain specifically to VR sound, revealing how consistent and embedded the underlying theme of integration is.

Audio Hardware

In 2009, Palmer Luckey took the decision to begin development of a VR prototype from the humble foundations of his parents' garage. Three years on and Luckey makes an acquaintance of *Doom* co-creator John Carmack, who promptly ports a build of *Doom 3* (Willitis et al. 2004) to Luckey's prototype and demonstrates the creation at the 2012 Electronic Entertainment Expo (E3). In the same year, Luckey founds the company Oculus VR and begins a crowdfunding campaign for the first *Rift* development kit (DK1). Across 2012 and 2013, media attention circulating the Rift increases exponentially and Oculus raises over $75 million for further prototype development. Then in the spring of 2014, Oculus makes the announcement that Facebook is to acquire the company, giving the Rift a massive increase in financial backing in addition to mainstream approval by way of its endorsement by a technology giant (Nelson 2014). Whilst other VR developments outside of Oculus occurred during this time, it is the story of the Rift that is credited as the first and most significant in contemporary VR, representing the beginning of the technology's second push to become mainstream and ubiquitous after stalling in the 1990s (Branda 2015). Various books already exist that continue this story to explore contemporary VR technology from a more general standpoint (Jerald 2015; Parisi 2015). Consequently, this section focusses primarily on cutting-edge VR technology as it pertains to sound. We begin with a review of the various VR systems currently available to the consumer market, looking specifically at how sound is treated in their overall designs, before progressing to examine new audio hardware technologies built specifically for VR applications. Subsequently, this section explores audio software, including the various sound engines and middleware solutions, before closing with a review of design techniques at the cutting edge of VR sound.

Outputs

At the time of writing, tethered (high end/PC-based) HMDs that include the HTC *Vive* and Oculus *Rift* are available to the general market and represent the cutting edge in consumer VR hardware. Offering greater accessibility and affordability in exchange for lower fidelity, mobile technology also accounts for the majority of well-known and widely available consumer VR, with notable systems including the Samsung *Gear VR* and Google's *Cardboard* platform and *Daydream* HMD. Finally, Sony's console-based solution, the *PlayStation VR* (PSVR), completes the mainstream VR line up, positioning itself between mobile and PC-based systems in terms of fidelity. In terms of hardware, consideration for sound across these platforms arguably highlights its continuing under-appreciation, as graphical output and kinaesthetic input enjoy substantial prioritisation. HTC may claim that 'audio is essential to bringing an immersive VR experience to life',[2] but their hardware doesn't quite reinforce this sentiment. The original

2015 Vive system did not include any audio hardware beyond a single mini-jack stereo output, requiring users to provide their own headphones. Integrated headphones were announced mid-2017 but even then, only as part of the *Deluxe Audio Strap*, a separate accessory for the headset that was primarily marketed as a more comfortable and flexible head support. PSVR reveals a similar attitude towards sound, with no bespoke hardware developed for the system, again requiring users to provide their own. Fortunately, however, and also like the Vive, this relatively poor treatment of sound is only in terms of audio hardware and not software, which we shall return to later in this chapter.

Mobile VR platforms echo this limited attention to audio hardware, with almost all HMDs requiring the user to either connect their own headphones to the smartphone stereo output or use the integrated speakers of the phone itself (several models of mobile HMDs obstruct the headphone output, limiting users to the latter option). That being said, independent start-up companies have sought to address this issue with the notable example being *Opto VR*,[3] a crowdfunded venture that presented a HMD for mobile VR with integrated stereo speakers and marketed this as the primary selling point of the product. As He (2016) rightly states, most headphones and earphones are typically optimised for listening to music. A VR experience, however, demands very different things from the hardware, such as noise cancellation that isolates the virtual soundscape to a flat frequency response that enhances 3D positional audio.

At present, Oculus arguably stands as the primary innovator of integrated audio hardware across the mainstream VR platforms. The 2016 consumer release of the Rift HMD featured integrated stereo speakers specifically designed and constructed to facilitate the *RealSpace 3D Audio*[4] sound engine (discussed later in this chapter). As James (2015) observes, the relationship between audio hardware and software in the development of the Rift was realised by way of a carefully constructed 'spectral neutrality' in the headphone design, in which the levels across the frequency spectrum are balanced to produce a crisp and transparent sound to accentuate the positional audio techniques instigated by the software. In addition, the connections between the headphone speakers and the audio source are proprietary (as opposed to the generic 3.5 mm mini-jack connection) and are optimised for conveying positional audio signals. These efforts made by Oculus have led to the device being praised specifically for its audio quality (Leadbetter 2016) and it is one of the only current HMDs to outwardly acknowledge the value of sound in VR at a hardware level.

Whilst integrated solutions such as Oculus' are rather uncommon, third-party headphones developed specifically for VR have started to make their presence known. Audeze's *iSINE VR*[5] headphones currently represent the high end of the market. Their design features clearly reflect a number of key requirements of optimal VR audio, including a relatively low weight (compared against sets using the same driver technology) for increased comfortability and planar magnetic driver technology[6] that produces a more precise sound as opposed to the more commonplace dynamic driver. The headset also boasts a frequency response range of 10 Hz–50 kHz which, although extending far beyond the standard human hearing range (20 Hz–20 kHz), has been associated with increased signal fidelity, though it should be noted that this claim has been disputed (see Attias 2013).

Another device worthy of mention here is Samsung's experimental *Entrim 4D* audio headset. The Entrim 4D is based on galvanic vestibular stimulation (GVS), a decidedly novel technology in consumer terms, but less so when considering the forerunning research that came before it. GVS describes a process of activating fibres of the vestibular nerve with a very minute electrical current (Cohen et al. 2011). Dating back to the work of its namesake, Luigi Galvani (1791), GVS stimulation encourages the human body to sway roughly in the direction of the stimulus (see Britton et al. 1993). Testing of GVS and exploration of its possible applications have featured in several academic research programmes throughout the later twentieth century, from being used to reduce right-hemisphere visuospatial neglect (difficulty in noting objects on your left side) in post-stroke patients (Rorsman et al. 1999), to creating a 'sensation interface' for enhancing musical experience by synchronising the GVS stimulation to tempo (Maeda et al. 2005). Back in the consumer domain, the Entrim 4D exploits GVS sensations for the purpose of enhancing user presence. Its design integrates GVS with the audio output by way of two stimulators built into the left and right sides of the headphones. When playing a racing simulator, for instance, as the player turns a right corner at speed, the corresponding gravitational force that would push them to the left is simulated by GVS from the left headphone. At present, the Entrim 4D remains an experimental device. It received a great deal of media attention in 2016 (Cragg 2016; Popper 2016) and was touted as a forthcoming feature of Samsung's *Gear VR 2*, but this never transpired and there is little recent information online regarding its progress or whether it will see a consumer release.

Regarding the most probable future of VR audio outputs, it would appear likely that progress in this area will focus upon refining and optimising the hardware specifically for VR experience. For example, efforts in minimising signal distortion and maximising frequency balance for the purpose of enhancing positional sound processing are highly likely. Beyond this, integration reveals itself once again as a trend in contemporary HMDs that appear to be moving increasingly towards integrated headphones. Experimental systems such as the Entrim 4D also point towards a prioritisation of integration, suggesting that future systems will likely bring together multiple technologies into HMD headphones. Potential examples of this could include biometric sensors (skin temperature and electroencephalography both good candidates for this) or even force feedback.

Inputs

In addition to audio output hardware, mainstream contemporary VR systems consistently incorporate some form of microphone to enable audio capture. The Vive, Rift and PSVR all include integrated microphones of comparable quality that are placed in similar locations within the HMD (towards the centre and near the nose of the headset). Voice command is the primary functionality of these microphones and this is reflected in their relatively basic design and quality. Audio capture is also prevalent in 360° video cameras and mobile VR, particularly as a hardware feature that enables users to create their own content.

Alongside integrated proprietary microphones, the industry has of late revealed several third-party devices that demonstrate how VR is decidedly reinvigorating consumer interest in binaural and ambisonic sound design. We return to the signal processing aspect of these techniques later in this chapter but, continuing the focus on hardware, both binaural and ambisonic principles have inspired a range of contemporary microphone designs with VR applications in mind. Boyce (2016), for example, outlines Sennheiser's *AMBEO* system, a cluster of four omnidirectional cardioid microphones that facilitate ambisonic recording. As Boyce notes, the reasoning behind production of such microphones has been to provide an auditory equivalent (and compliment) to the 360° video camera, with ambisonics (in very basic terms) creating a spherical representation of the soundscape that can be rotated across three axes of orientation. The resultant effect being that, just as 360° video capture enables the user to explore the visual scene by way of head rotation, ambisonic sound facilitates the same degree of interactivity, but with the soundscape.

Like ambisonics, binaural recording techniques also have an established history within research and niche consumer groups, but have only recently shown signs of becoming mainstream as a result of their great potential for VR. One notable example of this is 3DIO's *Free Space* and *Omni* series' of VR microphones.[7] Like most typical binaural microphone setups (see Rumsey and McCormick 2012), the Free Space system employs a pair of omnidirectional microphones (intended to pick up sound waves equally from all directions), embedded within prosthetic pinnae and ear canals so that the recorded audio signal is pretreated with head-related transfer function effects (again, see below). This gives sound objects within the captured soundscape a vertical position in addition to the horizontal and proximal axes (standard surround sound). 3DIO combines this binaural design with ambisonics principles in their Omni series. Here, the four-microphone array featured in the AMBEO system is doubled to present a four-by-two arrangement, four pairs of left/right ear microphones in an adjacent (north/south/east/west) formation. Both ambisonic and binaural sound production at the hardware level are intended primarily for fixed-point or 'on-the-rail' VR experiences such as 360° video, in which position and locomotion of the user are not interactive. However, this does not mean that such techniques are not applicable to other aspects of VR, as both have significant potential in such applications, but from more of a software foundation that we shall return to shortly.

For the purposes of capturing convincing three-dimensional soundscapes for computer-generated virtual worlds, it is unlikely that many novel and significant leaps forward will occur in the near future, as the technology is already functioning very well for its intended purpose. Consequently, ambisonic and binaural microphones are unlikely to stray from their current designs and development will instead focus upon general technology refinements (weight reduction, signal quality, affordability, etc.). That said, audio input technology is likely to see some notable changes, once again in terms of integration. At present, even professional models of 360° cameras continue to incorporate arrays of basic monophonic microphones that can only record surround sound. High-end products such as the *GoPro Odyssey*[8] continue to ship with an external microphone for enhanced sound capture, whilst the current industry-leading *Eye*[9] by 360 Designs is intended for capturing only video within the rig. This separation of audio and video capture, or low-quality audio within integrated systems, points to clear issues for future developments to address, and it is anticipated that future 360° VR cameras, inclusive of professional and consumer versions, shall steadily integrate higher quality positional audio recording technology into their core designs.

Audio Software

Whilst the hardware design and marketing of contemporary HMDs largely continue to deprioritise sound in VR, developments in audio processing software have shown subtle but significant progress. This section outlines the diverse range of tools currently available to sound designers for VR applications. The intention here is not to provide an in-depth analysis of these tools, as there is already a great deal of such information readily available online that can remain more up to date with the fast-moving changes in this area. Instead, this section identifies the key audio software systems for VR and broadly describes how they relate to one another and can be used collectively. Following on from this, we look to elucidate the themes, ambitions and techniques that constitute contemporary VR sound design.

A Wide Range of Options

For sound designers, there is now a relatively diverse range of options for creating and implementing audio for VR that includes audio source development kits (SDKs) produced by VR firms, third-party plug-ins that can work within game engines or be implemented within code, and middleware solutions that can create sound objects and environments separately before importing them into a game engine. The underlying design intention, consistent across all the above options, is for greater user presence by way of increased 'realism' which, in terms of these tools, refers to the perceived accuracy, fidelity and detail of the soundscape as it pertains the physics of the virtual environment. This intention is revealed when we consider the features of these tools, which are almost exclusively concerned with more accurately and efficiently modelling acoustic phenomena. Figure 8.1 broadly outlines the production pipeline of contemporary VR sound design, highlighting the various types of software that can be used and their relationship with one another and the final VR product.

In terms of SDKs that facilitate VR sound and are produced by firms within the VR industry itself (identified from this point as 'VR-SDKs'), Oculus once again positions themselves as industry pioneers in terms of VR audio technology with their *Audio SDK*. As Oculus' online documentation[10] describes, the Audio SDK bundles together various audio tools, plug-ins and libraries, also providing access to the source code, to give developers the opportunity to customise and build upon the given toolset. The features of the Oculus Audio SDK are highly integrative with regard to other commonly used VR-relevant design tools. For example, its plug-ins are compatible with popular game engines (such as *Unreal* and *Unity*), audio middleware (e.g. *FMOD*, *Wwise*) and desktop audio workstations (DAWs: e.g. *ProTools*, *Nuendo*, *Logic*). With software distributors Steam recently acquiring Impulsonic's third-party system *Phonon 3D* to create *Steam Audio*[11] (supporting the Vive HMD), Oculus is no longer the only VR-SDK publisher in town and whilst these systems highlight the industry's appreciation for sound at a software level in desktop-based (tethered) VR, the *Google VR*[12] (GVR)

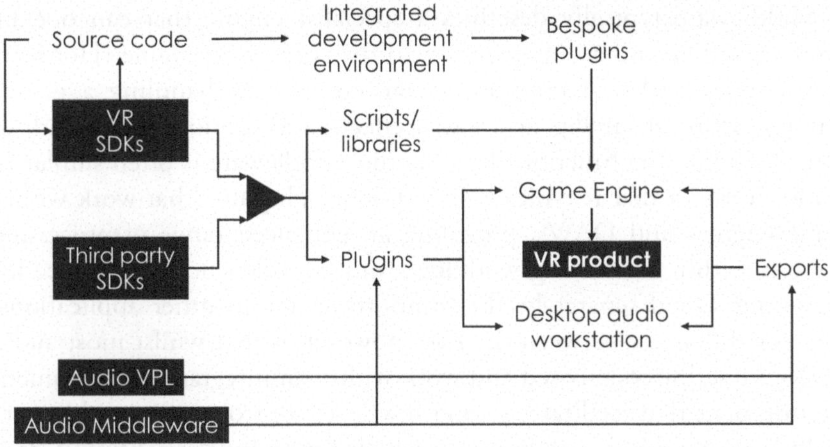

Fig. 8.1 Outline of various software tools for VR sound

SDK demonstrates an equivalent acknowledgement of sound in mobile VR. The GVR incorporates specialist audio tools within the wider SDK, with features that are mostly comparable with the tethered system SDKs. As with the Oculus and Steam systems, GVR plug-ins and libraries also work within game engines and middleware.

Outside of audio processing tools within the VR industry, numerous third-party plug-in developers constitute a large proportion of the overall VR sound market. As with VR-SDKs, third-party options typically bundle together various tools that operate as plug-ins within host applications. Devana (2015) outlines several current solutions of this type, including Two Big Ears' *3Dception*,[13] Astound Holdings' *AstoundSound*[14] and VisiSonics' *RealSpace 3D*.[15] These plug-ins commonly differ from the VR-SDKs in terms of depth of user-control, with the majority being closed systems, not providing access to the source code for low-level feature editing. Where they are notably similar, however, is in their integration. As with the VR-SDKs, most of the available third-party plug-ins work within DAWs, middleware and game engines, giving designers flexibility with regards to how they incorporate the plug-ins into their existing workflows.

Middleware typically describes a specialist engine that can operate both separately and as an extension to the core game engine (Horowitz and Looney 2014). Common operations include handling a game's physics, artificial intelligence, visual effects, full motion videos and, of course, audio. The functionality of audio middleware is often similar to audio SDKs in that it typically incorporates plug-ins (that work within game engines and DAWs, providing an enhanced range of processing options within an existing workflow) and an independent interface for designing sound separately, then importing it into other applications. The key difference between the two, however, is that whilst most audio SDKs are script/code-based and work within an integrated development environment that facilitates greater low-level control, audio middleware includes a graphical user interface which limits functionality but does not require coding skill to operate. This key difference highlights the primary intended appeal of the two alternative solutions, with SDKs typically preferable to programmers and middleware to designers.

One further type of software diverts significantly from those above, by presenting a markedly different aspect of sound design. Visual programming language (VPL) audio software typically focusses less upon signal processing of recorded audio samples and more upon generative audio that is produced synthetically from digital information. *Max*[16] by Cycling '74 and its open source counterpart *Pure Data*[17] have for many years represented VPL audio solutions for procedural generation of audio content. Although a quite different facet of sound design, VPL software also reinforces the theme of integration. Whilst currently requiring some technical know-how, both Max and Pure Data can be integrated into various DAWs and game engines. This opens up a huge range of design opportunities, with essentially every facet of the soundscape (as produced in the VPL) highly responsive to the any element of the virtual world. It is likely that this functionality shall increase and become more accessible to non-programmers in over the next few years. We shall return to the specifics of generative audio for VR later in this chapter.

It is unlikely that the diversification of audio solutions will change in the near future. As the processing techniques become more complex, greater specialist knowledge and skill are required to develop and implement them, thereby discouraging a single firm from attempting to produce a comprehensive VR sound solution. That said, VR sound design techniques do not contradict the integration theme. Whilst the existence of multiple specialist businesses is a likely continuation in future, the demand for increasing interoperability will dictate how the range of systems available must function in simpatico with one another. As was noted earlier in this section, the emerging theme of multiple businesses collaborating to produce DAWs, game engines, SDKs, middleware and third-party plug-ins that work together is almost certainly a continuing trend. Ultimately, it is likely that all possible functionality shall be made available as a customisable and modular system, all accessible from within a single interface. As a final note, one intriguing possibility returns us to the notion of VR as the central point of interaction with all digital content. At present, VR tools are available that enable users to create art (e.g. *Tiltbrush*[18]) and compose music (e.g. *LyraVR*[19]) from within a virtual environment. Both Unity (Takahashi 2016) and Unreal (Robertson 2016) are in the process of finalising 'in-VR' game development tools, with which users will soon be able to indulge in 'meta-virtuality', creating virtual environments from within virtual environments. These kinds of developments point to a possible future in which design and development across many (if not all) industries are conducted within a virtual space. With game design that can be done within VR itself now on the horizon, crafting and implementing soundscapes whilst standing in the same virtual space you are building for are a highly likely future scenario.

To summarise, SDKs, middleware and audio VPL solutions collectively offer sound designers a diverse range of tools with which to create realistic, dynamic and responsive sound for VR. Their alternative interface methods (i.e. text-based programming interface or graphics-based visual interface) cater to individuals from both compositional (design) and programming (development) backgrounds, and their increasingly integrative nature strongly reflects a central trend in VR technology. Following on from this, the next section goes into more depth regarding the specific features and functions across this range of software, to present a concise account of what VR sound can do at this precise moment, and what it may be able to do in future.

The Cutting Edge and Beyond in VR Sound

As posited by Oculus in their developer's documentation (see Note 10), there has been little need for consumers to demand anything more than basic surround sound in their media applications. This is because home cinema, television and traditional digital games continue to have their audio outputted by way of loudspeakers, either in stereo or 5.1/7.1 surround and so full three-dimensional processing would be lost at the point of output. VR changes this landscape significantly. With the functionality of head tracking coupled with the preference for headphones over loudspeakers and the need of the user to be able to orient their physical bodies within three-dimensional virtual space, VR has brought a range of sound design techniques back from obscurity and launched a renewed interest in developing these techniques further.

As one might expect, there is a great deal of overlap between sound design in VR and traditional digital games, particularly those utilising a first-person perspective. In the past few years, several excellent books have been published that specifically detail the principles and techniques of contemporary game audio programming (see Gouveia 2013; Marks 2012; Somberg 2016; Stevens and Raybould 2013). As such, an in-depth analysis of this area would certainly be reinventing the wheel. This section therefore serves as a brief overview of the key concepts that have the most relevance to VR. Namely, those that pertain to sound localisation, procedural generation and the modelling of physical environments as means to create more immersive and responsive virtual soundscapes that evoke deeper levels of presence for the user.

Ambisonics

The history of ambisonics dates back to the 1970s and the works of Peter Fellgett (1975) and Michael Gerzon (1975). The general principle of the term is that captured sound wave information can be positioned within a virtual sphere, by way of a particular configuration of microphones and a signal encoding and decoding process (Hu et al. 2012). The basic form of ambisonics is known as 'first order' or 'B format'. In simple terms, a B format recording setup consists of four microphones positioned closely together. A single omnidirectional microphone collects sound wave pressure information (the 'W-channel'), with three figure-8 microphones providing positional information along each respective axis ('X', 'Y' and 'Z-channels'). The unique polar pattern of the figure-8 microphones is such that they only detect sound waves positioned directly in front and behind the capsule. They are thereby able to capture sound wave information that indicates a source's position along a straight line. With the three figure-8 microphones oriented correctly, the combined WXYZ signal captures the source in three-dimensional space. It is the Z-channel, providing height information that distinguishes ambisonics from surround sound and, as Gerzon (1985) notes, B format can be decoded to enable playback on standard headphones and on almost any combination and configuration of loudspeakers, provided that the necessary information is incorporated into the decoding algorithm. The specifics of these processes are considerably complex and outside the scope of this book, but for a detailed explanation of the physics behind ambisonics see Ortolani (2015).

Where ambisonics stands apart from binaural recording techniques (and this is crucial for VR applications) is its responsiveness to head rotation. Unlike traditional binaural recordings, which require a fixed head position for the effect to work, B format audio positions sound objects within a virtual sphere. The tracked orientation information relative to the individual's head movement can be used to manipulate the XYZ signal in the B format recording. This enables the virtual sphere to rotate. As the user looks upward, adjusting the pitch of their head, the virtual sphere can move downward along the same orientation axis to create the illusion that the sound object has remained fixed in virtual space.

VR was identified as an exciting potential application of ambisonic recording techniques during the first generation of VR in the 1990s (Malham and Myatt 1995) and has re-emerged in more recent years to become a prominent approach to positional sound design in contemporary VR. Particular themes of ambisonics' application within VR research also show consistency with the preceding decades. The diverse applications range from helping the visually impaired construct spatial knowledge of a location before visiting it (Picinali et al. 2014) to facilitate an immersive telecommunications service (Nagata et al. 2017). Ambisonic recordings feature most prominently in 360° VR video, rather than graphically rendered virtual environments, due to the limited interactivity of a fixed soundscape (e.g. the user cannot manipulate specific sound objects, only the soundscape as a whole). This is not to say that such recordings have no use in virtual environments. The nature of an ambisonic recording does facilitate positional interaction as well as orientation, meaning a user can walk around an ambisonic sound object in addition to focussing upon it by way of their head movements. Creative sound design for virtual environments can further mitigate interactivity limitations by combining ambisonic sound with multiple other processing techniques. For instance, an ambisonic recording could be looped to create an immersive background/ambient soundscape, onto which numerous monophonic audio samples algorithmically converted into artificial positional sounds can be added to create a foreground layer with much greater interactivity (e.g. individual sound objects can be moved, distorted and muted).

Environmental Modelling

A very broad area, this section encompasses several audio processing techniques that can be grouped by their focus on producing virtual acoustic environments that mimic 'real-world' physical phenomena. Within this group, wave tracing is one notable aspect of environmental modelling that exemplifies the historical theme of first materialising in the 1990s before falling into relative obscurity, only to experience a resurgence in more recent years. Essentially the auditory cousin of graphical ray tracing (in which a graphical image is generated by tracing the path of light and simulating its effects on virtual objects), wave tracing describes the real-time processing of audio, relative to the user's virtual proximity (i.e. their avatar) to reflective surfaces (Ung 1998). At its most basic level, wave tracing can generate first-order reflections (the sound wave that reaches the ear after a single reflection) but can be extended to the Nth order, plus simultaneous calculation of reflections upon multiple surfaces, limited only by the available computing power.[20] In recent years, the term 'wave tracing' has been used less frequently, with the concept now overlapping with acoustic reflection processing and reverberation. Furthermore, it has now been absorbed into the overarching class of environmental modelling tools that also includes binaural rendering, occlusion and obstruction, global geometry reflections and environmental transitions. We take a look at these now.

Occlusion and obstruction both describe a physical effect of sound wave propagation in which some form of geometric intermediary (i.e. an 'obstacle'), located somewhere between the source and the listener, impacts upon the sound wave. Stevens and Raybould (2013) offer us one of the more comprehensive explanations of these effects and they also include a related third, exclusion. In their account, occlusion describes the obstacle muffling both the direct sound wave and its reflections (with either the entire sound wave or certain frequency components obscured), whilst obstruction denotes muffling of the direct sound only and exclusion refers to muffling of only the reflections. In practical terms, shutting a door that separates two rooms and no longer being able to hear the voice of the person in the other room (the listener would hear either nothing or the few low frequencies that penetrated the dividing wall) describes occlusion. By comparison, obstruction would be the effect of the listener leaving the door open, but moving to the other side of the room so no direct sound waves reached them (typically, some of the higher frequencies would be lost and the sound would lack clarity). Lastly, extraction would describe the effect of both listener and speaker being in the same room, but with a thick carpet and padded wallpaper, dampening the reflections so only the direct sound reached the listener (the received sound would likely be anechoic and 'deadened').

At its most basic level, virtual models of these phenomena can have a binary effect in which the direct sound wave, its reflections or both are either audible or muted. More sophisticated models, however, will instead attenuate the sound wave to varying degrees depending on the listener's precise location and the material properties of the obstacle. This can be taken a step further if we introduce multiple obstacles, or rather than attenuating the entire waveform, instead alter the intensity of specific spectral components of the sound wave. Occlusion effects, much like wave tracing and reverberation, suffer the same challenge of balancing physical accuracy and realism with computational efficiency. As a general rule, greater complexity of the virtual acoustic model bestows greater realism, but as the complexity increases so does the computational cost. Several of the VR audio packages documented earlier do not yet include occlusion effects for this exact reason, and many contemporary occlusion effects in VR are realised by way of more simple, parametric approaches, in which the occlusion values (parameters) are set by the designer rather than being computed from the virtual world geometry. There is also, however, a middle ground. Convolution reverb and global geometry reflections are two examples of artificial approaches that don't rely upon parametric values but rather simplified iterations of geometric modelling. The latter of these middle ground approaches is the more conceptually straightforward and is basically the auditory equivalent of a box collider in 3D collision modelling. Just as a box collider reduces computational expense by positioning a graphical object within a six-sided box (so that when other objects touch, there is the impression of collision without the system having to map the entire geometry of the object), global geometry reflections (often referred to as 'shoebox modelling') position the user-avatar within an invisible box, the six sides of which reflect any audio content within that defined space. In a virtual room, for instance, the shoebox would recreate the acoustic effects of the walls, floor and ceiling, but not any objects within the room. Whilst this technique is certainly efficient, it is not without problems. As anyone who has moved house recently will appreciate, there can be a significant acoustic difference between an empty and fully furnished room.

As an article by Nair (2014) elucidates, convolution is essentially the filtering of an audio sample by the spectral frequency of another. By extension, convolution reverb describes the use of an impulse response, captured from an actual location to filter a sound wave. Impulse responses are typically acquired by playing back a sine wave sweep (a single frequency tone that gradually increases and decreases in frequency over time) across the audible range of human hearing within the actual acoustic environment that is to be modelled. At the same time, a microphone records the sound waves following their acoustic interaction with the environment. The spectral composition of the recording can then be analysed against the sine wave source to accurately infer the acoustic properties of the space and produce the impulse response. Using a sine wave sweep generates information regarding how sound waves respond across the full spectral range, meaning that the impulse response can be applied to a separate audio sample of any frequency composition and, to the listener, it will appear as if that sound was captured in that same acoustic space. As Oculus themselves (see Note 10) assert, convolution reverb has great application for VR by way of accurately recreating physical spaces to create a realistic and detailed virtual soundscape. It is, however, not without limitations, with the prime issue being that impulse responses represent fixed acoustic spaces and therefore cannot simulate dynamic virtual geometry.

In terms of the conflict between accuracy/detail and computational resources, middle ground techniques such as convolution reverb and global geometry reflection exemplify efforts to address the great challenge for the sound designer in balancing this conflict from within a user-experience context. A perfectly accurate replication of a complex acoustic space would not only be hugely resource-intensive, but in all probability completely superfluous, as the listener would not be able to discern this 'perfect' replication from one that involved certain processing compromises. The trick is of course in knowing where these compromises can and cannot be afforded. It is therefore highly likely that future advances in this area will incorporate a great deal of user-experience research, working towards a framework for environmental modelling that is optimised for both effectiveness and efficiency.

So, from this, what can we infer regarding the likely future of environmental modelling for VR sound? Current trends in VR audio SDKs and plug-ins suggest that increased automation of environmental modelling is likely to feature in future systems, minimising the need for individuals working in audio to dedicate significant time to modelling the effects manually. Based on the identified limitations of convolution reverb, occlusion filtering and detailed geometric modelling of the virtual acoustic environment, the central avenue of progress is highly likely to be one of computational brute force. It has been observed that, with transistors reaching the limit of how small they can be, Moore's Law of exponentially increasing computing power could be reaching the end of its relevance (Simonite 2016). That said, manufacturers are making significant investments to facilitate 'another few turns of the exponential crank' (Cross 2016) whilst simultaneously making progress with quantum and diffusion methods of computation (Adamatzky 2009; Hirvensalo 2013). It therefore remains a safe bet that VR developers can count on sizeable increases in computational power at least over the next decade and will utilise this power to increase the complexity of their environmental modelling techniques. The need for optimisation will certainly remain, but by way of a combination of more power and better frameworks of environmental modelling, we can still expect virtual acoustic environments that increasingly reflect the detailed sonic properties of physical and (crucially) more dynamic spaces.

On a final note, more power and better frameworks are not independent strands of progress, as advances in computing power will also enable new methods of implementing environmental modelling into VR applications. For instance, whilst user locomotion (walking movement) will currently break the illusion of convolution reverb (as the impulse response is acquired from a single fixed point), increased processing power could enable multiple responses to be collected and used in combination with one another to enable the convolution effect to respond to movement. Lastly, it seems logical to expect that several pre-existing modelling effects, that have yet to be exploited within VR technologies due to optimisation issues, will follow the trajectory of most of the above techniques and be resurrected when the required power is made available. An example of this would be 'air effects' (soundscape responses to ambient temperature, air pressure and humidity) that originally featured in A3D technology which dates back to 1997 and Aureal's *Vortex ASD1* PC soundcard (see Note 20) and which currently adorns the 'coming soon' feature list of the *3Dception* audio engine.

Spatialisation

Spatial audio is, like several of the topics documented throughout this chapter, a vast research subject with several individual books exploring it exclusively (see He 2016; Rumsey 2012). As a result, this humble section cannot possibly hope to account for all of the relevant developments, even within the specific context of VR. That said, the intention here is for the following information to provide a brief glimpse into the main themes in spatialisation techniques as they pertain to current (and likely future) trends in VR sound design.

Across the various contemporary VR sound solutions, the term 'spatialisation' (albeit usually with an Americanised spelling) appears consistently at top billing upon the list of audio features. As Stevens and Raybould (2013) posit, sound spatialisation in first-person digital games 'is literally a matter of life and death [as] most things in the world are actually off-screen. Sound plays a vital role in describing this off-screen space for the player' (p. 301). The perceptual ambition for spatialisation is not dissimilar to that of ambisonics. In straightforward terms, it describes the ability of an audio system to position sound objects within a specific and clearly localisable point in three-dimensional space (see Note 10). The difference lies in the approach. Broadly speaking, spatialisation in virtual environments describes any audio processing method that gives sound some perceptual quality of spread or position across a virtual space and includes stereo panning, surround sound and 3D/positional audio (Garner 2016). This would of course apply to ambisonics and binaural recordings but whilst these techniques describe a process of capturing a soundscape inclusive of its spatial information (a pre-production process), spatialisation refers to means of artificially enhancing a non-spatial audio sample (a post-production process).

In terms of spatialisation techniques currently deemed most relevant to VR sound, head-related transfer function (HRTF) arguably stands as one of the most powerful approaches (see Note 10). HRTF denotes the Fourier transform of head-related impulse responses (HRIRs), sound wave signals that have been filtered by the acoustic effects that the head (including the pinnae and ear canal) and torso have upon incoming sound waves before they reach the middle-ear. HRTF and HRIR are commonly associated with binaural recordings, particularly dummy head recordings. The difference between HRIR capture and a typical binaural recording is that whilst the latter is intended to produce an actual sound recording, HRIR capture is used to facilitate HRTF filtering on separate audio samples. In this regard, it bears a notable resemblance to the process of acquiring impulse responses for convolution reverb. Whilst there are several alternative methods of HRIR capture, the 'direct method' (Gardner and Martin 1995) is arguably one of the most commonly implemented. In this method, a dummy head may be used, or an actual human may have a pair of microphones placed in their ears. An emitter (loudspeaker) is placed nearby, directing a source sound wave towards the head. The source sound is consequently altered by the geometry of the head and body before it reaches the microphone and this forms the HRIR. Subsequent to each capture, the position of the emitter is adjusted so that HRIRs can be recorded from multiple angles. As with convolution reverb, the complete set of impulse responses is then interpolated to form a single new data set. A further similarity is that HRIR capture is also most effective when using a source (emitted) sound wave that possesses a wide frequency range, such as a sweeping sine wave. Where HRIR capture differs significantly, however, is that it must take place in an anechoic chamber. This is because the impulse responses being collected need to identify acoustic effects from the head and body only, with environmental acoustics distorting the spectral frequency of the sound wave, thereby invalidating the impulse response (Völk et al. 2009).

As noted earlier, current VR audio SDKs and plug-ins predominantly feature HRTF processing as part of their core feature set, enabling VR sound designers to apply this powerful localisation technique without needing to conduct any of the HRIR capture or processing calculations themselves. As with environmental modelling features in these toolsets, automation of complex acoustic effects is a common thread in terms of their development, and it is highly likely that future iterations of these toolsets will continue to incorporate increasingly complex audio processing, with the power to edit and build upon them for those willing and able, whilst also offering the option for increased automation and presets that facilitate more efficient VR sound development.

Recent research on spatialisation pertaining to digital games and VR raises some notably interesting future possibilities. For example, progress in cross-talk cancellation, that for many years had only worked with a pair of stereo speakers, has seen the use of VR-based head tracking technology enabling the effect to work with 5.1 surround sound loudspeaker arrays. The result of this is that a greater range of loudspeaker setups can now be used to playback binaurally spatialised audio (see Kim et al. 2009). A review of more recent literature reveals that one of the most prominent themes of spatial audio is individualisation (also referred to as 'personalised HRTF'). For instance, Kim and colleagues (2015) observe that cultural background can have significant impact upon the preferred spatial properties of a VR system, indicating that future systems may incorporate culturally based spatial audio configurations, in much the same way as a game's language settings can be adjusted prior to play. Similarly, Meshram and colleagues (2014) present a framework for personalised HRTF that uses image-based 3D modelling to map a user's head and body with only a consumer-grade camera, then processes the captured data to compute a bespoke HRTF. In another example, Murphy and Neff (2010) take spatial audio personalisation a speculative step further by positing that future developments could see VR environments capable of adapting the entire soundscape in all its complexities based on the user's individual experience and emotional state. As a final note, this kind of flexibility in the soundscape would not be limited to spatialisation and could extend to any aspect of VR sound, opening up exponentially more complex prospects in which any aspect of the soundscape could be responsive to the user's physiology, the narrative of the game/experience, the amount of time spent playing and so on. The possibilities seem endless.

Distance Effects

Audio processing techniques that generate a perception of proximity between sound objects and the listener are well established in contemporary game engines, in relatively simplified forms such as minimum and maximum signal attenuation thresholds. The underlying principle is simple but effective; closer sound objects are louder, whilst more distant objects are quieter. The minimum threshold denotes a set proximity point at which the relative loudness value will begin to lower (i.e. the distance at which the sound will start to fade), whilst the maximum value indicates the point at which its loudness will reach zero. More complex techniques within this area of sound design include using HRTF processing to more realistically model the timbral effects of increased distance, rather than just a simple fade. Whilst both min/max attenuation thresholds and HRTF processing are relatively effective at differentiating mid-range and distant sound objects, neither are known for being particularly effective within 'nearfield ranges' (commonly defined as less than one virtual metre between sound object and listener—Boer 2003). This issue means that sound sources very close to the ear (a whispered voice, a buzzing insect, etc.) cannot evoke the equivalent perceptual 'intimacy' that actual nearfield sounds would have. This intimate effect has particular relevance to one of VR's fundamental principles, namely its aim to induce multisensory immersion by way of synaesthesia (stimuli of one modality that are so convincing they evoke sensations in another modality).

In research, an association has been established between nearfield sound and autonomous sensory meridian response (ASMR), which commonly refers to a tingling sensation across the neck and upper spine and is attributed to a mild, auditory-tactile form of synaesthesia (Barratt and Davis 2015). An approach such as nearfield sound that induced ASMR responses would therefore be a very powerful tool in producing deeply immersive and engaging VR experiences. In response to the nearfield problem, a particular approach emerged once again in early 2000s, in the form of Sensaura's *MacroFX*, a third-party extension of DirectSound 3D that divided the (one-metre radius) nearfield area into at least four separate regions. Each of these sub-regions was programed to alter the spectral composition of a sound wave differently, in a way that made the sound object appear extremely close to the ear without the intensity becoming uncomfortably high (Boer 2003). Since MacroFX first appeared in 2000, the technology has re-emerged as a feature of contemporary PC soundcards and whilst it is technically available for VR sound applications, nearfield audio is not explicitly referred to within any of the audio plug-ins or SDKs discussed earlier in this chapter. This identifies nearfield audio as one to keep an eye on with regards to likely future directions in VR sound. With many experiences in VR concentrating less on vast, expansive environments and more upon smaller, more intimate and highly detailed spaces, sounds that can get up close and personal with users would be of great value to the overall VR experience.

Volumetric Sound

In yet another instance of a highly consistent theme, the origin of volumetric sound can be traced back to the regrettably short-lived golden age of interactive audio in the late 1990s and early 2000s, with *ZoomFX*, another proprietary audio technology developed by Sensaura (Percy 2000). Volumetric sound describes an interesting solution to the 'point source problem' of positional audio. As Dvorak and colleagues (2004) explain, the bulk of positional audio technology is built around point sources (monophonic samples processed as surround sound or via HRTF that are perceived as existing at a specific point in the virtual space). This presents an issue when dealing with composite sounds (Dvorak et al. identify crowd cheers, flowing water and wind as notable examples) where the perceived source, in actuality, would not be localised to a point but rather spread across a region. In an approach that is similar to shoebox modelling, volumetric sound addresses the point source problem by way of a cuboid that defines the area in which the composite sound can be heard. When the listener is positioned outside of the designated area, volumetric sound has positional and distance attenuation qualities. By comparison, stepping into the area negates these effects to give the impression of being surrounded by the sound.

The current limitation of volumetric sound is primarily its static nature, whilst the listener is positioned within the area cuboid. Standing next to an actual waterfall would immerse an individual within a perceptually surrounding sound area, but changes in position and orientation within that area would ultimately still effect the quality of the sound, rather than it being entirely static and unresponsive. This issue presents us with another likely future attribute of VR sound. A recent paper by Schissler and colleagues (2016), all of whom are part of the audio development team at Oculus, presents an experimental means of integrating HRTF processing into volumetric audio, heavily hinting that forthcoming versions of the Oculus Audio SDK will include such a feature, and meaning that soon, composite sound will be just as responsive to our head and body movements as point sound.

Procedural Audio

Across the numerous elements of VR sound that have been discussed up to this point, one notably consistent point that groups these elements together is that they are all sample-based (i.e. methods of either pre-production or post-production signal processing upon complex, natural sound waves). However, virtual environments are certainly not limited to sample-based sound design and this brings us to our final topic of discussion in this chapter. Procedural audio is a modality-specific offshoot of a much broader method of computing, procedural generation, which describes data that has been created algorithmically rather than manually by a developer. In a digital games context, various methods of procedural generation form a collection commonly referred to as 'procedural content generation': 'the algorithmic creation of game content with limited or indirect user input' (Shaker et al. 2016, p. 1). As this quote implies, any content can arguably be traced back to some form of user input, but procedural methods describe circumstances in which the designer created the system, that in turn created the content; a sort of algorithmic intermediary if you will. Procedural content generation can be applied to seemingly any form of content, from graphical objects and textures to artificial intelligence and narrative. It can add elements of randomisation into a game (spawn locations, item properties, etc.), generate endless and evolving landscapes and facilitate adaptive difficulty in real-time during gameplay (Watkins 2016).

Procedural audio describes a family of sound design approaches built upon the above principle of algorithmically mediated content. Amongst its various applications for digital games (that include real-time speech synthesis for dialogue, algorithmic music composition and synthesis of sound effects), procedural audio is now well established and has received substantial research attention in recent years. As with other forms of procedurally generated content, the primary advantage and driving force of development in procedural audio is its capacity for producing theoretically limitless quantities of sound content within a single game or

VR experience (Pecino 2014). For Farnell (2007), 'procedural audio is sound qua process, as opposed to sound qua product [...] non-linear, often synthetic sound, created in real-time according to a set of programmatic rules and live input' (p. 1). Two elements of this statement stand out with regards to the importance of procedural audio for VR, namely 'real-time' and 'live input'. Nair (2014) asserts that increasingly reactive soundscapes (i.e. those that respond to live input in real-time) will contribute significantly to the creation of more convincing and engaging experiences. Essentially, the more a soundscape can respond directly and clearly to unpredictable user-behaviours, the more the user will take ownership of their avatar, evoking feelings of embodiment and presence.

What can be classified as procedural audio reveals another continuum based upon the ratio of 'process against product' (see the above quotation of Farnell) and the degree to which designer-control over the audio is direct or indirect. Engaging briefly in some light pedantry, almost all of the VR sound concepts that we have discussed throughout this chapter, from occlusion filtering to HRTF, are partially procedurally generated as they include significant algorithmic elements. Collins (2009) helps clarify this, citing Wooler and colleagues' (2005) distinction between 'transformational algorithms' and 'generative algorithms' to denote the difference in procedural game music between algorithms that transform characteristics of a composition (e.g. randomly altering a few notes in a phrase or chords in a progression) and algorithms that generate an entire piece of music from the most basic fundamental components. Considering procedural audio more generally, we can identify the filtering and editing methods of audio samples as primarily transformational, whilst a soundscape built in its entirety from algorithmically manipulated sine waves would be more generative. It is the latter that we are focussing upon in the remainder of this section.

What we understand to be procedural audio also highlights a substantial divide between research and industry. Böttcher (2013) explores this issue by way of detailed interviews with several eminent game sound designers, observing that the general quality of procedural models is currently too poor and that the necessary tools for producing such audio are not yet available as integrated solutions within game engines. Whilst this would indicate that the industry is not currently facilitating any procedural audio, it is arguably more accurate to suggest that research and industry, currently, understand it as different things. Whereas research tends towards meaning synthesised audio systems (generative audio), industry is largely referring to adding procedural elements to sample-based sound design techniques. For an example of the latter, Andersen (2016) outlines a process of creating a procedural explosion sound, using a series of samples that reflect components of a composite explosion. Each component has three variations that are connected to a gate and controlled by a randomisation generator. The resultant sound is consequently a random assortment of variations for each component, making the collective explosion sound different each time it is repeated (up to a point). Audio Middleware is notably geared towards this type of processing, enabling designers to import a small set of samples to produce a wealth of additional content. Commonplace examples include procedural transformation of footsteps, gunshots and user-interface earcons.

Research concerning generative forms of procedural audio is undoubtedly making some very interesting progress. Hoover and colleagues (2015), for instance, show that it is possible, with some restrictions, to procedurally generate audio content simultaneously to multiple other game facets (such as graphics and animations), using a series of interconnected artificial neural networks. In another recent study, Lopes and colleagues (2015) demonstrate how soundscapes within a survival horror game can be procedurally generated, based upon the player's intended emotional state as identified by the designer. Such techniques are yet to break through substantially into consumer VR or digital games, but instances of experimental games, such as *Muse*,[21] that feature procedurally generated audio are becoming steadily more common and we have also begun to see commercial releases, in games like *Proteus* (Key and Kanaga 2013).

The developments noted above indicate a likely convergence of generative and transformational audio progress between commercial and research interests. With the games industry developing it from the top-down and research from the bottom-up, the anticipated future of procedural audio as a VR technology is likely to be a race to the centre, as the quality of generative audio gradually begins to approach that of sampled recordings and transformative techniques imbue samples with ever greater flexibility and responsiveness. With regards to flexibility and responsiveness to user input, VR presents several highly significant opportunities for new approaches to procedural audio. To give an example, the head and body tracking interactions that are prominent in VR could act as a novel means of generating sound for the purpose of increasing the user's sense of embodiment with their avatar (Garner 2017b). If playing a robot, for example, movements in the head and arms could generate synthetic whirring noises. If playing a knight of the realm, looking around could instigate a gentle wave of clinks as the user's chainmail armour swayed. Other VR input methods present further opportunities. Integrated HMD microphones and voice recognition could be fed into a synthesiser to procedurally generate a characterised version of the user's speech, whilst biometrics such as respiration and heart rate sensors could be translated into rhythmic information, enabling the generative musical soundtrack to precisely adapt to the breathing and heartbeat of the user.

It has been said that 'the future of interactive entertainment is procedural' (Baker 2017). Virtual worlds may soon have the potential to fluctuate, extend and evolve infinitely, with the designers responsible not for any of the content, merely specifying the rules that generate and govern the content. In terms of audio, and in general, we are certainly not there yet and the distance between us and this ambition is unclear. Whether or not a complex, engaging and entirely procedural virtual world is achievable, the desire to attain it is reflected in the significant investment being made in both research and industry (even if they presently have differing priorities in this area). Consequently, it is a very safe assumption that procedural audio shall be a continued thread of VR development for the foreseeable future. Before such an ambition is realised, however, the sample processing techniques outlined above shall continue to dominate VR sound throughout the next few years at the very least. As noted earlier, the industry will not fully embrace procedural techniques until they are of a 'realism' quality that matches sample-based techniques. Consequently, it is highly probable that the future of VR sound is, once again, one of integration, with procedural and sample-based techniques used collectively within a single soundscape that is flexible, detailed, of a high-fidelity and matches users' many complex requirements for an ideal user experience.

Chapter Summary and References

In this penultimate chapter, integration is posited as the most significant underlying theme across VR, inclusive of the various aspects of audio technology and sound design. Our virtual lives are soon to be integrated with 'things' by way of IoT technology, whilst collaborative virtual environments and multiplayer VR as a new platform for social media integrates us with each other. In terms of VR sound, audio hardware arguably continues to be of secondary priority when compared to graphics. However, emerging developments in this technology, from more HMDs integrating audio outputs as standard to headphone designs that incorporate additional biometric technology, reveal that sound is not being forgotten. Audio software relevant to VR continues the integration theme. Expectations are that the diversification of options will continue as processes become more complex, but that systems will also become more interoperable, working together to give designers and developers more power, control and flexibility in their work. Sound processing techniques such as HRTF processing, procedural generation, ambisonics and spatialisation shall continue to be refined, consequently becoming more commonplace features across the range of VR sound solutions. In keeping with the overarching theme, these design approaches once again reiterate integration, with sample processing and procedural generation techniques highly likely to be utilised collectively within the virtual soundscapes of the future.

On a final note, the recurrence of integration is a constant reminder of the emergent nature of VR sound. Moreover, it emphasises the value of constructing such an understanding as technological developments are increasingly bringing diverse elements together. An emergent framework has value by way of acknowledging and keeping track of all the pieces of the puzzle. Beyond this, it has the potential to inspire further inquiry for those of us willing to go deeper, by suggesting connections and processes between the elements for us to discover, piece by piece making clearer the emergent visage of VR sound.

Notes

1. IAB-UK, Future VR. https://iabuk.net/resources/white-papers/future-trends-volume-16-the-future-of-virtual-reality.
2. HTC's *Deluxe Audio Strap* webpage. https://www.vive.com/uk/vive-deluxe-audio-strap/.
3. Opto VR website. https://optovr.com/.
4. RealSpace 3D Audio website. http://realspace3daudio.com/.
5. Audeze iSINE VR hesdphones. https://www.audeze.com/products/isine-series/isine-vr-ear-headphone.
6. In simple terms, the difference between the two is that planar magnetic drivers use a larger array of magnets, more evenly spread across the diaphragm (which disturbs the air to generate the sound wave). This spread reduces the subtle distortions that are prevalent in many dynamic headphones.
7. 3DIO binaural/ambisonic microphones. https://3diosound.com/collections/microphones.
8. GoPro Odyssey. https://shop.gopro.com/odyssey.
9. 360 Designs Eye. http://360designs.io/product/eye-vr-camera-full-3-axis-package/.
10. Oculus (2017). *Introduction to virtual reality audio*. Oculus Developer Guide. https://developer.oculus.com/documentation/audiosdk/latest/concepts/book-audio-intro/.
11. Steam Audio SDK: https://valvesoftware.github.io/steam-audio/.
12. Google VR-SDK. https://developers.google.com/vr/concepts/spatial-audio#atmospheric_sounds.
13. 3Dception. http://www.twobigears.com/spatworks/.
14. AstoundSound. http://www.astoundgaming.com/.
15. RealSpace 3D audio. http://realspace3daudio.com/.
16. Max. https://cycling74.com/products/max/.
17. Pure Data. https://puredata.info/
18. Tiltbrush. https://www.tiltbrush.com/.
19. LyraVR. http://lyravr.com/.
20. IXBT Labs (2003). *Modern audio technologies in games*. http://ixbtlabs.com/articles2/sound-technology/.
21. Muse (Game): http://musedev.net/.

References

Adamatzky, A. (2009). Reaction-diffusion computing. In *Encyclopedia of complexity and systems science* (pp. 7548–7565). New York: Springer.

Andersen, A. (2016). Why procedural game sound design is so useful. *A Sound Effect*. https://www.asoundeffect.com/procedural-game-sound-design/.

Attias, B. (2013). Subjectivity in the groove: Phonography, digitality and fidelity. In B. Attias, A. Gavanas, & H. Rietveld (Eds.), *DJ culture in the mix: Power, technology, and social change in electronic dance music*. New York: Bloomsbury Publishing USA.

Baker, E. (2017). What can you do with 3D sound that you can't do with 2D sound? *No Film School*. http://nofilmschool.com/2017/02/3D-audio-virtual-reality-sound-designer-viktor-phoenix-interview.

Barratt, E. L., & Davis, N. J. (2015). Autonomous Sensory Meridian Response (ASMR): A flow-like mental state. *PeerJ, 3*, e851.

Benford, S., Snowdon, D., Greenhalgh, C., Ingram, R., Knox, I., & Brown, C. (1995). VR-VIBE: A virtual environment for Co-operative information retrieval. In *Computer Graphics Forum* (Vol. 14, No. 3, pp. 349–360). Blackwell Science Ltd.

Bergel, P. (2015). The internet of sound. *Tech Crunch*. https://techcrunch.com/2015/07/26/the-internet-of-sound/.

Boer, J. R. (2003). *Game audio programming*. Hignham, USA: Charles River Media.

Böttcher, N. (2013). Current problems and future possibilities of procedural audio in computer games. *Journal of Gaming & Virtual Worlds, 5*(3), 215–234.

Boyce, K. (2016). Sennheiser virtual reality mic showcase at CES. *Visualise*. http://visualise.com/2016/01/sennheiser-virtual-reality-mic-showcase-at-ces.

Bradshaw, P. (2017). Carne y Arena review—Dazzling virtual reality exhibit offers a fresh look at the refugee crisis. *The Guardian*. https://www.theguardian.com/film/2017/may/22/carne-y-arena-review-inarritu-virtual-reality-refugee-cannes-2017.

Branda, E. (2015). Review: Oculus Rift. *Journal of the Society of Architectural Historians, 74*(4), 526–528.

Britton, T. C., Day, B. L., Brown, P., Rothwell, J. C., Thompson, P. D., & Marsden, C. D. (1993). Postural electromyographic responses in the arm and leg following galvanic vestibular stimulation in man. *Experimental Brain Research, 94*(1), 143–151.

Buyya, R., & Dastjerdi, A. V. (Eds.). (2016). *Internet of things: Principles and paradigms*. Amsterdam: Elsevier.

Chernyshov, G., Chen, J., Lai, Y., Noriyasu, V., & Kunze, K. (2016). Ambient rhythm: Melodic sonification of status information for IoT-enabled devices. In *Proceedings of the 6th International Conference on the Internet of Things* (pp. 1–6). ACM.

Churchill, E. F., Snowdon, D. N., & Munro, A. J. (Eds.). (2012). *Collaborative virtual environments: Digital places and spaces for interaction*. Berlin: Springer Science & Business Media.

Cohen, B., Yakushin, S. B., & Holstein, G. R. (2011). What does galvanic vestibular stimulation actually activate? *Frontiers in neurology, 2*.

Collins, K. (2009). An introduction to procedural music in video games. *Contemporary Music Review, 28*(1), 5–15.

Cragg, O. (2016). Samsung Entrim 4D headphones let you 'feel' virtual reality with your whole body. *International Business Times*. http://www.ibtimes.co.uk/samsung-entrim-4d-headphones-let-you-feel-virtual-reality-your-whole-body-1549641.

Cross, T. (2016). After Moore's Law. *Economist Technology Quarterly*. http://www.economist.com/technology-quarterly/2016-03-12/after-moores-law.

Devana, A. (2015). 3D Audio: Weighing the options. *Designing Sound*. http://designingsound.org/2015/04/3d-audio-weighing-the-options/.

Domingo, M. C. (2012). An overview of the internet of things for people with disabilities. *Journal of Network and Computer Applications, 35*(2), 584–596.

Drago, E. (2015). The effect of technology on face-to-face communication. *Elon Journal of Undergraduate Research in Communications, 6*(1).

Dredge, S. (2014). Facebook closes its $2bn Oculus Rift acquisition. What next? *The Guardian*. https://www.theguardian.com/technology/2014/jul/22/facebook-oculus-rift-acquisition-virtual-reality.

Dvorak, J., Pirillo, C., & Taylor, W. (2004). *Online!: The book*. New Jersey, USA: Prentice Hall Professional.

Farnell, A. (2007, September). An introduction to procedural audio and its application in computer games. In *Audio mostly conference* (pp. 1–31).

Fellgett, P. (1975). Ambisonics. Part one: General system description. *Studio Sound, 17*(8), 20–22.

Galvani, L. (1791). *De Viribus Electricitatis in Motu Musculari Commentarius*. Bologna: Institute of Sciences at Bologna.

Gardner, W. G., & Martin, K. D. (1995). HRTF measurements of a KEMAR. *The Journal of the Acoustical Society of America, 97*(6), 3907–3908.

Garner, T. A. (2016). from sinewaves to physiologically-adaptive soundscapes: The evolving relationship between sound and emotion in video games. In *Emotion in Games* (pp. 197–214). Springer International Publishing.

Garner, T. A. (2017a). Why connecting all the world's robots will drive 2017/s tops technology trends. *The Independent*. http://www.independent.co.uk/life-style/gadgets-and-tech/why-connecting-all-the-world-s-robots-will-drive-2017-s-top-technology-trends-a7505181.html.

Garner, T. A. (2017b). Virtual reality and new aesthetic opportunities for sound design. *The Soundtrack*. 9(1–2).

Gerzon, M. A. (1975). Ambisonics. Part two: Studio techniques. *Studio Sound, 17*(8), 24–26.

Gerzon, M. A. (1985). Ambisonics in multichannel broadcasting and video. *Journal of the Audio Engineering Society, 33*(11), 859–871.

Gouveia, D. (2013). *Getting started with C++audio programming for game development*. Birmingham: Packt Publishing Ltd.

Greengard, S. (2015). *The internet of things*. Cambridge: MIT Press.

He, F. (2016). Oculus earphones review and comparison with ultra high-end earbuds. *Road to VR*. http://www.roadtovr.com/oculus-earphones-review-rift-earbuds-comparison/.

He, J. (2016). *Spatial audio reproduction with primary ambient extraction*. Singapore: Springer.

Hermann, T., Nehls, A. V., Eitel, F., Barri, T., & Gammel, M. (2012). Tweetscapes-real-time sonification of twitter data streams for radio broadcasting. Atlanta: Georgia Institute of Technology.

Hirvensalo, M. (2013). *Quantum computing* (pp. 1922–1926). Netherlands: Springer.

Hoover, A. K., Cachia, W., Liapis, A., & Yannakakis, G. N. (2015, April). Audioinspace: Exploring the creative fusion of generative audio, visuals and gameplay. In *International Conference on Evolutionary and Biologically Inspired Music and Art* (pp. 101–112). Springer International Publishing.

Horrowitz, S., & Looney, S. (2014). Masterclass: Using game audio middleware. *Electronic Musician*. http://www.emusician.com/how-to/1334/masterclass-game-audio-middleware/48158.

Hu, R., Dong, S., Wang, H., et al. (2012). Perceptual characteristic and compression research in 3D audio technology. In *9th International Symposium CMMR 2012*. London, UK.

Ilif, R. (2016). 5 VR Trends to watch. *Inc*. https://www.inc.com/rebekah-iliff/5-virtual-reality-trends-for-2017-and-beyond.html.

James, P. (2015). Why the Oculus Rift's integrated headphones might replace your gaming headset. *Road to VR*. http://www.roadtovr.com/why-the-oculus-rifts-integrated-headphones-might-replace-your-gaming-headset/.

Jerald, J. (2015). *The VR book: Human-centered design for virtual reality*. San Rafael: Morgan & Claypool.

Key, E., & Kanaga, D. (2013). *Proteus*. London, UK: Curve Studios.

Kim, S., Ikeda, M., Takahashi, A., Ono, Y., & Martens, W. L. (2009). Virtual ceiling speaker: Elevating auditory imagery in a 5-channel reproduction. In *Audio Engineering Society Convention 127*. Audio Engineering Society.

Kim, S., King, R., & Kamekawa, T. (2015). A cross-cultural comparison of salient perceptual characteristics of height channels for a virtual auditory environment. *Virtual Reality, 19*(3–4), 149–160.

Kohlrausch, A., Braasch, J., Kolossa, D., & Blauert, J. (2013). An introduction to binaural processing. In *The technology of binaural listening* (pp. 1–32). Berlin and Heidelberg: Springer.

Leadbetter, R. (2016). Oculus Rift review. *EuroGamer*. http://www.eurogamer.net/articles/digitalfoundry-2016-oculus-rift-review.

Leppänen, T., Heikkinen, A., Karhu, A., Harjula, E., Riekki, J., & Koskela, T. (2014). Augmented reality web applications with mobile agents in the internet of things. In *Next Generation Mobile Apps, Services and Technologies (NGMAST), 2014 Eighth International Conference on* (pp. 54–59). IEEE.

Lockton, D., Bowden, F., Brass, C., & Gheerawo, R. (2014). Powerchord: Towards ambient appliance-level electricity use feedback through real-time sonification. In *International Conference on Ubiquitous Computing and Ambient Intelligence* (pp. 48–51). Springer International Publishing.

Lopes, P., Liapis, A., & Yannakakis, G. N. (2015). Targeting horror via level and soundscape generation. In *Eleventh Artificial Intelligence and Interactive Digital Entertainment Conference*.

Macias, J. A. G., Alvarez-Lozano, J., Estrada, P., & Lopez, E. A. (2011). Browsing the internet of things with sentient visors. *Computer, 44*(5), 46–52.

Maeda, T., Ando, H., Amemiya, T., Nagaya, N., Sugimoto, M., & Inami, M. (2005). Shaking the world: Galvanic vestibular stimulation as a novel sensation interface. In *ACM SIGGRAPH 2005 Emerging technologies* (p. 17). ACM.

Malham, D. G., & Myatt, A. (1995). 3-D sound spatialization using ambisonic techniques. *Computer Music Journal, 19*(4), 58–70.

Marks, A. (2012). *The complete guide to game audio: For composers, musicians, sound designers, game developers*. Boca Raton: CRC Press.

Martin, K. A., & Laviola, J. J. (2016). The transreality interaction platform: Enabling interaction across physical and virtual reality. In *Internet of Things (iThings) and IEEE Green Computing and Communications (GreenCom) and IEEE Cyber, Physical and Social Computing (CPSCom) and IEEE Smart Data (SmartData), 2016 IEEE International Conference on* (pp. 177–186). IEEE.

Matsuda, K. (2016). *Hyper-reality*. Columbia: Fractal.

Meshram, A., Mehra, R., Yang, H., Dunn, E., Franm, J. M., & Manocha, D. (2014). P-HRTF: Efficient personalized HRTF computation for high-fidelity spatial sound. In *Mixed and Augmented Reality (ISMAR), 2014 IEEE International Symposium on* (pp. 53–61). IEEE.
Murphy, D., & Neff, F. (2010). Spatial sound for computer games and virtual reality. In M. Grimshaw (Ed.), *Game sound technology and player interaction: Concepts and developments* (pp. 287–312).
Nagata, H., Mikami, D., Miyashita, H., Wakayama, K., & Takada, H. (2017). Virtual reality technologies in telecommunication services. *Journal of Information Processing, 25,* 142–152.
Nair, V. (2014). Audio and VR. *Designing Sound.* http://designingsound.org/2014/05/audio-and-vr/.
Nelson, F. (2014). The past, present and future of VR and AR: The pioneers speak. *Tom's Hardware.* http://www.tomshardware.co.uk/ar-vr-technology-discussion,review-32940-5.html.
Ortolani, F. (2015). *Introduction to Ambisonics.* Ironbridge Electronics.
Parisi, T. (2015). *Learning virtual reality: Developing immersive experiences and applications for desktop, web, and mobile.* Newton: O'Reilly Media, Inc.
Parkin, S. (2014). What Zuckerberg sees in Oculus Rift. *MIT Technology Review.* https://www.technologyreview.com/s/525881.
Pecino, I. (2014). Spatial and kinematic models for procedural audio in 3D virtual environments. In *International Computer Music Conference.*
Percy, M. (2000). Implementing ZoomFX: 3D objects in 3D space using DS3D. *Gamasutra features archive.* http://www.gamasutra.com/features/gdcarchive/2000/ZoomFX2.pdf.
Picinali, L., Afonso, A., Denis, M., & Katz, B. F. (2014). Exploration of architectural spaces by blind people using auditory virtual reality for the construction of spatial knowledge. *International Journal of Human-Computer Studies, 72*(4), 393–407.
Popper, B. (2016). Samsung's new Entrim 4D headphones add movement to VR by tricking your inner ear. *The Verge.* https://www.theverge.com/2016/3/14/11220836.
Porter, J. (2016). Star Trek: Bridge Crew proves that VR has a bright multiplayer future. *Tech Radar.* http://www.techradar.com/news/star-trek-bridge-crew-proves-that-vr-has-a-bright-multiplayer-future.
Riach, J. (2013). Virtual reality simulators could end England's penalty shootout woe. *The Guardian.* https://www.theguardian.com/football/2013/aug/23/virtual-reality-england-penalty-shootout.
Robertson, A. (2016). The Unreal Engine now lets you build games inside virtual reality. *The Verge.* https://www.theverge.com/2016/2/4/10908444.

Rorsman, I., Magnusson, M., & Johansson, B. B. (1999). Reduction of visuo-spatial neglect with vestibular galvanic stimulation. *Scandinavian Journal of Rehabilitation Medicine, 31*(2), 117–124.

Rosedale, P., et al. (2003). *Second life*. San Francisco, USA: Linden Lab.

Rumsey, F. (2012). *Spatial audio*. Boca Raton: CRC Press.

Rumsey, F., & McCormick, T. (2012). *Sound and recording: An introduction*. Boca Raton: CRC Press.

Schissler, C., Nicholls, A., & Mehra, R. (2016). Efficient HRTF-based spatial audio for area and volumetric sources. *IEEE Transactions on Visualization and Computer Graphics, 22*(4), 1356–1366.

Shaker, N., Togelius, J., & Nelson, M. (2016). *Procedural content generation in games*. Switzerland: Springer International Publishing.

Simonite, T. (2016). Moore's Law is dead. Now what? *MIT Technology Review*. https://www.technologyreview.com/s/601441/moores-law-is-dead-now-what/.

Somberg, G. (Ed.). (2016). *Game audio programming: Principles and practices*. Boca Raton: CRC Press.

Stevens, R., & Raybould, D. (2013). *The game audio tutorial: A practical guide to creating and implementing sound and music for interactive games*. Oxford: Taylor & Francis.

Takahashi, D. (2016). Unity demonstrates new way to develop games inside virtual reality. *VentureBeat*. https://venturebeat.com/2016/02/10/unity-demonstrates-new-way-to-develop-games-inside-virtual-reality/.

Ung, G. (1998, November). Vortex 2 and A3D 2.0. *Maximum PC*. Imagine.

Völk, F. (2009). Externalization in data-based binaural synthesis: Effects of impulse response length. In *Proceedings of International Conference on Acoustics NAG/DAGA* (pp. 1075–1078).

Waters, R. C., Anderson, D. B., Barrus, J. W., Brogan, D. C., Casey, M. A., McKeown, S. G., et al. (1997). Diamond park and spline: Social virtual reality with 3D animation, spoken interaction, and runtime extendability. *Presence: Teleoperators and Virtual Environments, 6*(4), 461–481.

Watkins, R. (2016). *Procedural content generation for unity game development*. Birmingham: Packt Publishing Ltd.

Williamson, R. (2017). The sound of the internet of things (and why it matters for brands). *Creative Review*. https://www.creativereview.co.uk/sound-internet-things-matters-brands/.

Willitis, T., Blackwell, M., Carmack, J., et al. (2004). *Doom 3*. USA: Id Software.

Wooller, R., Brown, A. R., Miranda, E., Berry, R., & Diederich, J. (2005). A framework for comparison of processes in algorithmic music systems. In *Generative arts practice* (pp. 109–124). Sydney: Creativity and Cognition Studios Press.

9
Applications of Virtual Reality

As has been alluded to at various points throughout this book, the ways in which VR technologies can be used extends far beyond recreational use and the consumer digital games market. Throughout the 1990s and the first generation of mainstream VR, research and development across numerous industries were inspired to explore how the technology could be utilised as a means of enhancing and extending their existing concepts and approaches; capitalising on the unique properties of VR to help overcome some of the limitations of existing solutions. However, in stark contrast to consumer VR, development into wider commercial applications of the technology continued to flourish throughout the 2000s. Consequently, with the recent resurgence in popular interest, the steadfast research progress made in the preceding years is beginning to merge with the narrative of consumer VR, with the wider applications themselves emerging from relative obscurity to become mainstream headliners.

With continuing attention paid to the role of sound, this chapter presents an introductory review of VR's commercial applications. Beginning with a look at representations of data in VR by way of visualisation and sonification, we proceed to examine the multipurpose

T.A. Garner, *Echoes of Other Worlds: Sound In Virtual Reality*,
Palgrave Studies in Sound, DOI 10.1007/978-3-319-65708-0_9

functionality of telepresence applications before addressing the potential of VR as a tool for learning and skills development. How VR can support creative processes follows, before the chapter then closes with a brief review of the technology's place in digital health.

Data Visualisation

As Emanuel Goldenweiser (1916) observed: 'Graphic methods of presenting statistical facts have been in use from times immemorial and are doubtless destined to become increasingly popular' (p. 205). More than a century later, Goldenweiser's prediction has proven to be exceedingly accurate, with graphical representations of information used for everything from comparing measles outbreaks in vaccinated and non-vaccinated communities (Harris et al. 2015) to displaying correlations between parental income and the likelihood of attending a higher education institution (Aisch 2015). Visual representations of data have been shown to discourage disparate (low-level) interpretations of data in favour of those that are more integrated (high-level), with recipients able to understand overarching systems and the relationship between components. This is largely in contrast with text, which encourages a greater focus on the individual components themselves (Cromley et al. 2010).

Data visualisation is widely accepted to be of substantial benefit to us as means of understanding and sharing information. However, as anyone who has spent hours attempting to decipher impossibly complex diagrams can attest, the visualisation of data as a static image spread across two dimensions has some rather distinct limitations. This section takes a look at some of these issues with a review of how contemporary visualisations are progressing, what the key features of these advances are, and how virtual technologies are poised to become the likely visualisation interface of the future. This discussion then leads into and exploration of sound, firstly looking at VR-relevant themes in sonification literature, before a brief analysis of the auditory aspects of VR visualisations. To try and minimise overlap between sections of this chapter, the following discussion on visualisation is restricted to that which pertains specifically to visualisation of data for analysis. It will not cover things such as virtual prototyping or visualisation for educational purposes. Both are discussed in later sections.

Relevant Issues in Data Visualisation

Kirk (2016) provides an elegantly concise definition of data visualisation, describing the term as: 'The representation and presentation of data to facilitate understanding' (p. 19). As the term inherently suggests, the primary modality of such representations is visual, with computer graphics now the most common form of visualisations. Friendly (2008) traces the history of data visualisation back to classical antiquity and the use of imagery to chart the positions of the stars to aid navigation. Modern statistical graphics, from bar and pie charts to scatterplots and line graphs, first appeared during the early nineteenth century (largely in the work of William Playfair—see Rimland et al. 2013). Official recognition and standardisation began to be introduced over subsequent decades. The latter years of the twentieth century saw visualisation emerge in its contemporary form, as an established yet innovative and multi-disciplinary research field. Friendly (2008) observes the primary requirements of contemporary data visualisation to be: the ability to manipulate data in its visual form; methods that facilitate visualisation of complex, multi-dimensional data; means of representing discrete and categorical data (e.g. race, gender and education level); and also that data visualisation progress be founded on increased attention to human cognition and perception. These points present us with a fundamental list of requirements that VR visualisations should be able to meet.

Moving closer to the present day, an article by Johnson (2004) presents numerous contemporary issues regarding visualisations, several of which have particular relevance to VR. These include a remaining incompleteness of understanding with regard to visual perception, and a lack of interactivity and dynamic, time-dependent content. It has also been suggested that, whilst visualisations have much to offer as a means of our consumption and understanding of information, they also have a great power to bias, manipulate and even deceive. In a study conducted by Idris and colleagues (2011), it is asserted that a carefully crafted colour scheme can have greater influence on a viewer's perception of credibility than the reputation and authority of the data source. This issue arguably becomes exponentially greater when visualisations are integrated with network technologies. As Burn-Murdoch (2013) notes, the networked globalisation of information has enabled the more engaging visualisations to spread rapidly to reach huge numbers of people, thanks largely to information sharing on social media. This technology not only facilitates the viral spread of visualisations, but simultaneously reinforces their perceived authenticity by being passed to the individual from within a space that they feel secure and comfortable in.

Making Sense of Data with VR

Referring back to the above issues, VR technology arguably has great potential to mitigate some of the current limitations of data visualisation, but also to exaggerate some of its other problems. Where VR presents a substantial opportunity for added value in data visualisation is with regard to immersion, interactivity and dimensionality. As has been discussed in previous chapters, enhanced immersion is a foundational advantage of VR. The immersive qualities of VR (particularly HMDs and CAVE systems) have been repeatedly identified as a benefit to data visualisation (Laha and Bowman 2012; Raja et al. 2004), likely due to its ability to minimise distraction, thereby supporting concentration, but also to present a more engaging experience that motivates the user to work with the data more productively.

By way of its inherent user experience and functionality, VR is also a significant asset to visualisations when considering Johnson's (2004) interlinked issues of interactivity and dimensionality. The interactivity offered by VR can empower visualisations to present multiple heterogeneous datasets within a single dynamic space. In many instances, the user can switch between multiple data sets, adjust the scale and orientation of the graphics, and apply filters with immediate feedback (see Reda et al. 2013). In 2016, Google created an interactive VR visualisation (using their *Cardboard* platform) of data pertaining to the UK's forthcoming (at the time) referendum on leaving the European Union. The experience enabled users to navigate a stylised virtual continent and select from various European countries. Doing so would reveal a ranked list of the most commonly asked questions each country had regarding the referendum, based upon the correlation of search engine queries. In this instance, a large volume of heterogeneous data, that could have otherwise looked intimidatingly incomprehensible as a static image, was made accessible by way of the ability to navigate the scene and interact with various elements to manipulate the data. Other recent visualisation applications have sought to combine VR environments with additional virtual technologies to add further, more naturalistic interaction to data visualisations. Marks and colleagues (2014), for example, combine the *Rift* HMD with marker-based motion capture to facilitate locomotion tracking (users can physically walk around the data) and interaction with individual components (user can select and manipulate data points with their hands) within 3D visualisations of neural networks.

It should be noted that modern data visualisations do not require a 'fully immersive' (HMD or CAVE) display to be dynamic or to facilitate interactivity. In fact, the first two 'semi immersive' (flat screen) examples noted in this chapter are both interactive and employ techniques for presenting complex data more accessibly. This presents us with a scale of dynamism and interactivity, with static visualisations positioned towards the far left and, at present, fully immersive VR stationed towards the far right. As observed at various points throughout this book, what VR technology *is* extends beyond just its display, to include various other devices and systems that sit between the extremes of this scale. For example, Saggio and Ferrari (2012) note that technologies such as stereoscopic 3D upon flat screen monitors can also provide some of the enhanced value offered by HMDs, most notably enhanced display of high-dimensional data (negative parallax for instance, as a means of drawing a virtual object outward from the screen so that its depth is more pronounced) and some limited immersion (via the 3D illusion, the position and orientation of virtual objects appears to shift as the viewer walks around the circumference of the monitor).

One way in which VR really comes into its own in terms of data visualisation is in systems that generate interactive virtual environments based on data outputs of a simulation. Griffon and colleagues (2011) present a good illustration of this with their *Seamless Landscape Explorer* prototype, a virtual environment generator that visualises landscape changes across large regions in response to a 'scenario simulation', such as the effect of further expanding an already large city upon its surrounding villages and countryside. Another significant advantage of VR visualisation lies in its capacity for multi-user, collaborative support. As Chen (2013) observes, the most profound effect of a convergence between visualisation and VR is a social one. Collaborative virtual environments (CVEs) bring with them the facility for multiple users to examine the same visualisation within a shared virtual space, enabling them to exchange interpretations from the same dataset in real time. Multi-user networked VR is itself a highly significant aspect of commercial VR and we shall be returning to CVEs at several points throughout this chapter.

Olshannikova and colleagues (2015) point to the technology as a logical means of presenting visualisations of 'big data' (i.e. data processing and interpretation that 'one can do at a large scale that cannot be done at a smaller one, to extract new insights or create new forms of value'—Mayer-Schönberger and Cukier 2013, p. 6). In their review of big data challenges, Olshannikova and colleagues discuss the '5 V of big data' that comprises volume (sheer quantity of data), velocity (the high speed at which new data is generated), variety (the wide range of forms the data can take), veracity (the great complexity of the data) and discovery[1] (difficulty in extracting quality data from the raw set). These issues are broadly condensed to an assertion that current visualisations of big data sets, typically generated by a computer, are time-consuming to interpret and lacking in nuance. VR presents a powerful solution to the 5 V problems. The capacity of a HMD can display large quantities of data within a single, surrounding scene, whilst head/body tracking enables real time, naturalistic interactions with the data. This empowers the user to be directly involved in the shaping of big data visualisations, making it much easier to interpret them and extract further meaning that simply would not be available in other visualisation forms.

Whilst VR arguably finds favour in visualisations of more abstract and numerically founded data forms (money, statistics, big data, etc.), its augmented cousin along the mixed reality continuum, AR, presents noteworthy value in more concrete applications. For instance, Schall and colleagues (2009) exploit AR technology to visualise data pertaining to underground infrastructure as virtual graphics overlaying actual outdoor environments. Aimed at supporting utilities operations, the system enables field workers to see precisely where (and what kind of) local cabling and pipelines are underneath the ground. Another interesting application of this kind is documented by Nóbrega and colleagues (2008), who convert rainfall and flooding data into virtual flood waters. Users can direct their phone across the edges of an actual body of water to observe likely water levels during a flood event. The system also overlays a virtual flood defence to indicate the size and shape required to successfully mitigate the rising water. Mixed reality applications such as these can provide most of the enhanced functionality of VR in terms of interactivity and three-dimensional imagery, but compromise somewhat on immersion in favour of actual-world contextualisation. Because the majority of AR visualisation tools exploit flat screen mobile technology (smartphone and tablet typically), immersion is not as extensive, compared to a HMD. However, drawing clear connections between the visualisations and actual objects or environments enables users to observe the visualisations within context, in a way makes the information that the visualisation is attempting to project, notably more situated and accessible.

With VR and social media integration very much on the horizon, the immersive, engaging and individualised quality of VR brings the power of visualisations to an unprecedented height. However, it should be acknowledged that VR is not the ultimate solution to all issues pertaining to data visualisation, and as a still emerging technology there is clearly further progress to be made. Furthermore, VR technology also has the potential to amplify existing data visualisation issues, particularly those concerning the perceptual and ethical implications of its usage. As Idris and colleagues (2011) observed how a flashy colour scheme could project false but convincing authenticity, VR presents a possible exponential increase of this problem. Bringing together the comfort and familiarity of social media platforms, the personalisation potential of big data, and the immersion and engagement of virtual environments could collectively be taking us on a path to a future eerily reminiscent of some of the dystopian futures discussed in Chap. 5, in which this technology could be exploited to manipulate receivers, blinding them with flash content and encouraging them to accept the information irrespective of truth. That would of course be a worst-case eventuality, but it does flag the importance of maintaining an awareness of the great potential of VR-based data visualisations to be used underhandedly to deceive and manipulate. Lastly, with regard to issues of limited perceptual understanding (noted by Johnson 2004), VR is highly likely to exaggerate this problem. Whereas standard visualisations (i.e. static graphics) highlight gaps in our understanding of visual perception, the multimodality and interactivity of VR-based visualisations reveal gaps in our understanding across multiple, interrelating modalities. This makes perception of VR visualisations a far more complex matter. Both of the above issues are interrelated and indicative of the need for future research to explore the perceptual aspects of VR visualisations across sensory modalities other than image, the first and foremost of which is most certainly sound.

Sonification Systems and Sound in Data Visualisation

Sound can feature in a variety of different ways with regard to the representation of data. Referring back to Kirk's (2016) description, the sensory modality of data visualisation is not restricted to (or even requiring any) image-based representation. Therefore, a representation composed entirely of sound content can still be classified as a form of data visualisation, albeit one more commonly bestowed with the alternate title of 'sonification'. Kramer and colleagues (1997) define sonification as 'the use of nonspeech audio to convey information' and 'the transformation of data relations into perceived relations in an acoustic signal for the purposes of facilitating communication or interpretation'. This definition clearly encapsulates the functionality of data visualisation and positions sonification as its auditory equivalent.

One of the fundamentals of sonification is the mapping of objective data into discrete sound wave parameters (Hermann 2008). Such approaches to data mapping characteristically involve measures of pitch (of which human hearing is capable of discerning roughly 400 discrete values), loudness (between 50 and 100 discernible values), spectral composition (timbre) and localisation (Kramer et al. 1997; Rimland et al. 2013). Due to such acoustic parameters being highly relevant to music, sonification also possesses a significant conceptual overlap with art. Vickers and colleagues (2017) note that a musical listening attitude embedded within Western culture (and our innate propensity for perceiving sound as a concreate, real-world entity rather than an abstraction) has caused sonification to become entangled with notions of music and art, thereby detracting from its profile as a tool for data representation.

Possibly, the most well-known example of sonification dates back to 1908 and the radiation detection principles that lead to the portable Gieger counter. This device converted each detected ionisation event into a single click, making a collective sound that was both distinct and highly recognisable. For Hermann (2008), sonification describes a relatively fledgling area of research and also a term that can be easily confounded with associates such as 'auditory display' and 'sound visualisation'. Hermann's own taxonomy does address this, defining an auditory display as a much broader concept that encompasses sonification but also can include dialogue systems and speech interface. Also, whilst a sonification system, by Hermann's definition, should represent only objective properties of the input data, an auditory display can be subjective, with the associations between audio and meaning crafted by a designer. Kaper and colleagues (1999) differentiate sonification from 'sound visualisation', identifying each as different sides of the same coin. Whilst sonification denotes a set of sound waves output to represent a dataset, sound visualisation essentially describes the opposite, sound-related information represented by way of a graphical output. A well-known example of sound visualisation would be the representation of ultrasound as a moving image for conducting medical diagnoses (see Nelson and Elvins 1993).

As it pertains to data representation, Rimland and colleagues (2013) posit that the primary function of sonification is 'exploratory data analysis' (identifying broader characteristics that summarise a dataset) and comparisons of expected and observed values. It is the fundamental differences between sound and image that present the latter as both a distinct alternative to visualisation and a significant means of enhancing it. In the previous chapter, a couple of sonification instances were detailed and it was observed that the temporal aspect of sound wave data is one of the primary features of sonification that enable it to realise many of its practical applications. This point is reinforced in various recent sonification projects, including Ballora and colleagues (2012), who combine stock market values with social media posts to create a sonification system that is sensitive to periodicities, patterns and sudden changes in the underlying data. Instances such as this neatly tie sonification to issues with representation of big data. As noted earlier in this chapter, big data is recognised as a significant challenge to data visualisation, due largely to its heterogeneity, generation rate and sheer volume. Dynamics and integration are acknowledged to be essential facets of data representations if we are to be able to draw accurate and valid inferences from them. This clearly points to sonification as a powerful contributor, particularly if utilised alongside visualisation within a powerful multimodal display.

Despite the clear advantages of sound content in this context, many instances of data visualisations, including those built for VR, continue to be silent. Of those that do utilise sound, most multimodal VR data visualisations reinforce the assertion made by Vickers and colleagues (2017), that sonification is intertwined with music and art. For example, a VR visualisation of Canada's current and projected energy demands[2] features various sounds that possess distinctly musical rhythms and timbres. This does not mean, however, that the use of sound within the visualisation is merely decorative. Its most noteworthy and consistent function is as user-feedback that reinforces the graphical changes that occur when the user makes a selection of manipulates some aspect of the scene. Other examples (such as Salesforce's *Rift*-based visualisation of data pertaining to their consumer relationship management company[3]) characteristically utilise sound sparingly and rely upon established earcon designs that although are unlikely to provide a substantial enhancement to the visualisation, are relatively safe implementations that minimise the risk of undermining the graphics.

Although the majority of VR visualisations, such as the above examples, are yet to fully recognise the value of sonification, contemporary research has increasingly utilised sonification systems for a variety of applications that have particular relevance to VR. This has been done by way of various technologies that are steadily being integrated into the broader family of virtual systems. For instance, Väljamäe and colleagues (2013) observe the progressing trend in sonification of electroencephalographic (EEG) data for applications such as neurofeedback, brain-computer interfacing and as a means of manipulating procedural music. This raises some fascinating possibilities for VR. For example, in line with the trend of hardware integration, EEG (or other biometric sensors) could be built into HMDs, acquiring physiological data before (by way of a sonification process) represent that data as audio content within the virtual environment.

A review by Kramer and colleagues (1997) identifies several key areas that warrant ongoing investigation, all of which continue to be relevant today. These include: (1) finding features of sound that can be reliably mapped to data; (2) uncovering ways in which sonification can provide unambiguous added value to data representation (in ways that visuals cannot); (3) developing more complex sonification tools that correspond to high-level perception and cognition; (4) bettering our understanding of the dynamic features of auditory perception; (5) investigating how sonification systems can be integrated with visualisation techniques to produce multimodal data representation and; (6) establishing more integrated and interdisciplinary research. These six points are all largely consistent with ongoing perspectives on sonification up to and including the present day. The third and fourth points, however, also reveals an interesting connection to some of the auditory perception concepts discussed in Chap. 3, including Sonic virtuality. For Kramer and colleagues (1997), the foundational research into perception and cognition should be high-level (i.e. should 'focus on dynamic sound perception, auditory scene analysis, multimodal interaction, and the role of memory and attention' p. 3). The assertion here is that our processing of auditory information does not occur in a perceptual vacuum. Broader perceptual processing must be taken into account in order to design an auditory display that reliably communicates the intended information to the listener, or that properly facilitates a listener's ability to extract new, unbiased information from a dataset. Without this understanding, the intentions of the sonification designer are in danger of being confounded by interpersonal (i.e. different perceptions from one listener to the next) and intersituational (i.e. different perceptions from one context to the next) differences.

At present, sonification and VR for data visualisation reflect two distinctly separate areas of study, despite their being a great deal of untapped potential in combining contemporary principles of sonification with the advanced audio processing toolkits that are now available to VR designers (see Chap. 8). Such tools provide significant opportunity, both in terms of enhancing more common sonification parameters (such as improved localisation potential by way of head-related transfer functions) and presenting new parameters (including occlusion and reverberation) to facilitate multimodal VR that is able to clearly communicate data, even as it becomes bigger and increasingly more complex. Integration is once again a recurrent theme, as future systems would arguably benefit from more sound content, provided that it was implemented with care and founded in the relevant theories of auditory perception.

Telepresence

Telepresence is arguably one of the most fundamental functions of all VR. The term refers to a process of exploiting the immersive and engaging qualities of virtual environments to make the user 'feel' as though they are there. Steuer (1992) goes so far as to assert that the very definition of VR is founded on concepts of presence and telepresence. Notions of presence and telepresence are discussed earlier in Chaps. 2 and 4, but are retuned to here with a focus less upon the underlying concepts and more on the practical applications, specifically those pertaining to VR and sound.

In a discourse featured in the journal *Science* several years ago, Riva and colleagues (2007) emphasise the experiential nature of VR and posit that the feeling of presence within a virtual environment is the defining user-experience characteristic of the technology. Reviewing relevant literature, Riva and colleagues define telepresence as a scenario in which the user experiences the illusion of non-mediation. They connect this to the theory of embodied cognition, asserting that telepresence requires the perception of intent being directly translated into action without any detection of a mediator. These notions are elucidated quite clearly in an ongoing point of contention between industry/popular literature and academia with regard to naturalistic interfaces and telepresence. The popular position repeatedly argues that such interfaces are more likely to break the telepresence illusion by drawing a player's attention towards the actual environment in which they are carrying out the physical actions, and away from the virtual environment in which they are supposed to be immersed (see Bulter 2014). In contrast, the academic position tends to assert that naturalistic interfaces enhance presence and increase enjoyment (see Williams 2014). The likely reason for this conflict is a difference in the underlying definition of presence, with the research-side describing a broader experience of play (i.e. a feeling of being present in a mixed reality space, between the virtual and the physical, and immersed in what you are doing rather than where you are) and the industry-side describing something more akin to feeling as if you are physically present within the virtual world. Either way, this contention raises the question of whether the naturalistic facets (primarily head/body tracking) of VR can actually provide tangible value for telepresence applications.

We shall return to this question later in this section, which outlines some of the key applications of VR for telepresence. We start by exploring how the integration between VR and network technologies has presented huge opportunities for communications, bringing people from across the globe closer together with increased multimodality and fidelity to the extent that we could almost be on the verge of 'virtual teleportation'. Following this look at how VR can bring us to people, the subsequent part of this section discusses how VR can bring us to places, with a review of the potential for this technology within broadcasting.

VR Communications and Social Telepresence

Telepresence reflects a wide range of practical applications for VR. When implemented well, the illusion of telepresence can transport the user to another location, almost akin to physical dissociation (out of body experiences). Telepresence also integrates readily with VR's penchant for multi-user interfaces and shared virtual environments; providing a virtual space in which individuals from around the (physical) world can come together, interacting as if they were only a few feet apart.

Communications technologies already transcend physical barriers to evoke telepresence and bring people together virtually. Examples include teleconferencing and videoconferencing systems but also more traditional actions such as reading a letter from a distant friend or listening to a recording of a live music performance (Steuer 1992). As methods of communication, however, such systems are not without limitations and VR presents a distinct opportunity for making various enhancements to teleconferencing, a single application that arguably represents one of the most substantial values of virtual technology for businesses across almost all commercial sectors. Being able to communicate in a manner more akin to a face-to-face conversation but with the freedom of a network, VR teleconferencing could support numerous interpersonal activities, including meetings, seminars and job interviews. The vision for VR teleconferencing is outlined by Edwards (2011): 'Ideally, participants are life-sized [and] every sound, gesture and facial expression are replicated [...] to make telepresence and in-person meetings virtually indistinguishable from each other' (p. 9). In recent years, Microsoft's AR technology, *Hololens*, has featured prominently in commercial prototype developments. Coining the neologism 'holoportation',[4] Microsoft's project aims to enable multiple users to interact by way of graphical representations that are rendered in real time and accurately mapped to their physical behaviours, which includes their mouth movements and facial expression. As an alternative, some systems utilise 360° video capture to stream live content from a host location to a remote user as a more straightforward means of teleconferencing.[5] More recent research developments in this area include work by Lan and colleagues (2016), who demonstrate a more mixed-methods approach, utilising an array of depth sensors to scan and reconstruct physical users as three-dimensional graphical avatars whilst also generating a 360° image of the host's conference venue.

As Zakrzewski (2016) observes, one of the most significant challenges VR teleconferencing faces is in the accurate and convincing portrayal of facial expression and gesture as they pertain to communicating affective information. Thies and colleagues (2016) present a good example of current virtual teleconferencing research that aims to address this issue. Their prototype system, *FaceVR*, combines facial capture and eye tracking with an AR conferencing system using the Oculus *Rift* headset. Whilst the user wears the Rift, their face and eye geometry and also their movements are captured and rendered graphically. Quick Response Code labels are then used to project the resultant image over the user's face, thereby cancelling the obfuscation caused by the headset. The result is that a user can participate in a teleconference using the Rift, but other individuals in the conversation observe the user as if they were not wearing a headset at all. Alternate approaches to VR teleconferencing include an interesting mixed reality study by Regenbrecht and colleagues (2015), in which VR and AR elements are combined. Here, users interact within a shared virtual (i.e. graphically rendered) environment with translucent webcam feeds of the physical users overlaid as a kind of heads-up display (essentially a kind of augmented reality simulator).

At present, VR teleconferencing is still very much in its experimental phase and the extent to which the technology can evoke the perception of teleporting into a shared space is yet to be established. Theoretically, whilst VR's naturalistic interfaces could have the potential to break (rather than evoke) the illusion of telepresence, this is not an issue inherent to physical movement as a whole, but rather with the technology's tendency to lack accuracy, reliability and contextualised feedback. These problems 'expose the wires' of the interface, revealing the presence of a mediator and, in accordance with Riva and colleagues' (2007) definition, compromising the illusion of telepresence. Therefore VR should possess significant potential for inducing telepresence provided the mechanisms are subtle, fully working and properly embedded within the virtual environment. This neatly raises one particularly important consideration for evoking telepresence in VR: sound.

One of the most surprising aspects of current developments in VR teleconferencing is that spatial/positional sound processing continues to be underutilised despite being a central issue in all forms of network-mediated conferencing. From a traditional conference telephone call to experimental mixed reality prototypes, multi-person conversations suffer most notably in their stilted dialogue, with each participant struggling to receive straightforward conversational cues and therefore unsure as to when they should speak. Long pauses abound, followed inevitably by multiple voices talking over one another. If the audio of a teleconference is presented mono/stereophonically, such moments of overlap typically require the conversation to reset, as the convolution of two voices into a single stream makes it near-impossible to decipher the speech of either voice. Positional sound could offer a powerful solution to this issue, with ambisonic and binaural capture as well as post-processing spatialisation techniques presenting all viable options. With each individual voice processed into a clearly discernible stream by way of its three-dimensional positioning in virtual space, the potential for a listener to comprehend a single voice from multiple simultaneous voices would be significantly increased. Furthermore, within a VR context, positional sound in tandem with head tracking would provide a substantial enhancement, with the user able to perceive multiple voices all around them and adjust their orientation to help focus on a particular speaker. Ultimately, this could lead to scenarios in which multiple conversations can be had simultaneously, all within the same virtual space.

That contemporary teleconferencing systems continue to use non-spatial forms of sound presentation becomes more perplexing when considering that the history of research into this precise issue dates back to the 1990s. For instance, a report by Evans and colleagues (1997) compares 'sound field simulation' (referring to spatial audio capture techniques such as ambisonics) against 'perceptual synthesis' (simulated localisation cues—HRTF processing), as alternate means of enhancing teleconferencing with spatial audio. The findings of this report note that both approaches have relative advantages and limitations; with ambisonics working particularly effectively with large speaker arrays and in fixed, regularly used physical spaces for which the B-format recording can be precisely calibrated, whilst HRTF processing favours headphones and more flexible environments.

Curiously echoing the gap between 1990s and contemporary consumer VR, research concerning spatial audio quietened significantly throughout the 2000s before resurfacing around 2010, where much of the published literature initially focussed upon revisiting established spatialisation concepts within a more contemporary context (see Hyder et al. 2012). However, as we get closer to the present day, a renewed interest in advancing spatial sound for VR teleconferencing becomes apparent. In a patent for a spatial sound conferencing server, Gardner (2013) observes that client-server teleconferencing architecture (i.e. each user has a bidirectional connection to a single server which acts as a central hub) is progressing in-part thanks to digital games, that are increasingly utilising the technology to facilitate 'chat' during online play. Gardner's patent was, in basic terms, a server-side 'multiplayer spatialisation' audio processor, the function of which was to allow participants in a networked virtual world to converse naturally as in a real-life situation. This was achieved using HRTFs, stereo reverb, occlusion, distant-relevant attenuation and a Doppler effect—all of which are key audio processing techniques already discussed in Chap. 8. In addition to digital gameplay, this technology has broader potential for bringing people together within various forms of virtual environments, in which they can communicate effectively through sound, with an experience almost indistinguishable from a face-to-face meeting.

VR Broadcasting and Telepresence in Physical Places

Being able to experience a live event, both as a recording and in real time by way of streaming, is a facet of contemporary VR that features heavily in marketing for the technology. Television advertisements for Samsung's *Gear VR,* for instance, include one in which a deeply uncomfortable but stalwart father stands amongst a sea of hysterically screaming adolescents, his arm aloft and clutching a 360° camera before the adverts cuts to a shot of his daughter watching the concert in her bedroom. Experiencing events in VR by way of real-time 'live-streaming', commonly known as 'live broadcast VR', is becoming very popular indeed. Live broadcast VR of music performances and sporting events, in particular, is increasing exponentially (Hayden 2016), with many partnerships forming between concert promoters and VR companies. The intention is to circumvent the physical limitations of audience numbers and potentially sell tickets to many multiples of a venue's actual capacity (della Cava 2016). The technology underpinning VR broadcasting is relatively straightforward and is essentially an integration of 360° video that responds in a HMD to the user's head rotations and lives streaming of audio-visual content. Extensions of this principle have begun to appear in experimental prototypes. For example, a recent patent by Huston (2016) describes a system and method for a 'virtual spectator' prototype that would enable the user to switch between multiple cameras, positioned along the perimeter of a sporting venue, and adjust (zoom, pan and tilt) the image from each camera.

Broadcast VR is of course, not limited to music and sporting contexts. According to Lawler (2015), the first live VR programme originated from California's Laguna Beach in January 2015, courtesy of the VR-specific broadcaster NextVR. Since then the technology has enabled users to remotely spectate live events as diverse as surgery (Volpicelli 2016) and news casting (Dredge 2015). With regard to VR news casting in particular, enabling the user to immediately visit the scene of a news story has significant potential value with regard to engaging more people in current affairs. As Rogers (2016) notes: 'the best journalism makes you feel like you are part of the story. You care what happens'. Furthermore, increasing direct access for viewers, delivering them more factual information and enabling them to be 'present' in the situation enable them to observe events for themselves and form their own opinions more confidently.

Whether viewing a sporting event, music concert, or even a political party conference (should one find such a thing appealing), the features of VR have the potential go beyond offering consumers a virtual equivalent of actual attendance. In much, the same way as embracing DVDs and Blu-Ray media rewarded the consumer with 'bonus content' (commentaries, deleted scenes, etc.), broadcast VR raises the possibility of presenting a wealth of additional, exclusive content to users. Whittle (2016) cites *Eon Sports VR*,[6] a broadcast VR platform that enables users to stream content from 360° cameras. In addition to the core content generated by cameras placed around the perimeter of the stadium, further cameras positioned in typically inaccessible areas (such as locker rooms) and affixed to the players themselves provide unique access that would not be available without the technology. This of course could extend to any other form of event, with bonus VR content including backstage access before and after a live music performance, live viewings of rehearsals and training sessions, or extended access to interviews and press conferences.

At present, broadcast VR is certainly not without issues. Image fidelity for instance is very rarely in high-definition due to the bandwidth constraints, a limitation that also affects the current capacity for broadcasting multiple camera feeds simultaneously to allow users to 'teleport' between perspectives, and the costs of acquiring the required proprietary hardware and software remain prohibitive.[7] Additionally, Low (2016) makes an interesting observation with regard to the intellectual property aspect of virtual concerts, asking if online piracy "has been the bane of the music industry [...] what happens when live performances are ported to a virtual medium that all of a sudden may be subject to piracy again?" (p. 425). With VR broadcasting extending beyond music events, the opportunity for illegitimate accessing of content is even greater and may have far-reaching implications for multiple industries.

Returning briefly to the more optimistic side of VR broadcasting, this section has so far focused upon VR as it pertains to telepresence and the virtual experiencing (and re-experiencing) of actual events. VR technology does however extend its functionality into scenarios in which the user is physically present at the actual event. Mixed reality systems, in particular, provide noteworthy value in a live event context. Recent AR prototypes utilise the technology to present digital information as a graphical overlay to the physical action. Bielli and Harris (2015), for example, demonstrate how various statistics and background information can be superimposed over the spectator's view, enabling them to observe everything within the same visual frame rather than requiring them to turn away from the action (as would be the case if viewing on a flat screen).

As primarily a capture of the physical world, the priority for sound in broadcast VR is arguably tied to relevant signal processing techniques, of which the foremost is positional audio. Dysonics[8] lay claim to being the first instance of a live, audio-based VR programme, in which a bespoke array of eight omnidirectional microphones captured the ambient soundscape of a roomful of people. In a separate demonstration space of similar dimensions, users adorned headphones that tracked their movements, enabling them to walk around the broadcasted soundscape. Unlike VR teleconferencing, broadcast VR appears to be more considerate of the auditory aspects of the technology. In addition to its support of live-streamed 360° video, YouTube incorporated support for streaming of binaural recordings (Kelion 2016). As it is asserted in an ongoing research and development project at the BBC,[9] the increasing ubiquity of smartphones and mobile hardware has resulted in many more people consuming sound content by way of headphones, thereby providing significant impetus for broadcasters to incorporate spatial audio in their programming.

However, there is further opportunity for creative audio processing as more than a way of complementing the immersive quality of the visual display. A good example of this is 'augmented audio'. Augmented audio describes several applications that can be broadly separated into two categories. The first is the auditory equivalent of the AR systems mentioned above and comprises the overlaying of a user's natural soundscape with additional live audio content. The second (and more interesting) involves the capture, manipulation and immediate playback of incoming audio signals in real time. This allows the listener to experience a live augmentation of a physical soundscape (see Moustakas et al. 2016), functionality that has notable application for live events. As O'Kane (2016) notes, consumer-grade AR earbuds by way of connection to a smartphone can enable a user to filter the spectral composition of their immediate soundscape. During a live music concert for example, a user can deploy the earbuds to customise the EQ (frequency components) of a live performance to their individual liking.

The earlier discussion on VR data visualisation highlights several ways in which sound can be of value by way of enhancing functionality through positional audio, environmentally modelled sound and sonification systems. VR for telepresence reveals additional important applications of VR sound but with a focus that is slightly less upon function and more upon experience. Once again, positional audio and environmental modelling are identified as vital elements but, in this context, the intention is to present an experience that meets a user's expectations for realism, thereby evoking telepresence and drawing them into the virtual world. This context also raises the issue of user-preference and individualised audio, in which the VR environment enables the user to control various aspects of a physical soundscape. Whether the user wishes to amplify the bass frequencies of a rap concert, or attenuate low-intensity environmental noise during a play to increase the perceived fidelity of the actors' voices, VR sound can oblige.

Education and Skills Training

As a further application of virtual technologies, education and skills training encapsulates a vast range of possibilities. Although still an emerging pedagogic tool, schools have begun to consider the potential of VR as an asset to their curriculum delivery, with some teachers already implementing HMDs in their lessons (Lawrie 2017). Technology companies have also been quick to highlight and support the educational potential of VR in an attempt to dispel presumptions that VR is primarily for playing games. Online communities such as *Virtual Reality for Education*[10] and *Unimersiv*[11] have been established to help raise awareness of VR for education and also to consolidate multiple forms of content that includes 360° videos, mobile VR apps and online tools for helping users create their own VR materials.

In this section, we take a brief look at VR as it pertains to education, including a brief history on how this application of VR has developed in recent years and also a review of the most salient issues that feature in related research. We examine how virtual technology can enhance the learning of abstract concepts and practical skills, both in terms of mainstream schooling and as part of various forms of skills training and professional development. The structure of this section begins more broadly with the theoretical foundations of constructivism and the 'extended classroom' concept. Within this concept, its computer-mediated/virtual component 'eLearning' is discussed. This then leads us to Virtual Learning Environments before the role of VR-specific technology is considered and we consider the role and value of sound as a means of enhancing and extending VR for education.

The Extended Classroom, Virtual Learning Environments and eLearning

During the first generation of consumer VR as it developed throughout the 1990s, the potential of VR to support learning was acknowledged repeatedly (Helsel 1992; Psotka 1995). As Seidel and Chatelier (2013) assert, pedagogic practice of the time was embracing 'the extended classroom' concept, an idea that laid the foundation for VR as an educational tool. The extended classroom refers to methods of facilitating learning opportunities outside of the classroom environment, supplementing traditional teaching to enhance the overall effectiveness of a learning programme. The concept is itself built from the constructivist philosophy of education founded by Jean Piaget, the central tenet of which states that humans establish meaning from interactions between ideas and experience (see Piaget 1955). Approaches to the extended classroom typically utilise either physical locations within the wider natural world, such as field trips, or by way of a computer to extend into the wider virtual world (also known as 'eLearning'). As its name implies, eLearning refers to any method of educating that utilises some form of electronic technology. The practices and skills of eLearning extend from Bloom's (1956) taxonomy of educational objectives to include searching and selecting, exploration and promotion, testing theories, analysis and synthesis of information, collaboration and discussion, greater understanding, the ability to apply knowledge and finally fostering creativity (Holmes and Gardner 2006). As we shall get to shortly, these objectives elucidate why associations between VR and constructivist eLearning principles have been so easily and frequently drawn. The eLearning aspect of the extended classroom is encompassed significantly by the Internet as a means of accessing, creating and sharing information (Schneider 1998). It is, therefore, unsurprising that the majority of applications discussed below utilise the Internet to varying degrees, and why many of the educational VR tools embrace networks and collaborative multi-user systems.

As Mallon and colleagues (2012) observe, sound in eLearning has struggled to make progress due to the lack of a theoretical framework, leaving designers unsure as to how to implement their audio effectively. Their response highlights the numerous contributions well-designed audio can have in an eLearning context, including: decorative function and increased aesthetic appeal, better representation/realism (i.e. auditory icons: causal and environmental sounds that reinforce the visual representation); greater engagement and presence by way of enhanced user-feedback; increased embodiment by way of relational sound (e.g. positional audio that gives the user a relative sense of place in the virtual world); transformational understanding (using the dynamics of sound to indicate state changes); and easier interpretation by way of earcons and musical cues that can clearly present a perspective, thought or intended meaning from the designer to the user.

As defined by Dillenbourg and colleagues (2002), the term 'Virtual Learning Environment' (VLE) describes an important subdivision of the eLearning approach to extended classroom learning. More specifically, a VLE is an informational and social space, within which multiple users can interact and also contribute to the construction of the environment. Dalgarno and Lee (2010) identify five key learning benefits afforded by a well-designed and implemented VLE, namely: (1) the opportunity to represent spatial knowledge (i.e. understanding relative measurements, navigation/map reading, etc.); (2) more experiential learning (learning through experience and role-playing); (3) less abstract learning that can be more readily understood in its context; (4) greater potential for learning collaboratively and; (5) improvements to learner engagement.

With regard to the requirements for developing an effective VLE, immersive experience and interactivity are pivotal characteristics that draw clear associations with constructivism. This point is underlined by

Mikropoulos and Natsis (2011), who posit that the theoretical framework behind the majority of VLEs is fundamentally based on constructivist pedagogy. In their report that presents a general framework of VLEs, Dalgarno and Lee (2010) assert that 'representational fidelity' (i.e. the quality of the output—this could include image resolution, object behaviours consistent with user expectation, reliable feedback to user's actions and spatial audio) and 'learner interaction' (the extent to which user's actions are evoke embodiment by being meaningful and involved) are the central requirements of a functioning, fully immersive VLE. In addition to these requirements, Stiles (2000) asserts that learning is, amongst other things, an inherently social process. Consequently, in order for VLEs to work effectively within a curriculum, social processes must be an integral component of their functionality. The expectation for a VLE is ultimately to deliver an educational experience that enriches classroom teaching and facilitates the extended classroom approach by integrating with distance learning. Adding to this list, a VLE is also expected to facilitate multiple pedagogic approaches and this kind of flexibility is also noted by Dillenbourg and colleagues (2002) to be an important feature for both teachers and learners, enabling teachers to have freedom over how they design and distribute the materials, whilst allowing learners to access and interact with these materials in different and creative ways, thereby reflecting the heterogeneity of learning and appreciating the uniqueness of the individual.

In practical terms, VLE most commonly refers to an online two-dimensional toolset that enables educators to upload various types of resources (video lectures, course guides, presentation slides, etc.), schedule programmes of learning and setup online assessments. For the learners, VLEs provide remote access to these resources but also enable direct messaging between the learner, their tutor and their peers: facilitating communication and collaborative working. The software known as *Moodle* (Dougiamas 2002) is currently one of the most prominent open source examples of a VLE, with an architecture that enables the functionality of the software to be infinitely customised and extended by its users. Moodle exemplifies a specific form of VLE commonly referred to as 'learning management software'. If we consider this functionality against the theoretical requirements outlined above, learning management software arguably meets these requirements in quantified and purely functional terms. However, questions regarding its quality emerge when we consider this type of VLE in terms of user experience, in particular when we compare it against in-person education. Text-based messaging lacks the embodied experience of face-to-face communication. Learning materials are more abstract and their contexts are more difficult to understand. Feedback from peers and tutors is rarely in real time, and the VLE itself provides greater access to learning materials but their presentation lacks any immersive quality. Of course, a VLE such as Moodle is typically provided as a supplement to in-person teaching. Learning management software is however also used in some distance learning courses, a context in which the above points would present difficulties.

With regard to sound, the majority of learning management software tools are almost entirely silent. Sound content is typically only present as user-uploaded material and the main interface of these VLEs, much like the majority of browser-based interfaces, relies completely upon visual information and cues. From an eLearning perspective, this distinct lack of audio highlights part of the qualitative issue with such types of VLE. As Blazey (2015) observes, sound offers several benefits to learning, from increasing the potential to engage with the learner on an emotional level, to providing important cues without relying solely on visuals. Avoidance of sound in VLEs presents additional issues with regard to accessibility, with blind or partially sighted individuals unable to make use of a graphics-only system. This restriction also applies to some extent to individuals whose preferred learning styles are auditory as opposed to visual (see McNutt and Brennan 2005). Whilst learning management tools and other two-dimensional VLE forms benefit less from the spatialisation aspects of VR sound, various other auditory features would still be of considerable value. Well-designed earcons in particular are essential components of a good user interface. They are also powerful means of providing users with regular formative feedback that is far less likely than visual cues to distract their attention away from the task at hand.

Virtual Reality Learning Environments

Whilst Virtual Learning Environments undoubtedly present significant and continuing value in their two-dimensional form, a wealth of further opportunities are presented when we consider VLEs in three-dimensions. Meredith Bricken (1991) was one of the first researchers to publish the term 'Virtual Reality Learning Environments' (VRLE) in her review of the possibilities and challenges of VR technology for education. As Bricken observes, 'VR is experiential' (p. 178); a spatial, interactive and multisensory environment within which the user is physically and perceptually involved. The VRLE concept of the early 1990s was, much like in recreational VR, embodied by the HMD and glove-based tactile interfaces. In its ambitions, it supported more naturalistic interactions with facts and ideas, which could be more readily contextualised by way of high fidelity audio-visual representations. In principle, VRLEs had the potential to satisfy all of the constructivist requirements for an extended classroom experience, but to a greater qualitative degree than a traditional VLE. Despite this great promise, Bricken (1991) also acknowledges several challenges posed by the technology of the time. These included prohibitive costs of both the hardware and the software (particularly in a classroom context that would necessitate multiple units), difficulty in setting up and using the technology, and also several fears and ethical concerns relating to VR use (both in the classroom and in general—see conservative fears in Chap. 2). As the 1990s progressed and consumer interest in the technology faded, use of VRLEs failed to become ubiquitous and eventually diminished into obscurity. However, the fundamental concepts underpinning the technology remained and, in the years that followed, an alternative virtual technology emerged that would both keep the VRLE alive and aid its development: the humble digital game.

Several years prior to more recent developments, the VRLE concept embraced various characteristics of digital games as a means of realising the benefits of the computer-mediated extended classroom, whilst avoiding some of the pitfalls that HMD-based systems had presented. More accessible software packages, operating upon a standard personal computer platform, presented what Fowler (2015) describes as 'semi-immersive' VRLEs, sacrificing 'full immersion' (as attributed to CAVE/HMD systems) for cost effectiveness, accessibility and greater ease of use. Such semi-immersive VRLEs have now become a well-established educational tool, used throughout the last two decades and continuing to be popular today.

Whilst many semi-immersive VRLEs are built specifically for educational purposes, some of the most recognised and highly regarded instances were not built with education exclusively in mind. Possibly, one of the most noteworthy instances of this is Philip Rosedale's massively multiplayer online (MMO) title *Second Life*. Commonly described as a three-dimensional means of interfacing with the Internet, the central mechanics involve exploration and multiplayer interaction within a vast virtual world, the content for which is almost entirely created and uploaded by the game's community of users. This content can theoretically be anything that can be modelled graphically or as audio and includes the design of the user's own avatar. As an MMO, Second Life comprises a very large community of subscribers who can interact either by direct speech using VoIP (voice over internet protocol) or through gestural animations built into their avatar. Users can also share and exchange content, providing the Second Life world with a functioning virtual economy (content can be bought and sold for actual currency) and opening up even more applications for the platform. As a recognised and accredited higher education institution that specialises in distance learning, the Open University became one of the first in the UK to establish an educational platform within Second Life[12] and in 2008, it was estimated that approximately three-quarters of UK universities were using Second Life in some capacity (Kirriemuir 2008). Warburton (2009) identifies numerous specific educational functions of Second Life that include (but are not limited to): role-playing and simulations, data visualisations, exhibits, historical recreations and of course, the development of skill relating to the generation of graphical and audio content for Second Life itself. Warburton's review of Second Life substantially reflects the learning affordances of VLEs in general (see earlier discussion on points raised by Dalgarno and Lee 2010) by asserting that Second Life presents opportunities for more deeply contextualised learning. Specifically, learning that enables collaborations by way of a community presence, and also the immersion of users in a way that evokes an affective and memorable experience to enhance users' motivation for learning and their retention of knowledge.

To give another key example of game-based VRLEs; a few short months following the release of the sandbox survival and creation game *Minecraft* (Persson et al. 2011), a flurry of academic interest appeared, largely advocating the use of the game as an educational tool. Minecraft's core functionality involves the exploration of a procedurally generated cubist environment. Its environment includes night and day cycles and the player has the ability to smash blocks for resources that are then used for building. A deceptively simple game and in some ways comparable to Second Life, Minecraft presents distinctly open gameplay in which the player operates with a great deal of freedom. This freedom contributes significantly to the pedagogic potential of the game. Short (2012) asserts that Minecraft's 'ecological base' (being a simplified simulation of earth and its processes) imbues the game with numerous educational applications that include the teaching of biology (creating maps of the human body and its functions), ecology (geography, climate, ecosystems, etc.), electronics (blocks can be used as logic gates to construct circuits) and chemistry (various block 'element' types can be splited, mixed and bonded). As with Second Life, the key facets of Minecraft's user experience, provided the game was utilised appropriately in conjunction with classroom teaching, resonated strongly with the constructivist principles of VLEs.

Sound features significantly in both Second Life and Minecraft. The user interface of Second Life incorporates various earcons to enable users to easily navigate the menus and options, whilst the three-dimensional world of both titles feature various environmental and weather sounds. Positional audio appears in both titles and various actions of the users' avatars are also fed back with audio cues that help to enhance embodiment. Second Life also presents users with the ability to upload their own audio content and attribute it to various actions (known as 'gestures') such as attaching a sample of the user speaking 'hello' to the waving of the avatar's hand.

As we approach the present day, semi-immersive VRLEs such as Minecraft and Second Life have arguably laid the groundwork for the return of fully immersive VRLEs, a point highlighted by both of these titles now featuring HMD-compatible versions (Metz 2017). With contemporary HMDs now approaching the stage at which its costs, usability and fidelity are on par with semi-immersive VRLEs, the education industry has renewed its interest in the technology's potential. With more accurate head and body tracking now afforded by contemporary HMD systems, the potential for reinforcing learning through telepresence is becoming more accessible. Fully immersive headsets obscure distractions, putting the focus squarely on the contents of the virtual world. High fidelity and detailed multisensory environments establish more concrete representations of the information that is to be learned. Naturalistic interfacing with hand and head movements coupled with a multisensory environment that surrounds the user evokes presence and reduces disconnect between the learner and the VRLE. This does not mean, however, that contemporary VRLEs can be encapsulated by a digital environment, being experienced through a head mounted display. As has been asserted on several occasions throughout this book, what we think of as VR should extend beyond HMDs and include wider aspects of human–computer interaction and mixed reality systems. More recent research concerning VRLE approaches is supportive of this view, with various studies examining the value of, amongst others, haptic and force feedback interfaces for interacting with semi-immersive virtual environments (de Boer et al. 2017; Fiard et al. 2014) and electroencephalographic biofeedback mechanisms that provide learners with real-time cognitive and affective state information whilst interacting with a VRLE (Hubbard et al. 2017). This reinforces the notion that, as with the various other wider applications of VR, the headset provides value as a means of greater immersion, but in conjunction with further VR technologies, also becomes a central hub to a much broader repertoire of functionalities.

As three-dimensional environments with a user-experience prioritisation for immersion, clear representation of the learning context and affect-led engagement, VRLEs have a great deal to gain from good sound design and implementation. For learning contexts that relate to actual physical environments, detailed and high fidelity audio modelling is essential. For example, experiencing a VR recreation of a historical battle that visually surrounds the user but is presented in unresponsive stereophonic sound will almost certainly undermine any sense of presence. With regard to the wider virtual technologies that have applications for education, the potential value of sound within these applications also increases significantly when such technologies are introduced. The integration of biofeedback and full body tracking for example raise some very interesting possibilities; from sonification of a user's actual heartbeat or respiration rate as a real-time indicator of stress, to physically modelled audio cues that respond accurately and in real time as the user slowly scrapes the virtual chalk across the virtual board.

Skills Training and Augmented Reality Applications

The many applications within the field of VR for education are incredibly diverse and can theoretically be developed for any industry that has a requirement for engaging either its customers or employees in learning activity. VRLEs, in particular, present their wide-ranging value when we consider the two broad categories of learning, namely academics and skills. To elucidate on this point, Gemma (2014) explains that knowledge can be differentiated into knowledge that is propositional and explicit (*of* something) and knowledge that is procedural and tacit (*how to do* something). Gemma also observes that whilst the former of these types can be relatively easily communicated in abstract terms by static means (such as a textbook), tacit knowledge, is particularly difficult to express and typically realised solely by way of one-to-one training, provided by an expert (or 'Master'). As being able to communicate tacit and procedural knowledge is fundamental to developing practical skills, the value of this ability is self-evident and, as a result, the potential of contemporary VRLEs to facilitate this form of learning by way of an automated and network-accessible system has arguably become of great interest to both industry and academia. For instance, the VRLE as a means of developing surgical skills has featured in numerous studies, particularly those that in actuality are conducted with a camera and visual display, such as endoscopic and laparoscopic (keyhole surgery) procedures (Aggarwal et al. 2007; Kühnapfel et al. 2000; Larsen et al. 2009). VRLEs also have noteworthy value for those who work in more dangerous environments. Grabowski and Jankowski (2015), for example, present a prototype VRLE for training underground coal miners, in which users progress through various exercises representative of common actions in the actual work, from drilling blasting holes to preparing and detonating explosives safely.

In an evaluation by Gavish and colleagues (2015), learning tools that utilise augmented reality (AR) present a different set of pedagogical advantages. With the ability to superimpose digital information upon actual physical environments, what could be described as 'Augmented Reality Learning Environments' (ARLEs) essentially bridge the gap between the natural extended classroom and the technological. For Gavish and colleagues, ARLEs are of significant value to any practical skills that are to be utilised in the physical world, citing industrial maintenance (from repairing a circuit board to performing checks on an aircraft) and military training as two key examples.

When considering the vast range of possibilities for this application of VR and its current status as an emerging and experimental technology, it is clear that there is much progress to be made. As Salzman and colleagues (1999) posit, how VR works in a learning context is a highly complex question, with none of the key elements (such as the features of the VR hardware/software being used, the concept that the learner is attempting to engage with, the subjective nature of the interaction and learning experience, and of course traits of the individual learner themselves) operating in isolation. They are instead interrelated, making it very difficult to predict with certainty, how well a learner or group will respond to a specific VRLE configuration. Another noteworthy concern relates to the various side effects that are characteristic of VR experience, particularly with prolonged use (see Chap. 2). Issues such as simulator sickness, fatigue and eye/neck strain present an ongoing challenge to developers and progress in this area is certainly required.

Returning once again to consider sound, the benefits and approaches to sound design in VRLEs for procedural skills training largely resonate with those already discussed above. Well-designed audio presents benefits in terms of aesthetics, emotional engagement of the learner, reinforcement of VRLE representations, and greater accessibility to auditory learners and individuals with visual impairments. In ARLEs, sound has the advantage over visuals of being less distracting, to the extent that good ALRE design should arguably prioritise auditory overlays and utilise graphics more minimally as a reinforcement. For instance, if directing a learner on repairing a circuit board, overlaying text instructions and graphical feedback on each step would very quickly obscure the overall visual field. Alternatively, speech-based instructions and earcons as performance feedback, coupled with minimal graphics that precisely direct the user, would avoid visual overload and present a better learning experience.

Speech sound certainly poses a significant value, both in pre-recorded forms attached to non-player avatars and in real-time VoIP that enables multiple users to communicate and collaborate easily and naturally. Gigly asserts that non-speech auditory information within VRLEs has less function as a deliverer of explicit information and more as a 'situation monitor' (p. 118). Whilst this perspective does align with aspects of our earlier discussions, it should not be interpreted to mean that such sound is of limited value. As Mallon and colleagues (2012) assert, the dynamics of sound can be of vital importance in communicating transitions and changes of state. Furthermore, when considering our earlier discussions on sonification as a means of representing data, sound presents a distinct advantage in terms of demonstrating information. As a dynamic modality that can utilise numerous fluctuating parameters (loudness, reverberation, attenuation, spectral composition, 3D positon, etc.), sound is far more able that graphics to represent information and concepts that are bigger, broader and more complex. This has significant potential benefit to VRLEs and indeed all eLearning as a means of teaching overarching concepts where the priority is for the learner to appreciate the wood and not the trees.

Virtual Creativity and Artistic Applications

The connection between VR and creativity is profoundly expressed by Thomas Hohstadt (2013), who asserts that 'VR is the prototype of all creativity. It trumpets the very spirit of creativity' (p. 34). Virtual worlds are 'created worlds', crafted with intent to match a design drawn from the imagination. With regard to the wider practical applications of VR technology however, contemporary VR presents us with several innovative means of creating; from painting and sculpture to musical composition and architectural design. This section begins with a brief review of several of the most prominent areas of VR's creative applications, commencing with more artistic and expressionistic endeavours before examining how VR can benefit product design by way of rapid virtual prototyping. This section then contextualises VR for creativity with a discussion on various theoretical frameworks of creativity. As with the other sections of this chapter, the current status of sound within this context is reviewed and the section closes with a look at how the quality of VR for creativity can be enhanced in future systems by way of sound.

Creative Applications in VR

VR as a means of artistic expression has begun to successfully gather recognition and acceptance within the arts community, with the technology beginning to feature in popular exhibits within prominent galleries. VR has even been credited with contributing to the sustaining of the art market.[13] Both VR and AR have an established presence in the arts. In the majority of cases however, creative content is produced *for* VR/AR, not *within* VR/AR. This section focussed primarily on the latter, examining how contemporary virtual technology can facilitate and innovate the creative process.

Published by Google in 2016, *Tilt Brush*[14] is possibly the most renowned creative VR application presently available to the general public. Operating on both the HTC *Vive* and the Oculus *Rift*, the concept of Tilt Brush is relatively straightforward. Users don the HMD and use motion controllers to 'paint' with various brushes and effects, all within a three-dimensional environment. The most significant aspect of the experience relates to the immersion of VR, with users able to paint around themselves, essentially creating art from within the artwork itself. Of course, the creative applications of VR are certainly not limited to painting. Tools such as *Oculus Medium*[15] enable users to conduct 3D modelling from within a virtual space. This tool provides standard geometric modelling, sculpting and painting of virtual objects (or even entire environments) that can be implemented within a game engine to work as part of a VR/digital game; creating worlds from within worlds.

Although the first thought in VR for art is typically in the visual modality, sound is nevertheless quite well represented. For instance, *SoundStage*[16] is a VR 'sandbox' music performance and sound design tool in which users can create, configure and play bespoke synthesised instruments (from electronic drums, to keyboard synths and even a Theremin for good measure) from within a fully immersive virtual space. The timbral qualities of each instrument can be customised and the VR environment can be filled with multiple pieces of virtual music equipment. This surrounds the user, positioning them at the centre of a virtual music studio of their own design. A similar product both in name and function, *Soundscape* (see Pangburn 2015) balances performance and compositional aspects of music creation, again all from within a virtual world. Users can edit the timbral parameters of a VR synthesiser but also generate melodic phrases by 'drawing' sequences of notes. Soundscape also facilitates collaborative creation, with multiple users able to share the same virtual space and contribute to the music.

Product design and prototyping is arguably where VR as a means of facilitating creative processes is most established and embedded within multiple industries. As Kerttula and colleagues (1997) assert, immersive multimodal experience encompasses the primary advantage of virtual technologies for prototyping, within which both designers and customers can 'see, touch, hear and operate a future product before its physical implementation with lower cost and effort' (p. 2). For Drogemuller (2009), virtual prototyping describes 'the use of computer-based tools to model a proposed construction and to run a series of analyses against the model […] before starting to physically construct the proposed facility'. Here, virtual prototyping is characteristically interactive, with users able to influence numerous aspects of the design and its context to enable analysis and evaluation. 'VR-prototyping', largely refers to a subset of virtual prototyping, the defining characteristic of which is an immersive display; this can include HMDs and CAVE systems but also extended flat-screen systems such as Powerwalls (a matrix of displays that present a very large overall screen size without stretching the resolution—as would be the case with a large projector screen).

Remaining consistent with the historical theme that has emerged throughout this chapter, the implementation of VR technologies for prototyping and product design has roots in 1990s research and innovation, with many of the key themes reappearing in more recent years For example, Jayaram and colleagues (1997) assert the value of VR as an extension of Computer-Aided Design, specifically with regard to assembly processes (which accounts for the majority of costs across many industries); an aspect of VR-prototyping that also appears repeatedly in more recent research (Gavish et al. 2015; Seth et al. 2011).

Like some of the other applications discussed throughout this chapter, VR-prototyping primarily serves as an extension of an existing virtual technology; utilising the distinct experiential and interactivity features of VR to both enhance and extend existing functionality. These enhancements can be relevant across the various stages of product development, meaning that VR-prototyping can improve aspects such as ergonomics/usability refinement, constructability (i.e. the virtual assembly line) and of course, aesthetics (Craig et al. 2009). A recent review by Berg and Vance (2017) reinforces these benefits and adds 'storytelling' (prototypes within virtual environments that can be operated within a virtual narrative relating to an actual-world scenario), 'packaging' (relative positioning of objects in virtual space to give the evaluator a sense of perspective) and abstract data visualisation. In terms of the specific ways in which VR-prototyping offers significant value, Berg and Vance also identify the key industrial domains as: aerospace, agriculture, automotive, construction, consumer goods, energy and the military. To expand a little, VR-prototyping and construction can be brought together by way of architectural walkthroughs. In this instance, the technology can enable users to become fully immersed in a not-yet constructed building, as a means of more comprehensively communicating the architect's intention to the various stakeholders. As Craig and colleagues (2009) state, VR in this area has application reaching both into the future and the past, with walkthroughs able to present virtual recreations of historical architecture as a further means of informing contemporary design.

Creativity Theory and Its Relevance to VR

In an article concerning the implications of integrating technology with the creative design process, Fairchild and colleagues (2011) chart the history pertaining to our understanding of our own creativity. Initially perceived as an unknowable entity defiant of scientific observation, study of creativity focussed upon examining the effects that environmental and situational variables had upon creative outputs, with the intermediary process itself a mystery. During the early twentieth century, multi-stage models appeared, including that of Wallas (1926), who explained the creative process as 'preparation' (conscious analysis of the problem), leading to 'incubation' (subconscious synthesis and testing of ideas) before emerging as 'inspiration' (the idea translates from the subconscious to the conscious mind) and finally 'verification' (conscious checking of the idea). In more recent years, several additions have been made to this four-stage process. Goleman and colleagues (1992) posit the existence of an additional 'frustration' stage, in which cognitive overload during preparation forces the mind to cease examining the problem consciously, thereby engaging the incubation process. A further 'imple mentation'/'communication' phase has also been suggested, in which the individual utilises social support to develop ideas prior to inspiration (Amabile 1996). As Lubart (2001) observes, various observational studies throughout the twentieth century failed to produce significant evidence to corroborate the existence of a linear multi-stage creative process. Instead, theorists asserted that creativity was fundamentally a more dynamic and fluid phenomenon with co-occurring components. Citing Getzels and Csikszentmihalyi (1976), Lubart (2001) notes that even the separation of 'analysing the problem' from 'considering the solution' was shown to be contradicted in observations, with evidence implying that the creative process characteristically involved repeated revisiting and reconceptualising of the problem at the same time as ideas were being formulated and verified.

More contemporary models on creativity have attempted to further unpack the 'black box' of creative processing. Santanen and colleagues (2000) present the Cognitive Network Model, in which working memory mentally disperses ideas and then bridges multiple connections between them, from which emergent processes (see Chap. 3 and the discussion on emergent perception) actualise an idea. Alternatively, Amabile (1996) explains creativity to be an amalgamation of expertise (technical and intellectual knowledge, and also declarative knowledge pertaining to procedures), motivational factors (including rewards and influences from the working environment) and psychological characteristics directly pertaining to creative thinking (procedural skills, flexibility and imagination). A further framework, the Systems Model of Creativity (Csikszentmihalyi 2014), identifies creativity as an emergent property of interrelations between the individual (inclusive of personal background), the 'field' (i.e. environmental, social factors, etc.) and the 'domain' (culture and contextual factors).

In a recent review, article that considers the creative process from a more linguistic perspective, Jordanous and Keller (2016) reveal several key factors that contribute to creative thought, including: active and sustained involvement, domain competence, general intellect, independence and freedom, social interaction and communication, spontaneity, emotional investment and variety. These points not only resonate significantly with the models of Amabile (1996) and Csikszentmihalyi (2014), but also greatly elucidate Hohstadt's (2013) assertion that VR, for all intents and purposes, *is* creativity. Elements such as active involvement and emotional investment are fundamentally characteristic of VR by way of its immersive, novel and interactive design. Within a collaborative virtual environment context, VR inherently facilitates naturalistic social interaction and communications. Furthermore, aspects such as variety, spontaneity and freedom highlight both the function and dramatic appeal of virtual worlds such as Minecraft and Second Life, in which the foundations of the world are established but the mechanics and user-interactions are designed specifically to support free, varied and spontaneous actions. Tying back to our earlier discussion on education, VR also readily accounts for Jordanous and Keller's (2016) knowledge-related requirements for creativity. As a tool for learning both declarative knowledge and procedural skills, VR has the potential to support both general intellect and domain competence. Software such as Tilt Brush and SoundStage possess inherent learning content. The user does not just paint in VR, they learn *how* to pain in VR. Referring back to Csikszentmihalyi's (2014) assertion (that creativity emerges from the individual, the field and the domain), VR presents yet more significant value with regard to its bespoke and flexible virtual space. Unlike a physical environment, which is limited in it capacity for customisation, a virtual creative space can be transformed to the user's liking. This means that Csikszentmihalyi's 'field' can be adjusted to best align with the 'individual' and the 'domain', thereby optimising overall creative potential. If you find your creativity is at its peak whilst on a boat on a river, with tangerine trees and marmalade skies, VR can oblige.

VR and creativity most certainly make good bedfellows and the question is not about how VR can support creativity, but how it can better support creativity. This section closes with a speculative review of this question, with specific regard to sound design.

Likely Future Developments and the Potential of Sound

In the late 1990s, Jayaram and colleagues (1997) observed that, despite its great potential, the successful implementation of enhanced VR technologies for prototyping raised issues that required attention. Difficulties integrating VR systems with the various closed-source software already being used at the time was one notable problem. Another was that the costs associated with the technology, particularly for small/medium enterprises, were prohibitively expensive. Whilst contemporary VR has arguably made substantial progress in addressing both of these issues, several challenges still remain. Referred to earlier in this chapter, Berg and Vance's (2017) survey of contemporary VR-prototyping also includes a concise outline of the key research challenges associated with this domain. These include improving graphics (including increased brightness, general fidelity and mapping quality of visual image to user-tracking), incorporating more detailed and physically realistic environmental simulation, and providing more accurate haptic feedback. These challenges are fundamental to VR user experience and arguably of equal relevance to the other creative applications of art, modelling/sculpture and music production. Therefore, it is highly likely that future developments in this domain will focus upon addressing these issues. There is also a safe assumption that the increasing availability, affordability and usability of VR hardware and software will lead to continuing growth in both the number of VR tools being developed for facilitating creative processes and the number of industries utilising such tools at various stages in their production pipelines.

Once again, with the exception of VR-based music production tools (see *Soundscape* above), there is a tendency for sound in this domain to be overlooked or at least deprioritised. Yet again, this presents us with a significant opportunity for improving such tools by using informed audio design, the inherent qualities of which can further enhance the quality of VR for creative processes. For example, immersion and affective engagement are central requirements of creativity (see the earlier descriptions of models by Amabile [1996] and Csikszentmihalyi [2014]) and both are intrinsically linked to not only VR experience, but also to sound. Furthermore, as noted earlier within this chapter, sound presents opportunities for enhancing the learning potential of VR, itself also a significant contributing factor to creativity.

Despite sound not being a priority feature in HMD/CAVE systems, acoustic prototyping is not a non-existent entity. Recent research documents this with regard to various contexts in which sound features as a significant aspect of the product. This includes loudspeaker development (Salvatti 2010) and automotive design (Van der Auweraer et al. 2007). Such instances are however, not fully immersive VR-prototyping methods and there remains the potential for the principles of acoustic prototyping and those of VR-prototyping to be brought together. Referring back to Berg and Vance's (2017) benefits of VR, packaging and storytelling describe two ways in which the comprehensive evaluation of a prototype requires it to be examined within a simulation of its actual environment. With one of the central research challenges in this area being how to increase the realism of such environments, the potential for positional and environmentally modelled audio is significant. Indeed, assessing the auditory performance of theoretically any design would arguably require an accurately modelled virtual soundscape, within which the prototype could be tested. Architectural walkthroughs, for example, highlight this potential, with realistic auditory environments able to simulate how a structure affects its surrounding soundscape. An architect could, for instance, situate the stakeholder within their VR-prototype home and enable them to compare the comparative levels of street-noise occlusion between double and triple-glazed windows. Attention to sound bestows further benefit when considering VR creative tools more generally. As the above creativity frameworks assert, being creative requires domain competence and procedural ability. Therefore, any VR-based creativity tool is required to present an accessible systems design and positive user experience, without which creativity would likely be undermined as the frustrated user grapples with indecipherable controls and menus. Also, as noted earlier in this section, the unlimited potential for customisation in a virtual creative space presents opportunity for the user to adapt their environment to suit their creative preferences. Sound is once again a crucial component here for enabling VR to fully support the requirements of a user. This could including both naturalistic and fantastic spaces. Should the user feel at their most creative when they hear waves gently crashing on a beach, wind blowing through the trees or the muffled hum of a busy city occluded by closed windows, soundscape options would mean that VR could support this.

As both an enhancement and extension of the semi-immersive virtual technologies that laid its foundations, creative tools are now beginning to utilise the substantial advantages of contemporary VR. At almost every turn, attention to sound offers further enhancements and new opportunities within this context. From real-time collaboration with individuals from across the globe and integrated opportunities for learning and skills development, to endlessly customisable virtual environments that immerse and emotionally engage, sound contributes the very same benefits as VR adds. VR for creativity may be a more fledgling application, but it is one with great promise that should most certainly not underestimate the value of good sound.

Health and Well-Being

To comprehensively review the health and well-being-related applications of VR is a book (or more likely a series)-sized undertaking. Therefore, as with every form of commercial VR applications that we have discussed throughout the chapter, this section again serves as more of an introductory overview of the field; comprising key aspects of VR for both physiological health (inclusive of pain distraction techniques, phantom limb pain treatments and motor rehabilitation) and psychological health (phobias, anxieties that include post-traumatic stress disorder and social/communication difficulties).

Pain Distraction

The experience of pain has a significant basis in psychological processes, with responses to pain provocation modulated by factors that include fear of injury, reduced emotional stability and conscious attention upon the locality of the pain (Main and Watson 1999). Hoffman and colleagues (2000) work from this principle to assert the value of VR as a distraction method during physiological rehabilitation (e.g. during the stretching of grafted skin). Specifically developed for individuals suffering severe burns, their application, *SnowWorld*,[17] presents users with a VR rail-shooter game, in which they throw snow balls at various targets within an arctic-style environment. Reinforcing another common trend that has been observed throughout this book, VR for pain distraction also has origins in semi-immersive virtual technologies such as digital games (see Das et al. 2005). Where VR presents significant advantage over semi-immersive approaches is consistent with the benefits of VR discussed throughout this chapter, namely increased immersion, deeper engagement and better embodiment of the virtual avatar. Using functional magnetic resonance image (fMRI) to generate a neural correlate to subjective user-feedback, testing of the SnowWorld application revealed both a substantial reduction in pain measures when compared to a semi-immersive digital game. This effect included both very intense and milder pain cases. In a study by Hoffman and colleagues (2003), a significant correlation between VR's analgesic effects and increased subjective ratings for presence. Wiederhold and Wiederhold (2007) reinforce this, noting that '[w]hen immersion is high, much of the user's attention is focussed on the virtual environment, leaving little attention left to focus on other things, such as pain' (p. 182). Because the principles underpinning VR for pain distraction can be generalised across various forms and sources of pain, the applications of VR in this domain vary significantly, with existing literature examining its potential in varied contexts such as dental pain control (Hoffman et al. 2001) and distraction for young children during intravenous placement (Gold et al. 2006).

With a priority on presence, embodiment and engagement, VR for pain distraction can clearly benefit from good VR sound design, primarily with regard to telepresence techniques. Immersive soundscapes featuring detailed spatialisation as a means of drawing the user deeper into the virtual world has obvious implications for pain distraction. Furthermore, the use of more novel and possibly hyperreal sounds could improve engagement, thereby contributing to presence for the user away from their physical body and environment.

Phantom Limb Pain and Virtual Reality Mirror Therapy

Positioned in centre of the overlap between physiological and psychological pain, Phantom Limb Pain (PLP) dates back to the nineteenth century, in which the term was first coined by Silas Weir Mitchell (see Louis and York 2006). The first medical account of PLP was provided by the French military surgeon Ambroise Paré. Paré noted patients' complaints of severe pain following amputation. What was most unexpected for Paré, was that the patients' perceived location of their pain was not at the point of amputation, but rather within the missing limb (Nikolajsen and Jensen 2001). In a review by Nikolajsen and Jensen (2001) various characteristics of PLP are outlined, including its onset (PLP typically first occurs within days of amputation but has been known, on rarer occasions, to develop years afterwards), duration (it is common for PLP to decrease in perceived intensity overtime but less likely to disappear completely; many cases persist for many years), frequency (intermittent in most cases but instances of constant pain have been reported) and localisation (PLP is primarily reported in distal regions [limbs, fingers, palms, toes, etc.]). As with pain in general, PLP is a particularly difficult phenomenon to precisely characterise and Nikolajsen and Jensen's definition does not address all ambiguities (Crawford 2014), making PLP something for which the underlying theory is continuing to develop and various preventative and curative treatments continue to be prescribed. Flor (2002) outlines several of these, grouping treatments into pharmacological (opiods, muscle relaxants, etc.), surgical (e.g. removal of pain-relevant nerves), anaesthetics (receptor blockers), psychological (mirror therapy, hypnosis, sensory discrimination training, etc.) and 'other' (which includes ultrasound, acupuncture and transcutaneous nerve stimulation [TENS]). Of this very wide range of treatments, mirror therapy is the approach of most interest here, being as it is the foundation for VR-based PLP therapy.

Mirror therapy for PLP is built from the theory of 'mirror-neurons'. Rizzolatti and Craighero (2004) unveiled this concept of a mirror-neuron system based on observations of monkeys. Their findings that revealed certain neurons would fire both when the animal performs an action and when it observes the same action performed by another. This discovery was applied to PLP in a study that revealed human amputees would experience tactile sensations of their phantom limbs when an associated virtual image of that limb was 'touched' (Ramachandran and Rogers-Ramachandran 2008). Mirror therapy relates to the mirror-neuron theory and, in physical (rather than virtual) form, typically comprises an amputee observing a reflection of their intact limb moving in a mirror. They then use this image to practice moving their phantom limb, stretching and releasing perceived strain and muscle tension, thereby reducing pain.

Extending from its physical progenitor, Virtual Reality Mirror Therapy (VRMT) describes a comparable approach but within a VR/AR environment in which the user is presented with an animated graphical representation of their phantom limb (at present, this is almost always an arm). The representation can be presented within a fully immersive VR experience (by way of a HMD) or semi-immersive VR or AR experience (e.g. a flat-screen displaying an augmented camera feed of the user, with their virtual limb overlaying their physical body). In most cases, electromyography (muscle activity) sensors are then attached to the stump of the arm and the acquired signal data is fed into the system as a means of controlling the virtual limb (see Ortiz-Catalan et al. 2014). Recent research has argued that a vital condition for PLP relief is 'body ownership' (tied to embodiment: the extent to which an individual feels that a virtual body, or body part, is their own—see Imaizumi et al. 2014). Recent studies have posited that various factors can influence ownership. These include: synchronous action (the virtual body's movements accurately reflecting their physical, or their intended, movements, see González-Franco et al. 2010), temporal synchrony (no delay between virtual and phantom motor activity, Imaizumi et al. 2014), tactile stimulation of the physical limb synchronised with relative visible stimulation of the virtual limb (Slater et al. 2009) and connectivity of the virtual limb to the rest of the virtual body (Perez-Marcos et al. 2012).

From the above we can draw relatively safe inferences regarding the potential value of VRMT. With embodiment highlighted as a principle requirement of successful mirror therapy, the immersive qualities of VR offer us self-evident value. The flexibility of virtual environments raises the possibility for presenting the user with a virtual limb that is more accurately representative of their own body, with features such as skin tone, age, markings and hair all taken into account. As with pain distraction, and arguably a common threat throughout VR for health and well-being, very little research has been conducted regarding the potential benefits of sound. This reveals yet another opportunity to enhance VR that is yet to be seized. The priority for VRMT is for the user to embody their virtual limb. Therefore, approaches to sound design that can contribute to embodiment, such as real-time sonification of avatar movement, are of particular interest. For example, a VRMT environment could enable the user to move their arm across a shallow pool of water, with the ripples across the water's surface presented by both visuals and audio that were accurately modelled on the physical action and its effects.

Motor Rehabilitation

As Lucca (2009) observes, VR for motor rehabilitation is built upon similar principles to PLP treatments; that 'some functional re-arrangement of the damaged motor cortex can be activated with the mediation of mirror neurones or through the subject's motor imagery' (p. 1006). Levac and Sveistrup (2014) identify four alternative approaches. In the first, the user is presented with a virtual environment and representation of their body. The virtual body is then animated to carry out the desired motor action, enabling the individual to observe the action repeatedly within an immersive and embodying environment. The second approach also utilises body tracking but feeds back the movements by way of a separated avatar (comparable to looking in a mirror). In the third, the user observes and attempts to duplicate the actions of a separate avatar. The fourth is based more upon motor imagery, with the VR environment more abstractly encourages the user to mentally simulate a given action.

Motor rehabilitation can incorporate various actions, apply to numerous regions of the body, and relate to several alternative causes and conditions. The specific contexts in which VR-based techniques feature most prominently is, at present, upper and lower limb (reaching and walking) rehabilitation for post-stroke patients. Articles published during the first few years of the 2000s acknowledged VR for such motor rehabilitation to be an emerging approach not requiring ongoing clinical evaluations and interdisciplinary collaborations, particularly with regard to any assertions the outcomes of VR procedures being significantly greater than standard rehabilitation (see Rizzo and Kim 2005). Further research published around this time also raised concerns regarding the ecological validity of VR-based approaches (i.e. 'transfer of training': the extent to which practicing the action in a virtual environment carries over into the physical world). More recent research, however, has shown progress in this area. For example, Corbetta and colleagues (2015) compare standard and VR-based walking rehabilitation approaches in post-stroke patients by way of gait, balance and mobility measures. Their findings reveal small but significant differences that explicitly point to VR as the marginally better approach. Their conclusions assert that progress made in such VR systems, and in the evaluative research conducted upon them, has reached a point where we can more confidently assert the ecological validity of VR and its ability to support better transition between therapeutic and real-world activities. Of course, there remains a great deal of progress to be made here, with specific points of interest including identification of precisely which characteristics of VR are most relevant to producing the desired outcomes, and also establishing the long-term benefits of the technology (Laver et al. 2015).

In many cases, physiological rehabilitation is an arduous, painful and repetitive journey. As a result, increased engagement evoked by a novel and variable medium such as VR is of clear benefit to patients who are struggling to maintain motivation (Sveistrup 2004). Further explanations for the increased value of VR-based rehabilitation therapies include its capacity for delivering multisensory (audio-visual and haptic) feedback and that it can be easily customised and repeated over a prolonged period of time. It is also relatively cost-effective and its wide availability and portability means that VR rehabilitation can be accessed by more people, potentially used in their own homes.

Based on these points, one of the principle values of VR in a motor rehabilitation context is arguably its capacity to engage and motivate. This immediately highlights the value of sound. As Wolfson and Case (2000) assert, when properly implemented in conjunction with visual design, audio can significantly and substantially increase arousal and engagement. A review by Weller and Baker (2011) notes that music in particular has been shown to significantly enhance the outcomes of physical rehabilitation programmes, with outcomes specifically linked to increased patient engagement and motivation.

At present, there is very little evidence of research bringing together VR with sound to explore specifically how the two should be optimally utilised together. As with other physiological applications, motor rehabilitation has been shown to benefit from increased user-presence and embodiment of their avatar. In addition, the transfer of training issue raises the importance of the virtual environment's capacity to accurately represent the physical world. This once again highlights the potential added value of environmentally modelled audio, spatialisation and an auditory feedback system that accurately reflects users' actions. Lastly, a graphics-centred system makes VR for motor rehabilitation largely inaccessible to the large number of people who are visually impaired (Rizzo and Kim 2005), emphasising the need for audio-focussed equivalent VR therapies to ensure more people are able to benefit.

Virtual Reality Exposure Therapy

Traditional 'in vivo' exposure therapy encapsulates various methods of attenuating the adverse effects of an anxiety response by way of controlled exposure of the patient to the source of that anxiety. In vivo exposure is typically delivered either by way of flooding (rapid exposure to intense stimuli/situations) or systematic desensitisation (progressive exposure—gradually increasing the intensity to desensitise the patient over a longer period). As with the various other applications discussed above, VRET was an extension of a physical precursor that researchers believed could be improved upon by way of virtual technology. In a review, article that considers VRET for phobic treatment, Gregg and Tarrier (2007) observe that research within this field has an established history dating back to the early 1990s, in which early HMD and CAVE systems were frequently employed as an alternative means of treating phobias, the most common sources of which included heights, enclosed and open spaces, flying, driving, and spiders. In addition to specific phobias such as these, VRET is also commonly utilised as a means of addressing post-traumatic stress disorder (PTSD). Various studies dating back many years have examined VR-based therapies for PTSD, with patients seeking the therapy to address often quite debilitating anxieties resulting from life threatening or otherwise catastrophic events (Sanchez-Vives and Slater 2005), frequently affecting individuals who have returned from active military service (Rothbaum et al. 2001).

In terms of the research conducted up to 2007, Gregg and Tarrier note that findings consistently assert the value of VRET as a viable (and often preferable) alternative to in vivo approaches. In terms of clinical efficacy, research evaluating VRET does not favour it over in vivo approaches, but it does however reveal comparable effect sizes (Powers and Emmelkamp 2008) and observes that VRET has equivalent power in terms of ecological validity (transfer of trainingMorina et al. 2015). With VRET matching in vivo approaches in terms of efficacy and ecological validity, the further advantages of VR are brought to the forefront. These include greater situational immersion and more focussed attention upon the stimuli, more comprehensive and detailed recreations of actual anxiety-evoking scenarios, and also more direct control for the therapist over the parameters of the exposure (Motraghi et al. 2014).

Regarding sound in VRET; grading of the exposure intensity is predominantly achieved by way of graphics (e.g. in an acrophobia context, the user could stand within a glass-bottomed elevator and observe the distance from the ground increasing) with no change to sound content; if sound is even used at all (Opdyke et al. 1995). Related research supports acoustic parameters (loudness, frequency, periodicity, etc.) as both powerful amplifiers (Ekman and Kajastila 2009) and attenuators (Prabhakar et al. 2007) of anxiety. This suggests that grading the sound content of a VRET could not only enhance efficacy in general but, as noted above with regard to motor rehabilitation, help address the problem of therapies for partially sighted and blind patients. Furthermore, audio-centric VRET designs would have notable benefit for contexts in which the phobic stimuli are not primarily visual (e.g. during a dental procedure).

Social and Communications Difficulties

The use of VR as a support tool for individuals diagnosed on the autistic spectrum dates back to the mid-1990s. During this time researchers such as Strickland and colleagues (1996) advocated the technology, asserting its advantages with regard to tackling sensory problems (i.e. the versatility and control a therapist can have over a virtual environment, with which they can isolate specific stimuli and avoid sensory overload), providing individualised support and offering greater degrees of user-engagement. The underlying principle of VR for autistic spectrum condition (ASC) is that the controlled and finite nature of the virtual environment can provide a safe space that immerses the user in simulations of actual social scenarios and enables them to practice their interactions (Parsons and Cobb 2011). Such scenarios can include attending a job interview (Smith et al. 2014) or navigating a busy supermarket (Herrera et al. 2008). VR can also be used in this context as a means of reducing repetitive and overtly idiosyncratic behaviours (Bellani et al. 2011). Work in this field is also relatively broad with regard to the different forms ASC, with recent research also examining the potential of VR to support individuals with high functioning autism and attention deficit hyperactivity disorder (Kandalaft et al. 2013).

Sound has featured prominently across ASC research, typically examining the correlates between ASC and auditory processing, but also with regard to methods of using sound as a means of supporting individuals with ASC. For example, Kandalaft and colleagues (2013) utilised real-time audio cues to improve the performance of adolescents during a social task; their results supportive of the assertion that clear and prompt auditory guidance and feedback supports individuals with ASC in their daily social interactions. A study by Rosenhall and colleagues (1999) revealed mild and moderate hearing loss to be more pronounced in children and adolescents with ASC than those classified as 'neurotypical'. An even larger proportion of those with ASC also experienced hyperacusis (a substantially increased sensitivity to sound). Exploring a similar issue, Gage and colleagues (2003) used magnetoencephalography to measure brain activity in response to audio. They discovered that the neural patterns in children with ASC differed to a neurotypical control group when responding to various audio frequencies. Difficulty with integrating auditory and visual information has also been associated with ASC (see Williams et al. 2004). Such findings highlight the fact that individuals with ASC are likely to experience sound very differently to those without the condition, thereby emphasising the need for sound to be much more carefully considered in VR design, both in terms of as a VR-based tool for developing social and communications skills, and in general.

Continuing another theme that has become apparent throughout this chapter, research that concerns VR for ASC and that which looks at the nature of sound in this context appear to be kept almost entirely separate. With auditory perception for individuals with ASC something of a continuing unknown it is certainly vital that VR development in this area considers sound, both as a way of accommodating the unique auditory requirements of individuals with ASC, but also to utilise the specific benefits sound could return.

Chapter Summary and References

The wider applications of VR are great in number. With ever-improving technology and designs they are continuing to expand. A virtual environment can visualise datasets with dynamics and interactivity, to help us make sense of that which is too complex to understand in a static image. Though its immersive qualities and its power to evoke telepresence, VR for telecommunications and broadcasting interests can effectively beam users anywhere on Earth, from boardrooms to Glastonbury. With detailed representations and applicable interactions, VR for education has the potential to impart both declarative knowledge and procedural skills in almost any context. As a creative tool with the ability to centre the user within their work, VR presents an entirely new way of pursuing artistic endeavours. By way of virtual embodiment, the technology even has the potential to partially dissociate a patient from their physical body, giving them respite from pain. This account of VR applications is certainly not complete. Even if it were, the rate of development and expansion is such that, by the time this book were published, many new applications would probably be in use. That said, the above types of wider application represent the primary areas in which VR is finding value outside of recreational use. Moreover, the intention of this chapter is not simply to elucidate these applications, but to assert the consistent value of sound as a means of both enhancing quality and extending functionality. Sound is an essential component, from offering a unique route to the presentation of complex data in VR by way of sonification, to enabling visually impaired users the opportunity to engage with VR therapeutic applications from which they would otherwise be excluded. Unfortunately, sound remains a component of VR that continues to be overlooked, but this should be seen as an opportunity.

VR can already do great things without the added value of considered VR sound design. With sound, imagine what else it could do.

This chapter completes our little adventure. There are certainly more pieces of the emergent puzzle to be uncovered, but the topics reviewed throughout this book will hopefully have encouraged a more comprehensive understanding than before and an appreciation for VR sound from an emergent perspective. With regard to this, the book now closes by drawing together the many elements discussed, to form an emergent framework of VR sound.

Notes

1. Presumably they present the 'v' in 'discovery' in upper-case to try and make the '5 V' consistent.
2. VR visualisation of Canada energy demands. Nirvaniq Labs: https://nirvaniq.com.
3. Salesforce VR visualisation: https://www.salesforce.com/video/192746.
4. Microsoft's Holoportation project: https://www.microsoft.com/en-us/research/project/holoportation-3.
5. AR Works *LiveVR 360* system: http://www.arworks.com/en/live-vr-360-broadcasting-app-system-by-arworks.
6. Eon sports VR: https://eonsportsvr.com.
7. Limitations of current VR broadcasting: http://www.rgbbroadcasting.com/2016/11/23/virtual-reality-broadcasting.
8. First audio-based VR broadcast: https://dysonics.com/360-degree-audio-live-broadcast-for-virtual-reality.
9. Binaural sound. BBC R&D: http://www.bbc.co.uk/rd/projects/binaural-broadcasting.
10. VR for education: https://virtualrealityforeducation.com.
11. Unimersiv: https://unimersiv.com.
12. The Open University and Second Life: http://www.open.ac.uk/virtual worlds/content/ou-second-life.
13. Wall Street Journal: http://www.wsj.com/video/vr-art-brings-hope-to-struggling-art-market.
14. Tilt Brush main site: https://www.tiltbrush.com.
15. Oculus Medium main site: https://www.oculus.com/medium.

16. Sound Stage VR press page: http://www.soundstagevr.com/press.
17. Snow World research: http://www.vrpain.com/.

References

Aggarwal, R., Ward, J., Balasundaram, I., Sains, P., Athanasiou, T., & Darzi, A. (2007). Proving the effectiveness of virtual reality simulation for training in laparoscopic surgery. *Annals of Surgery, 246*(5), 771–779.

Aisch, G., Cox, A. & Quealy, K. (2015). You draw it: Family income predicts children's college chances. *The New York Times*. https://www.nytimes.com/interactive/2015/05/28/upshot/you-draw-it-how-family-income-affects-childrens-college-chances.html.

Amabile, T. (1996). *Creativity in context*. Boulder, USA: Westview press.

Ballora, M., Cole, R. J., Kruesi, H., Greene, H., Monahan, G., & Hall, D. L. (2012, May). Use of sonification in the detection of anomalous events. In *SPIE Defense, Security, and Sensing* (pp. 84070S–84070S). International Society for Optics and Photonics.

Bellani, M., Fornasari, L., Chittaro, L., & Brambilla, P. (2011). Virtual reality in autism: State of the art. *Epidemiology and Psychiatric Sciences, 20*(03), 235–238.

Berg, L. P., & Vance, J. M. (2017). Industry use of virtual reality in product design and manufacturing: a survey. *Virtual Reality, 21*(1), 1.

Bielli, S., & Harris, C. G. (2015, March). A mobile augmented reality system to enhance live sporting events. In *Proceedings of the 6th Augmented Human International Conference* (pp. 141–144). ACM.

Blazey, M. (2015). *The power of sound and vision in eLearning*. Bray Leino Learning. http://brayleinolearning.co.uk/blog/2015/august/06/the-power-of-sound-and-vision-in-elearning.

Bloom, B. S. (1956). Taxonomy of educational objectives. Vol. 1: Cognitive domain. *New York: McKay*, 20–24.

Bricken, M. (1991). Virtual reality learning environments: potentials and challenges. *ACM SIGGRAPH Computer Graphics, 25*(3), 178–184.

Burn-Murdoch, J. (2013). *Why you should never trust a data visualisation*. The Guardian. https://www.theguardian.com/news/datablog/2013/jul/24/why-you-should-never-trust-a-data-visualisation.

Butler, M. (2014). *Kinect doesn't boost immersion – it kills it.* Wow247. http://www.wow247.co.uk/2014/04/09/game-on-kinect-doesnt-boost-immersion-it-kills-it-46122/.

Chen, C. (2013). *Information visualisation and virtual environments.* Springer Science & Business Media.

Corbetta, D., Imeri, F., & Gatti, R. (2015). Rehabilitation that incorporates virtual reality is more effective than standard rehabilitation for improving walking speed, balance and mobility after stroke: a systematic review. *Journal of physiotherapy, 61*(3), 117–124.

Craig, A. B., Sherman, W. R., & Will, J. D. (2009). *Developing virtual reality applications: Foundations of effective design.* Morgan Kaufmann.

Crawford, C. S. (2014). *Phantom limb: Amputation, embodiment, and prosthetic technology.* NYU Press.

Cromley, J. G., Snyder-Hogan, L. E., & Luciw-Dubas, U. A. (2010). Cognitive activities in complex science text and diagrams. *Contemporary Educational Psychology, 35*(1), 59–74.

Csikszentmihalyi, M. (2014). *Society, culture, and person: A systems view of creativity* (pp. 47–61). Netherlands: Springer.

Dalgarno, B., & Lee, M. J. (2010). What are the learning affordances of 3-D virtual environments? *British Journal of Educational Technology, 41*(1), 10–32.

Das, D. A., Grimmer, K. A., Sparnon, A. L., McRae, S. E., & Thomas, B. H. (2005). The efficacy of playing a virtual reality game in modulating pain for children with acute burn injuries: A randomized controlled trial [ISRCTN87413556]. *BMC Pediatrics, 5*(1), 1.

de Boer, I. R., Lagerweij, M. D., de Vries, M. W., Wesselink, P. R., & Vervoorn, J. M. (2017). The effect of force feedback in a virtual learning environment on the performance and satisfaction of dental students. *Simulation in Healthcare, 12*(2), 83–90.

della Cava, M. (2016). Live Nation wants you to hit that next big concert in VR. *USA Today.* https://www.usatoday.com/story/tech/news/2016/09/28/91225372.

Dillenbourg, P., Schneider, D., & Synteta, P. (2002). Virtual learning environments. In *3rd Hellenic Conference "Information & Communication Technologies in Education"* (pp. 3–18). Kastaniotis Editions, Greece.

Dougiamas, M. (2002). *Moodle.* Dougiamas.

Dredge, S. (2015). Vice News debuts 'virtual reality news broadcast' of US Millions March. *The Guardian.* https://www.theguardian.com/technology/2015/jan/23/vice-news-virtual-reality-news.

Drogemuller, R. (2009). Virtual prototyping from need to preconstruction. In P. S. Brandon & T. Kocatürk (Eds.), *Virtual futures for design, construction and procurement*. Wiley.

Edwards, J. (2011). Telepresence: Virtual reality in the real world [special reports]. *IEEE Signal Processing Magazine, 28*(6), 9–142.

Ekman, I., & Kajastila, R. (2009, February). Localization cues affect emotional judgments–results from a user study on scary sound. In *Audio Engineering Society Conference: 35th International Conference: Audio for Games*. Audio Engineering Society.

Evans, M. J., Tew, A. I., & Angus, J. A. (1997). Spatial audio teleconferencing-Which way is better. Georgia Institute of Technology.

Fairchild, J., Cassidy, S., Cushenbery, L., & Hunter, S. (2011). Integrating technology with the creative design process. *Technology for creativity and innovation: Tools, techniques and applications*, 26–51.

Fiard, G., Selmi, S. Y., Promayon, E., Vadcard, L., Descotes, J. L., & Troccaz, J. (2014). Initial validation of a virtual-reality learning environment for prostate biopsies: Realism matters! *Journal of Endourology, 28*(4), 453–458.

Flor, H. (2002). Phantom-limb pain: Characteristics, causes, and treatment. *The Lancet Neurology, 1*(3), 182–189.

Fowler, C. (2015). Virtual reality and learning: Where is the pedagogy? *British Journal of Educational Technology, 46*(2), 412–422.

Friendly, M. (2008). A brief history of data visualization. *Handbook of data visualization*, 15–56.

Gage, N. M., Siegel, B., Callen, M., & Roberts, T. P. (2003). Cortical sound processing in children with autism disorder: An MEG investigation. *NeuroReport, 14*(16), 2047–2051.

Gardner, W. G. (2013). *U.S. Patent No. 8,559,646*. Washington, DC: U.S. Patent and Trademark Office.

Gavish, N., Gutiérrez, T., Webel, S., Rodríguez, J., Peveri, M., Bockholt, U., et al. (2015). Evaluating virtual reality and augmented reality training for industrial maintenance and assembly tasks. *Interactive Learning Environments, 23*(6), 778–798.

Gemma, W. (2014). *The 6 types of knowledge: From a priori to procedural*. Udemy. https://blog.udemycom/types-of-knowledge.

Getzels, J., & Csikszentmihalyi, M. (1976). *The creative vision: A longitudinal study of problem finding in art*. New York: Wiley.

Gold, J. I., Kim, S. H., Kant, A. J., Joseph, M. H., & Rizzo, A. S. (2006). Effectiveness of virtual reality for pediatric pain distraction during IV placement. *CyberPsychology & Behavior, 9*(2), 207–212.

Goldenweiser, E. A. (1916). Classification and limitations of statistical graphics. *Quarterly Publications of the American Statistical Association, 15*(114), 205–209.

Goleman, D., Kaufman, P., & Ray, M. (1992). The art of creativity. *Psychology Today, 25*(2), 40–47.

González-Franco, M., Pérez-Marcos, D., Spanlang, B., & Slater, M. (2010, March). The contribution of real-time mirror reflections of motor actions on virtual body ownership in an immersive virtual environment. In *Virtual Reality Conference (VR), 2010 IEEE* (pp. 111–114). IEEE.

Grabowski, A., & Jankowski, J. (2015). Virtual reality-based pilot training for underground coal miners. *Safety Science, 72,* 310–314.

Gregg, L., & Tarrier, N. (2007). Virtual reality in mental health. *Social Psychiatry and Psychiatric Epidemiology, 42*(5), 343–354.

Griffon, S., Nespoulous, A., Cheylan, J. P., Marty, P., & Auclair, D. (2011). Virtual reality for cultural landscape visualization. *Virtual Reality, 15*(4), 279–294.

Harris, R., Popovich, N. & Powell, K. (2015). Watch how the measles outbreak spreads when kids get vaccinated – and when they don't. *The Guardian.* https://www.theguardian.com/society/ng-interactive/2015/feb/05/-sp-watch-how-measles-outbreak-spreads-when-kids-get-vaccinated.

Hayden, S. (2016). *Live Nation and NextVR to broadcast hundreds of live performances in VR.* Road to VR http://www.roadtovr.com/live-nation-nextvr-broadcast-hundreds-live-performances-vr.

Helsel, S. (1992). Virtual reality and education. *Educational Technology, 32*(5), 38–42.

Hermann, T. (2008). Taxonomy and definitions for sonification and auditory display. In *Proceedings of the 14th International Conference on Auditory Display (ICAD 2008).*

Herrera, G., Alcantud, F., Jordan, R., Blanquer, A., Labajo, G., & De Pablo, C. (2008). Development of symbolic play through the use of virtual reality tools in children with autistic spectrum disorders: Two case studies. *Autism, 12*(2), 143–157.

Hoffman, H. G., Patterson, D. R., & Carrougher, G. J. (2000). Use of virtual reality for adjunctive treatment of adult burn pain during physical therapy: a controlled study. *The Clinical Journal of Pain, 16*(3), 244–250.

Hoffman, H. G., Garcia-Palacios, A., Patterson, D. R., Jensen, M., Furness, III, T., & Ammons, W. F., Jr. (2001). The effectiveness of virtual reality for dental pain control: A case study. *CyberPsychology & Behavior, 4*(4), 527–535.

Hoffman, H. G., Richards, T., Coda, B., Richards, A., & Sharar, S. R. (2003). The illusion of presence in immersive virtual reality during an fMRI brain scan. *CyberPsychology & Behavior, 6*(2), 127–131.

Hohstadt, T. (2013). *The age of virtual reality*. Lulu.com.

Holmes, B., & Gardner, J. (2006). *E-learning: Concepts and practice*. Sage.

Hubbard, R., Sipolins, A., & Zhou, L. (2017, March). Enhancing learning through virtual reality and neurofeedback: A first step. In *Proceedings of the Seventh International Learning Analytics & Knowledge Conference* (pp. 398–403). ACM.

Huston, C. D. (2016). *U.S. Patent No. 9,445,225*. Washington, DC: U.S. Patent and Trademark Office.

Hyder, M., Das, G., Qureshi, I., Memon, M., Jalbani, A., & Saba, E. (2012). On the Requirements of 3D Audio Telephone and Teleconferencing Calls: A Mini Review on Background of 3D Audio–From Human Hearing toTechnology. *Sindh University Research Journal-SURJ (Science Series), 44*(3).

Idris, N. H., Jackson, M. J., & Abrahart, R. J. (2011, April). Map mashups: what looks good must be good. In *Conference on GIS Research U* (pp. 137–143).

Imaizumi, S., Asai, T., Kanayama, N., Kawamura, M., & Koyama, S. (2014). Agency over a phantom limb and electromyographic activity on the stump depend on visuomotor synchrony: A case study. *Frontiers in human neuroscience, 8*.

Jayaram, S., Connacher, H. I., & Lyons, K. W. (1997). Virtual assembly using virtual reality techniques. *Computer-Aided Design, 29*(8), 575–584.

Johnson, C. (2004). Top scientific visualization research problems. *IEEE Computer Graphics and Applications, 24*(4), 13–17.

Jordanous, A., & Keller, B. (2016). Modelling Creativity: Identifying Key Components through a Corpus-Based Approach. *PloS One, 11*(10), e0162959.

Kandalaft, M. R., Didehbani, N., Krawczyk, D. C., Allen, T. T., & Chapman, S. B. (2013). Virtual reality social cognition training for young adults with high-functioning autism. *Journal of Autism and Developmental Disorders, 43*(1), 34–44.

Kaper, H. G., Wiebel, E., & Tipei, S. (1999). Data sonification and sound visualization. *Computing in Science & Engineering, 1*(4), 48–58.

Kelion, L. (2016). YouTube live-streams in virtual reality and adds 3D sound. *BBC News*. http://www.bbc.co.uk/news/technology-36073009.

Kerttula, M., Salmela, M., & Heikkinen, M. (1997, June). Virtual reality prototyping-a framework for the development of electronics and telecommunication products. In *Rapid System Prototyping, 1997. Shortening the Path from Specification to Prototype. Proceedings., 8th IEEE International Workshop on* (pp. 2–11). IEEE.

Kirk, A. (2016). *Data visualisation: A handbook for data driven design*. Sage.

Kirriemuir, J. (2008). A spring 2008 "snapshot" of UK Higher and Further Education developments in Second Life. *Online at* http://www.eduserv.org.uk/foundation/sl/uksnapshot052008.

Kramer, G., Walker, B., Bonebright, T., Cook, P., Flowers, J., Miner, N. et al. (1997). Sonification Report: Status of the Field and Research Agenda, prepared for the National Science Foundation by members of the International Community for Auditory Display, 1997.

Kühnapfel, U., Cakmak, H. K., & Maaß, H. (2000). Endoscopic surgery training using virtual reality and deformable tissue simulation. *Computers & graphics, 24*(5), 671–682.

Laha, B., & Bowman, D. A. (2012). Identifying the benefits of immersion in virtual reality for volume data visualization. In *Immersive visualization revisited workshop of the IEEE VR conference* (pp. 1–2).

Lan, G., Luo, Z., & Hao, Q. (2016, October). Development of a virtual reality teleconference system using distributed depth sensors. In *Computer and Communications (ICCC), 2016 2nd IEEE International Conference on* (pp. 975–978). IEEE.

Larsen, C. R., Soerensen, J. L., Grantcharov, T. P., Dalsgaard, T., Schouenborg, L., Ottosen, C., … & Ottesen, B. S. (2009). Effect of virtual reality training on laparoscopic surgery: Randomised controlled trial. *British Medical Journal, 338*, b1802.

Laver, K., George, S., Thomas, S., Deutsch, J. E., & Crotty, M. (2015). Virtual reality for stroke rehabilitation: An abridged version of a Cochrane review. *European journal of Physical and Rehabilitation Medicine, 51*(4), 497–506.

Lawler, R. (2015). The first live broadcast brought the beach to my backyard. *Engadget*. UK. https://www.engadget.com/2015/01/26/first-live-vr-broadcast-next-vr.

Lawrie, G. (2017). How our school is using virtual reality to prepare pupils for a future dominated by technology. *The Telegraph*. http://www.telegraph.co.uk/education/2017/01/23/school-using-virtual-reality-prepare-pupils-future-dominated.

Levac, D. E., & Sveistrup, H. (2014). Motor learning and virtual reality. In *Virtual Reality for Physical and Motor Rehabilitation* (pp. 25–46). New York: Springer.

Louis, E. D., & York, G. K. (2006). Weir Mitchell's observations on sensory localization and their influence on Jacksonian neurology. *Neurology, 66*(8), 1241–1244.

Low, C. H. (2016). Assessing the Future IP Landscape of Music's Cash Cow: What Happens When the Live Concert Goes Virtual. *New York University Law Review, 91,* 425.

Lubart, T. I. (2001). Models of the creative process: Past, present and future. *Creativity Research Journal, 13*(3–4), 295–308.

Lucca, L. F. (2009). Virtual reality and motor rehabilitation of the upper limb after stroke: A generation of progress? *Journal of Rehabilitation Medicine, 41*(12), 1003–1006.

Main, C. J., & Watson, P. J. (1999). Psychological aspects of pain. *Manual therapy, 4*(4), 203–215.

Mallon, B., Quinlan, J. J., & Nolan, K. (2012). eLearning modalities: A framework for selecting audio. *Proceedings of the 11th European Confernece on e-Learning: ECEL*, 319.

Marks, S., Estevez, J. E., & Connor, A. M. (2014, November). Towards the Holodeck: fully immersive virtual reality visualisation of scientific and engineering data. In *Proceedings of the 29th International Conference on Image and Vision Computing New Zealand* (pp. 42–47). ACM.

Mayer-Schönberger, V., & Cukier, K. (2013). *Big data: A revolution that will transform how we live, work, and think*. Houghton Mifflin Harcourt.

McNutt, L., & Brennan, M. (2005, October). Work in Progress–Learning Styles and elearning, what is the Connection? In *Frontiers in Education, 2005. FIE'05. Proceedings 35th Annual Conference* (pp. F1H-F1H). IEEE.

Metz, R. (2017). Second Life is back for a third life, this time in virtual reality. *MIT Technology Review*. https://www.technologyreview.com/s/603422/second-life-is-back-for-a-third-life-this-time-in-virtual-reality/.

Mikropoulos, T. A., & Natsis, A. (2011). Educational virtual environments: A ten-year review of empirical research (1999–2009). *Computers & Education, 56*(3), 769–780.

Morina, N., Ijntema, H., Meyerbröker, K., & Emmelkamp, P. M. (2015). Can virtual reality exposure therapy gains be generalized to real-life? A meta-analysis of studies applying behavioral assessments. *Behaviour Research and Therapy, 74,* 18–24.

Motraghi, T. E., Seim, R. W., Meyer, E. C., & Morissette, S. B. (2014). Virtual reality exposure therapy for the treatment of posttraumatic stress disorder: A methodological review using CONSORT guidelines. *Journal of Clinical Psychology, 70*(3), 197–208.

Moustakas, N., Floros, A., & Kapralos, B. (2016, September). An Augmented Reality Audio Live Network for Live Electroacoustic Music Concerts. In *Audio Engineering Society Conference: 2016 AES International Conference on Audio for Virtual and Augmented Reality*. Audio Engineering Society.

Nelson, T. R., & Elvins, T. T. (1993). Visualization of 3D ultrasound data. *IEEE Computer Graphics and Applications, 13*(6), 50–57.

Nikolajsen, L., & Jensen, T. S. (2001). Phantom limb pain. *British Journal of Anaesthesia, 87*(1), 107–116.

Nóbrega, R., Sabino, A., Rodrigues, A., & Correia, N. (2008, September). Flood emergency interaction and visualization system. In *International Conference on Advances in Visual Information Systems* (pp. 68–79). Berlin and Heidelberg: Springer.

O'Kane, S. (2016). *Doppler Labs' new augmented reality earbuds can also stream audio from your phone*. The Verge https://www.theverge.com/circuitbreaker/2016/6/28/12043946.

Olshannikova, E., Ometov, A., Koucheryavy, Y., & Olsson, T. (2015). Visualizing Big Data with augmented and virtual reality: Challenges and research agenda. *Journal of Big Data, 2*(1), 22.

Opdyke, D., Williford, J. S., & North, M. (1995). Effectiveness of computer-generated (virtual reality) graded exposure in the treatment of acrophobia. *American Journal of Psychiatry, 1*(52).

Ortiz-Catalan, M., Sander, N., Kristoffersen, M. B., Håkansson, B., & Brånemark, R. (2014). Treatment of phantom limb pain (PLP) based on augmented reality and gaming controlled by myoelectric pattern recognition: A case study of a chronic PLP patient.

Pangburn, D. (2015). Now you can make beats in virtual reality. *Creators*. https://creators.vice.com/en_uk/article/now-you-can-make-beats-in-virtual-reality.

Parsons, S., & Cobb, S. (2011). State-of-the-art of virtual reality technologies for children on the autism spectrum. *European Journal of Special Needs Education, 26*(3), 355–366.

Perez-Marcos, D., Sanchez-Vives, M. V., & Slater, M. (2012). Is my hand connected to my body? The impact of body continuity and arm alignment on the virtual hand illusion. *Cognitive Neurodynamics, 6*(4), 295–305.
Persson, M., Bergensten, J., Toivonen, M., et al. (2011). *Minecraft.* Sweden: Mojang.
Piaget, J. (1955). The construction of reality in the child. *Journal of Consulting Psychology, 19*(1), 77.
Powers, M. B., & Emmelkamp, P. M. (2008). Virtual reality exposure therapy for anxiety disorders: A meta-analysis. *Journal of Anxiety Disorders, 22*(3), 561–569.
Prabhakar, A. R., Marwah, N., & Raju, O. S. (2007). A comparison between audio and audiovisual distraction techniques in managing anxious pediatric dental patients. *Journal of Indian Society of Pedodontics and Preventive Dentistry, 25*(4), 177.
Psotka, J. (1995). Immersive training systems: Virtual reality and education and training. *Instructional Science, 23*(5), 405–431.
Raja, D., Bowman, D., Lucas, J., & North, C. (2004, May). Exploring the benefits of immersion in abstract information visualization. In *Proc. Immersive Projection Technology Workshop.*
Ramachandran, V. S., & Rogers-Ramachandran, D. (2008). Sensations referred to a patient's phantom arm from another subject's intact arm: perceptual correlates of mirror neurons. *Medical Hypotheses, 70,* 1233–1234.
Reda, K., Febretti, A., Knoll, A., Aurisano, J., Leigh, J., Johnson, A., ... & Hereld, M. (2013). Visualizing large, heterogeneous data in hybrid-reality environments. *IEEE Computer Graphics and Applications, 33*(4), 38–48.
Regenbrecht, H., Alghamdi, M., Hoermann, S., Langlotz, T., Goodwin, M., & Aldridge, C. (2015, March). Social presence with virtual glass. In *Virtual Reality (VR), 2015 IEEE* (pp. 269–270). IEEE.
Rimland, J., Ballora, M., & Shumaker, W. (2013). Beyond visualization of big data: A multi-stage data exploration approach using visualization, sonification, and storification. *SPIE Defense, Secur. Sens, 8758,* 87580K.
Riva, G., Leggenhager, B., Tadi, T., Matzinger, T., Blanke, O., & Ehrsson, H. (2007). Virtual reality and telepresence. *Science, 318*(5854), 1240–1242.
Rizzo, A., & Kim, G. J. (2005). A SWOT analysis of the field of virtual reality rehabilitation and therapy. *Presence: Teleoperators and Virtual Environments, 14*(2), 119–146.
Rizzolatti, G., & Craighero, L. (2004). The mirror-neuron system. *Annual Review of Neuroscience, 27,* 169–192.

Rogers, S. (2016). *How we made a VR visualisation*. Medium. https://medium.com/google-news-lab/how-we-made-a-vr-data-visualization-998d8dcfdad0.

Rosenhall, U., Nordin, V., Sandström, M., Ahlsen, G., & Gillberg, C. (1999). Autism and hearing loss. *Journal of Autism and Developmental Disorders, 29*(5), 349–357.

Rothbaum, B. O., Hodges, L. F., Ready, D., Graap, K., & Alarcon, R. D. (2001). Virtual reality exposure therapy for Vietnam veterans with posttraumatic stress disorder. *The Journal of clinical psychiatry*.

Saggio, G., & Ferrari, M. (2012). *New trends in virtual reality visualization of 3D scenarios* (pp. 3–20). INTECH Open Access Publisher.

Salvatti, A. (2010, November). Virtual Acoustic Prototyping—Practical Applications for Loudspeaker Development. In *Audio Engineering Society Convention 129*. Audio Engineering Society.

Salzman, M. C., Dede, C., Loftin, R. B., & Chen, J. (1999). A model for understanding how virtual reality aids complex conceptual learning. *Presence: Teleoperators and Virtual Environments, 8*(3), 293–316.

Sanchez-Vives, M. V., & Slater, M. (2005). From presence to consciousness through virtual reality. *Nature Reviews Neuroscience, 6*(4), 332–339.

Santanen, E. L., Briggs, R. O., & De Vreede, G. J. (2000, January). The cognitive network model of creativity: A new causal model of creativity and a new brainstorming technique. In *System Sciences, 2000. Proceedings of the 33rd Annual Hawaii International Conference on* (pp. 10-pp). IEEE.

Schall, G., Mendez, E., Kruijff, E., Veas, E., Junghanns, S., Reitinger, B., et al. (2009). Handheld augmented reality for underground infrastructure visualization. *Personal and Ubiquitous Computing, 13*(4), 281–291.

Schneider, A. (1998). Sociology: The Internet as an extended classroom. *Social Science Computer Review, 16*(1), 53–57.

Seidel, R. J., & Chatelier, P. R. (Eds.). (2013). *Virtual reality, training's future?: Perspectives on virtual reality and related emerging technologies* (Vol. 6). Springer Science & Business Media.

Seth, A., Vance, J. M., & Oliver, J. H. (2011). Virtual reality for assembly methods prototyping: A review. *Virtual Reality, 15*(1), 5–20.

Short, D. (2012). Teaching scientific concepts using a virtual world—Minecraft. *Teaching Science-the Journal of the Australian Science Teachers Association, 58*(3), 55.

Slater, M., Perez-Marcos, D., Ehrsson, H. H., & Sanchez-Vives, M. V. (2009). Inducing illusory ownership of a virtual body. *Frontiers in Neuroscience, 3*(2), 214.

Smith, M. J., Ginger, E. J., Wright, K., Wright, M. A., Taylor, J. L., Humm, L. B., … & Fleming, M. F. (2014). Virtual reality job interview training in adults with autism spectrum disorder. *Journal of Autism and Developmental Disorders, 44*(10), 2450–2463.

Steuer, J. (1992). Defining virtual reality: Dimensions determining telepresence. *Journal of Communication, 42*(4), 73–93.

Stiles, M. J. (2000). Effective learning and the virtual learning environment. In *EUNIS 2000: Towards Virtual Universities: Proceedings of the European University Information System 2000 Conference held at INFOSYSTEM 2000, Poznan*. Poland, Poznan: Instytut Informatyki Politechniki Poznanskiej, pp. 171–180.

Strickland, D., Marcus, L. M., Mesibov, G. B., & Hogan, K. (1996). Brief report: Two case studies using virtual reality as a learning tool for autistic children. *Journal of Autism and Developmental Disorders, 26*(6), 651–659.

Sveistrup, H. (2004). Motor rehabilitation using virtual reality. *Journal of neuroengineering and rehabilitation, 1*(1), 10.

Thies, J., Zollhöfer, M., Stamminger, M., Theobalt, C., & Nießner, M. (2016). FaceVR: Real-time facial reenactment and eye gaze control in virtual reality. *arXiv preprint* arXiv:1610.03151.

Väljamäe, A., Steffert, T., Holland, S., Marimon, X., Benitez, R., Mealla, S., … & Jordà, S. (2013). A review of real-time EEG sonification research. Georgia Institute of Technology.

Van der Auweraer, H., Janssens, K., de Oliveira, L., Da Silva, M., & Desmet, W. (2007). Virtual prototyping for sound quality design of automobiles. *Sound and Vibration, 41*(4), 26.

Vickers, P., Hogg, B., & Worrall, D. (2017). Aesthetics of sonification: Taking the subject-position. In C. Wöllner (Ed.). (2017). *Body, sound and space in music and beyond: Multimodal explorations*. New York: Routledge.

Volpicelli, G. (2016). What's next for virtual reality surgery? *Wired*. http://www.wired.co.uk/article/wired-health-virtual-reality-surgery-shafi-ahmed.

Wallas, G. (1926). *The art of thought*. New York: Harcourt Brace.

Warburton, S. (2009). Second Life in higher education: Assessing the potential for and the barriers to deploying virtual worlds in learning and teaching. *British Journal of Educational Technology, 40*(3), 414–426.

Weller, C. M., & Baker, F. A. (2011). The role of music therapy in physical rehabilitation: A systematic literature review. *Nordic Journal of Music Therapy, 20*(1), 43–61.

Whittle, M. (2016). Virtual reality and sport: Breaking ground on and off the pitch. *The Guardian*. https://www.theguardian.com/media-network/2016/oct/21/virtual-reality-changing-sport-game-nfl-nba.

Wiederhold, M. D., & Wiederhold, B. K. (2007). Virtual reality and interactive simulation for pain distraction. *Pain Medicine, 8*(s3), S182–S188.

Williams, K. D. (2014). The effects of dissociation, game controllers, and 3D versus 2D on presence and enjoyment. *Computers in Human Behavior, 38,* 142–150.

Williams, J. H., Massaro, D. W., Peel, N. J., Bosseler, A., & Suddendorf, T. (2004). Visual–auditory integration during speech imitation in autism. *Research in Developmental Disabilities, 25*(6), 559–575.

Wolfson, S., & Case, G. (2000). The effects of sound and colour on responses to a computer game. *Interacting with Computers, 13*(2), 183–192.

Zakrzewski, C. (2016). Virutal reality takes on the videoconference. *The Wall Street Journal.* https://www.wsj.com/articles/virtual-reality-takes-on-the-videoconference-1474250761.

10

Conclusion

The reality of VR is one of great interdisciplinarity. From the system side, it is an aspect of computer science. From the user side, it is most certainly relevant to psychology. The virtual worlds themselves resonate with art and design, whilst the underlying concepts of virtuality draw upon organisational theory and philosophy. This significant blend of disciplines is still only half the story, as the complexity increases further still when we consider the applications of this technology. Rehabilitation and exercise management adds health care to the mix. Teleconferencing presents us with matters of communications technology, whilst VR learning and skills training connect us to seemingly every other discipline in existence. VR sound is no less complicated and remains connected to all the above elements whilst also bringing forth additional issues in auditory perception, acoustic ecology, sound design practice and audio technologies.

The emergent frameworks of VR and VR sound presented below are attempts at representing the virtual aggregates from which these actualisations spring forth. As observed in Chap. 3, to know the aggregate is not equal to knowing the actualisation. Therefore, it is strongly asserted that to better understand VR sound is to both study it and experience it

T.A. Garner, *Echoes of Other Worlds: Sound In Virtual Reality*,
Palgrave Studies in Sound, DOI 10.1007/978-3-319-65708-0_10

in equal measure. As with the Allegory of the Cave, if we look too long at the shadows, we will lose all sense of the objects that cast them. But to look too long at the objects is to surrender our ability to perceive the shadows.

Although we cannot elucidate every component and process of VR sound, this book has hopefully provided a glimpse into its complex, interrelating nature as an emergent concept. Additionally, it is hoped that the argument for acknowledging and working with this perspective is a convincing one.

So then, What Is…

The Virtual

As a component of reality, the virtual encapsulates everything in existence that is not actual. Any object, quality, idea, etc. can emerge as an actualisation, the requirement for which is quite simply that it is consciously perceived. The term 'consciously perceived' specifies that the actualisation is a singular conceptual object of the perception that which is focussed upon, at one particular moment in space and time.

VR

The organisational model of the virtual essentially incorporates everything in existence and VR is no exception. Understood from a holistic, ecological and emergent perspective akin to Sonic virtuality, VR is an emergent phenomenon actualised by experience. Effectively meaningless without a user to interact with, the technological-virtual (VR system: hardware, design, materials, etc.) forms only part of the whole. Figure 10.1 consolidates the information that we have discussed throughout this book to give an idea of the scale of complexity that is revealed when we acknowledge VR as an emergent and holistic concept. This illustration is by no means a comprehensive model of all relevant entities within the emergent aggregate, nor does it account for

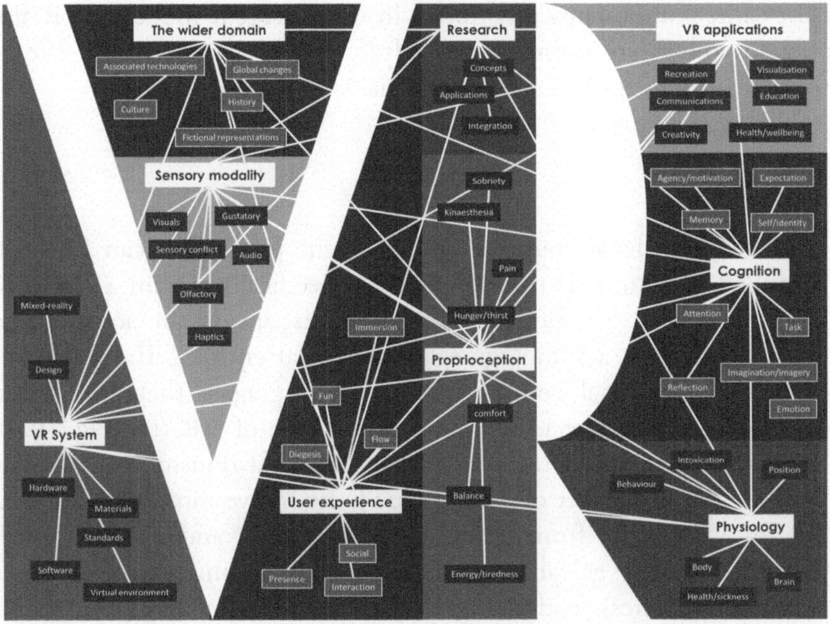

Fig. 10.1 A broad illustration of VR as an emergent concept

the nature of the many connections that have been drawn between each component. Instead, it serves to demonstrate how considering VR from solely a system-side perspective risks missing important factors and confounding a designer's expectations for user experience.

Sound

The most commonplace definition of sound is as an acoustic wave, but this is a mere fraction of the full phenomenon. This reveals a notable consistency with VR, as both continue to be thought of in largely empirical terms when they are in fact, deeply connected to a vast array of additional elements. As with our experience of VR, sound is not a static, physical entity and its qualities fluctuate significantly, both for a single listener over time and between multiple listeners. How we experience, interpret and draw meaning through listening reveals how many of the same components mentioned in the emergent model of VR also apply to sound, including wider cultural aspects, multimodal effects, proprioception, cognition and emotional state.

VR Sound

The fundamentals of sound as an emergent perception can be readily applied to sound that is experienced specifically within a VR context. Any sound, irrespective of context, is always an actualisation that emerges from a sonic aggregate of virtual entities. The difference lies in the additional components of the aggregate that arise when we bring together aspects of sound with those of VR (see Fig. 10.2). Developments in wider technology (wireless data transmission, mobile technology, the Internet of Things [IoT], etc.) have numerous implications for VR sound, from changing the nature of audio output hardware (such as the galvanic vestibular stimulator integrated within a pair of headphones) to defining its primary function (e.g. as a sonification system in support of IoT). Fictional representations have dramatically affected our expectations and overall experiences of VR sound in the past, and there is little reason to believe that this will change. Also, highly relevant are the broader cultural issues that contribute to the nature of VR sound overall, from the corporatisation effect that pushes us towards particular brands, to increasing individualisation that requires VR sound to become increasingly customisable and responsive to the precise characteristics and desires of the individual. Further advances in audio hardware and software, from enhanced real-time HRTF processing to high fidelity procedural generation, grant developers and designers ever-increasing opportunities to both improve and extend the quality and functionality of VR sound. Through research, we are continuing to learn more about ourselves, with ongoing progress being made in understanding auditory perception (modes of listening, acoustic ecology, Sonic virtuality, etc.), VR user experience, (from attaining a perfect and ongoing flow to mitigating cybersickness), human physiology (body shape, neurology, health, etc.), cognition (memory, attention, emotion), proprioception (balance, comfort, pain)

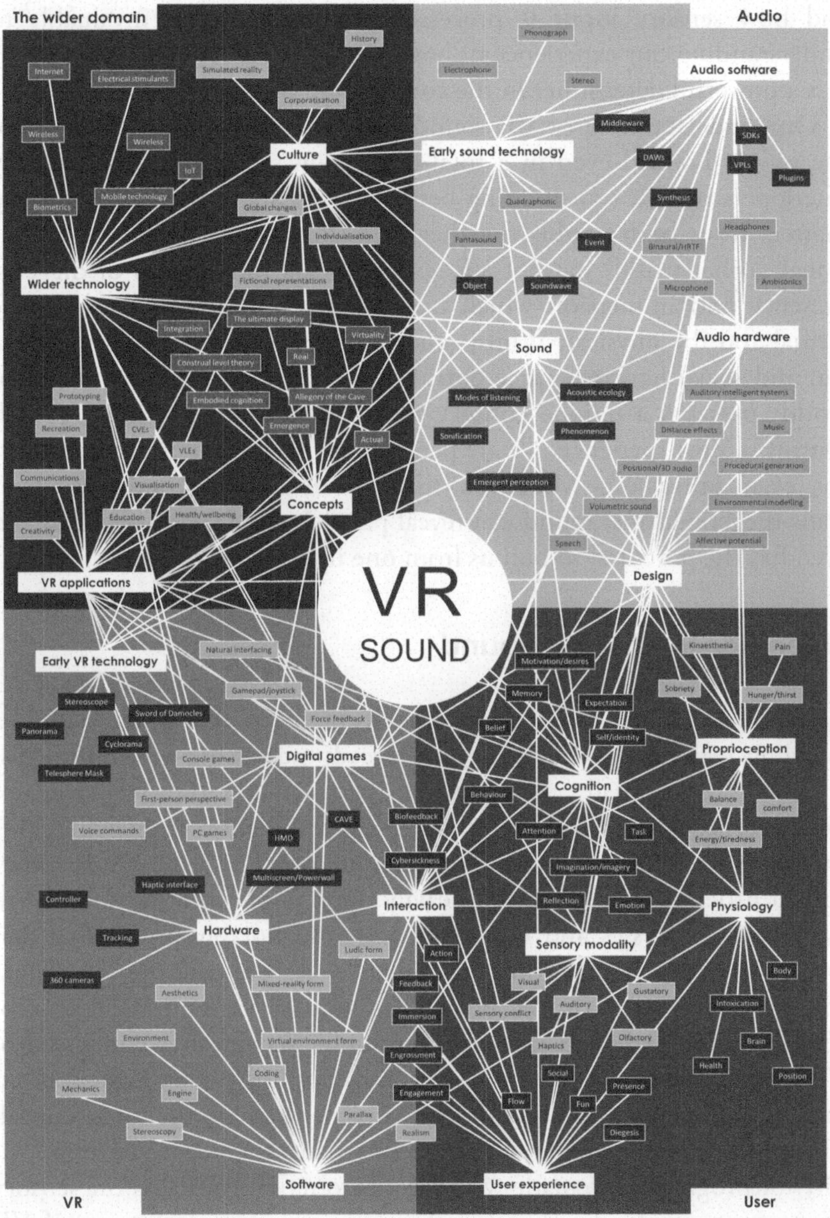

Fig. 10.2 Mapping VR sound as an emergent concept

and how sensory input is processed. Such research is invaluable to understanding our experience of both VR and auditory perception, and it is a powerful determiner in the constitution of VR sound.

Our understanding of VR sound must also incorporate elements of the physical world even in a total-immersion scenario, as we are only interlopers in VR, transitioning between worlds and therefore bringing matters from both worlds together in a single experience. The transitional, 'other-world' state of VR sound brings forth issues of diegesis, presence and immersion which have different meanings between the virtual and the physical. Whilst other interactive elements such as flow, fun and social have relevance outside VR, our relationship with virtual worlds presents a unique context within which we experience sound. During a VR experience, we are perceptually straddling multiple worlds (or even universes). Moreover, through diegetic, extra-diegetic and trans-diegetic forms, sound can reveal parallel worlds within worlds and even has the potential to pull us from one into the next.

Open Season for VR Sound

As has been stated more than once throughout this book, VR is a fledgling technology. This is despite already possessing a colourful history that conceptually dates back to early civilisation. We have not until recently possessed the technological foundation from which VR can flourish. As a result, the current nature of VR is exploratory, excitable, a little arrogant, focussed heavily upon application as opposed to underlying processes and methodologies, and notably prioritising that with which it can enrapture an audience. At present, the 'wow' factor revolves around visuals, but this is arguably a temporary state of affairs. As the hype subsides and the public become less convinced of VR's value when only presented with novel prototypes and ostentatious marketing, quality and function will become dramatically more crucial. This point in time is rapidly approaching, and there is a vast opportunity for progress to be made with regards to refining VR. As the sensory counterpart to visuals, sound is the underappreciated element of VR that has the most substantial potential for improving overall quality and functionality, from how sound can enhance VR data visualisations and make VR therapies available to the visually impaired to how it can aid information retention in VR learning and make VR games more engaging, immersive and fun. These questions and many more are yet to be fully answered. On the proviso that we are prepared to approach VR from a broader, emergent and more holistic perspective, great leaps forward are potentially just over the horizon. Now is most certainly a very exciting time for sound in VR.

Index

T.A. Garner, *Echoes of Other Worlds: Sound In Virtual Reality*, Palgrave Studies in Sound, DOI 10.1007/978-3-319-65708-0

The manufacturer's authorised representative in the EU is Springer Nature Customer Service Centre GmbH, Europaplatz 3, 69115 Heidelberg, Germany. If you have any concerns regarding our products, please contact ProductSafety@springernature.com

Printed and bound by CPI Group (UK) Ltd, Croydon, CR0 4YY
30/09/2026
02225597-0010